Brilliant Discourse

Brilliant Discourse

Pictures and Readers
in Early Modern Rome

Evelyn Lincoln

YALE UNIVERSITY PRESS

NEW HAVEN AND LONDON

Designed by Charlotte Grievson

Printed in China

Library of Congress Cataloging-in-Publication Data

Lincoln, Evelyn.

Brilliant discourse : pictures and readers in early modern Rome / Evelyn Lincoln.

pages cm

Includes bibliographical references and index.

ISBN 978-0-300-20419-3 (cl : alk. paper)

1. Illustrated books–Italy–Rome–History–16th century. 2. Books and reading–Italy–Rome–History–16th century. I. Title.

Z1023.L75 2014

096'.10945632–dc23

2013033954

A catalogue record for this book is available from The British Library

Frontispiece: *Allegory of Agrippa with Nature and the Muses*, Camillo Agrippa (detail of fig. 37).

Contents

Preface

But wise men pierce this rotten diction and fasten words again to visible things; so that picturesque language is at once a commanding certificate that he who employs it, is a man in alliance with truth and God. The moment our discourse rises above the ground line of familiar facts, and is inflamed with passion or exalted by thought, it clothes itself in images. A man, conversing in earnest, if he watch his intellectual processes, will find that always a material image, more or less luminous, arises in his mind, co[n]temporaneous with every thought, which furnishes the vestment of the thought. Hence, good writing and brilliant discourse are perpetual allegories.

Ralph Waldo Emerson, *Nature*, 1886

The urge to clothe words in images as a way to attest the character of authors resulted in an unusual category of printed books in the late sixteenth century. Early modern publishing provided opportunities for regular people to become authors, with immediate notoriety and an authorial reach they could not previously have imagined. As authors they greeted the world from the stage provided by the fact of publication. To issue a book with pictures was to endow the words of the speaking author with more specific clarity, aiming them with an accuracy that words alone failed to achieve. It also displayed a level of skill in orchestrating the separate arms of the material world of publishing that was in itself proof of technical, financial, and administrative ability.

This book is about sixteenth-century Italian treatises that engaged their audiences through the purposeful use of printed pictures. When books were read out loud, pictures added a visual dimension to the vocalization of the text. To silent readers, images provided a parallel experience to reading. This could be limiting in some ways; in other ways images opened out meaning onto other pictorial enterprises with which the book could ally itself through similarities of style. Illustrations in books give us a way to pair pictorial style with prose style, and we get a more vivid, visceral idea of where the books were intended to end up. Sometimes a carefully articulated

OPPOSITE PAGE Aliprando Capriolo after Bernardino Passeri, Plate 26, *Vita et miracula sanctissimi Patris Benedicti*, Rome, 1579, engraving, detail of fig. 16.

author appears in the pictures; other times the picture effects a connection between the reader, the book, and, importantly, the circumstances of its making. This was clearly meant to impress and delight, but not necessarily to instruct a reader in occupations like precise classification or the fine points of the techniques under discussion. These illustrated books are the residue of, and catalyst for, conversations; they are scenarios of an ideal relationship between someone with something to impart and a curious, knowledgeable audience who lacks only the information the author willingly shares. And yet, these books are not, in general, only about this knowledge, or that technique. They stand in for entire relational worlds, put into play through a discourse launched in words and images.

To access these worlds – and in particular to trace the access offered through pictures – I approached each book as a project undertaken through the collaboration of individuals with specific ambitions on several fronts at once. I chose the books I discuss here because they are picture books, the pictures make important visual arguments that lead to an understanding of what it was that publishers and authors thought of pictures as being able to do, and because in discussing them I could engage expectations for pictures on the part of different sorts of readers, writers, and patrons. The successful publication of a book meant a high level of consensus and communication on the part of its producers, and I was interested in the social networks formed by these commercial associations. The commercial interest that books enabled was not always directly related to the sale of books, since early modern books were players in all sorts of enterprises. Social capital was an important byproduct of this kind of authorship, and in these books we also see writers managing the creation and literary construction of a credible author. Books by little-known authors that were written on the fly, or were private commissions, could be very sparsely documented, but in fact, it was just such a notion of publishing as an opportunity for gaining authority in and among a describable culture that is the main subject of this book. The culture here is a class of readers and writers in Rome made up of clerics and merchants, artisans, and mathematicians, who took an interest in the natural, spiritual, and commercial realms ordered, as Roger Chartier put it, by books.

Years ago I became interested in why Italian Renaissance prints looked so different from the paintings they often claimed to reproduce. They differed not only in point-to-point comparison, but in those aspects of Renaissance painting that we have come to value as visual innovation: the brilliance and softness allowed by the use of oil paint, the emphasis on portraying motion in persuasive ways, the illusionistic qualities of depicting space through perspective and exploiting the properties of different media to trick the eye into perceiving depth instead of flatness. Few Italian prints trafficked in these visual pleasures. Books that were conceived of as ostensive objects, written to explain techniques, for example, would depend the most on the didactic work of images, and for these illustrations all the visual stops would be pulled out. Depth and movement in particular would seem to be important aspects of pictures designed to teach a technique, as anyone who has tried to roll out pasta by hand from the drawings in cookbooks will understand. As with most things, the story proved to be far more complicated, and each book opened out readily – too readily – onto a surprising number of uses that quickly began to seem limitless. In spite of that, the prints – as we will see – often had very little to offer by way of explaining the subtleties of the techniques described in the texts.

In this book I therefore did my best to stick with the books as projects in themselves, and to resist the temptation to take up the histories of the subjects they offered more than I needed to in order to discuss the work of the printed pictures included in them. Everywhere there is more that could have been said – about the Spanish Benedictines in Rome, for example, or Jews at Renaissance courts or in the silk trade, or whether or not bleeding people really worked, or if the Vatican obelisk might have been safely moved with one of several different methods. All of these topics cry out for further discussion and almost all of them have been discussed elsewhere. Here, I wanted to bring attention to the difference that pictures, as a kind of other speaking, made to a book: what kind of discussions, plans, hopes, and ambitions went into the production of an illustrated book, and what happened when a reader opened such a book and became immersed in its stories as they unfolded in images and text. It was not easy to keep focused on the readers and producers of the books in the moments of encounter, and this book is full of digressions that ended up being both useful and delightful to pursue.

Ideally, readers of this book would also be holding in their hands copies of the books I discuss, with their wrinkled vellum covers and elegant letters stamped into the creamy pages, with their decorative woodblock initials and chapters diminishing into the v-shaped typographic farewell that came to constitute a dramatic visual ending, no matter how inconclusive the textual footwork. Interspersed among the pages of type, readers (and viewers) would encounter the woodcuts and engravings I discuss here, exactly where they were originally set and meant to be found, marked by other readers and smudged by hundreds of years of page turning. In the absence of this, in some cases (the first edition of Magni, or Agrippa's fencing treatise, for example) the books are, as I write, available online. I urge readers to look for them there, at least, with the important caveat that few copies of sixteenth-century books are, in the end, exactly alike. But I hope this book makes its readers want to take them in hand in one of the rare-book rooms where they have been preserved for people like us to find them.

I was able to spend a lot of time with these old books in my own hands in some of the world's most splendid libraries. At the Vatican Library, Barbara Jatta helped me in researching printed pictures and the provenance of collections of illustrated books; at the Biblioteca Angelica, Daniela Scialanga has been particularly helpful. I was warmly treated at the Biblioteca Casanatense, and the Biblioteca Alessandrina in Rome, as well as the Biblioteca Nazionale di Firenze, and the Biblioteca Nazionale di Napoli. In the United States I worked at the James Ford Bell Library at the University of Minnesota, the Folger Shakespeare Library, the Avery Library at Columbia University, and the New York Public Library. At the Houghton Library at Harvard University I would like to thank Caroline Duroselle-Melish and Hope Mayo; at the John Hay Library at Brown University, Peter Harrington and Ann Dodge have been generous with their knowledge of and access to old books. The Brown University Library has supported my obsession with its rare books beyond the usual amount, for which I thank Harriette Hemmasi, and especially Karen Bouchard, Ann Caldwell, and Elli Mylonas. At the Archivio di Stato di Roma, I would especially like to thank Anna Lia Bonella and Orietta Verdi.

This research began at what was then the Bunting Institute, and is now the Radcliffe Institute, where the intense conversation between women working in the arts, humanities, and especially the sciences led me to think about Renaissance technical imagery, publishing, and work in uncanonical ways. Conversations in Cambridge in that period with Mario Biagioli, Katharine Park, Lisa Pon, and Abby Zanger began then, and have continued to this day to my great benefit. A fellowship at the Clark Art Institute and conversations there, especially with Sam Edgerton, Michael Ann Holly, Adrian Randolph, Davide Stimilli, and Jonathan Weinberg, significantly changed aspects of my approach to this material, much for the better. Discussions in various venues, formal and informal, in Rome and the United States with Renata Ago, Svetlana Alpers, Leonard Barkan, Cristelle Baskins, Harry Berger, Daniel Brownstein, Whitney Chadwick, Sandro Corradini, Susan Dackerman, Marina d'Amelia, Daniel Jütte, J. H. Heilbron, Maggie Hennefeld, Jane Ginsburg, Angela Groppi, David Kennedy, Elizabeth King, Medina Lasansky, Pamela Long, Ken Mondschein, Loren Partridge, Patricia Reilly, Pascale Rihouet, Marco Ruffini, Rose Marie San Juan, Suzanne Scanlan, Pat Simons, Pamela Smith, Barbara Sparti, Randolph Starn, Claudia Swan, Nicò Wey-Gomez, and Rebecca Zorach were important at one point or another in leading me to think about different aspects of Renaissance literature, history, natural knowledge, and printing, and I thank them for their very real intellectual generosity. During the time I was researching this book I benefited from the support of the Cogut Humanities Institute and the Pembroke Institute at Brown University, as well as the generosity of my colleagues at Brown. Maggie Bickford, Sheila Bonde, Caroline Castiglione, Hal Cook, David Kertzer, Ron Martinez, Rebecca Molholt, Jeffrey Muller, Dietrich Neumann, Doug Nickel, Tara Nummedal, Amy Remensnyder, Massimo Riva, Cristina Abbona-Snyder, Suzanne Stewart-Steinberg, Adam Teller, and most especially Dian Kriz have answered my questions, asked questions, read or listened to versions of this, shared their own work, and made critical suggestions that have helped me to make this book stronger. My work with early modern prints has been nourished by the generosity of my colleagues at Print Council of America and at the Museum of Art, Rhode Island School of Design, and in particular Cliff Ackley, Suzanne Boorsch, Francesca Consagra, Jan Howard, Emily Peters, Andrew Raftery, Sue Reed, and Joan Wright. Vincent J. Buonanno has shared his enthusiasm for Rome and delight in old pictures in ways that continue to be both materially important and spiritually grounding. Over the years that I have been working on this material, conversations about art, life, and prints with Brown students remain important to my writing and thinking, especially Matilde Attolico, Matthew Leuders, Oded Rabinovitch, Anna Sermonti, and Felipe Valencia. My father, Richard Lincoln, read the whole manuscript with his editorial eye and his writer's ear. Michael Baxandall made suggestions, and he warned that the book would take a long time. He was right – I wish I had finished it in time to show it to him. At Yale, I am grateful to Charlotte Grievson for her patience, advice and expertise in designing this book. Gillian Malpass's support and suggestions were as usual fundamental, and were always in my mind as I wrote about Renaissance publishers.

Robert Haddad, Roy Tishler, and Rosella Zarkin deserve grateful mention for having seen to it that I am still here to write this book, and Brian Shure continually makes me so glad that they did. He read every page that follows, and his shared love of reading and printing, in all its inky, thoughtful splendor, colored and shaded this book from library to print room to computer to your hands.

SPECVLVM ROMANAE MAGNIFICEN=TIAE·

OMNIA FERE QVAECVNQ
IN VRBE MONVMENTA
EXTANT· PARTIM IVXTA
ANTIQVAM PARTIM IVXTA
HODIERNAM FORMAM
ACCVRATISS· DELINEATA
REPRAESENTANS·

Accesserunt non paucæ, tum antiquarum, tum modernarum
rerum Vrbis figuræ nunquam
antehac æditæ·

Roma tenet proprys monumenta sepulta ruinis
Plurima, quæ profert hic rediuiua liber·
Hunc igitur lector scrutare benigne; docebit
Vrbis maiestas pristina quanta fuit

Antonius Lafreri exc Rome

1

Pictures and Readers in Early Modern Rome

To publish a book in early modern Rome demanded money, shrewdness, diplomacy, and connections in equal parts, which could mean involving many participants. It also required patience, a stomach for controversy, and nerves of steel; and this was in addition to possessing the certainty of having something useful to communicate to the world. Publishers and authors had to worry about the piracy of books, images, and ideas; the premature death of a patron or author; changing restrictions resulting from Counter-Reformation scrutiny of printed material of any kind; and peer challenges to the author's right to write. Printers worried about the excessive pickiness of patrons, the instability of their workforce, the liquidity of funds, misunderstandings between printers and clients, and the limited availability of good materials. Priests contracted with booksellers, surgeons with engravers, inventors with popes, printers, authors, and especially bankers: the dedicated historian may find out why only when the deals have gone bad and the names appear again in transcriptions from the courts of law.

Pictures added a whole other set of actors, specialized equipment, preferences, and contractual difficulties to the enterprise of printing a book. Illustrated books were without doubt made for the pleasure and instruction of children, women, and the illiterate. However, in the half-century between the papacies of Julius III (1550–5) and Urban VIII (1623–44) illustrated books were a common offering from Roman presses, often aimed through lengthy dedications and richness of illustration to the pleasure of a wealthier and educated set of readers. These books commonly claimed to be as useful as they were delightful. There were print shops that specialized in pictorial work, and a cadre of artists in the city who were primed to prepare pictures for the books that required them. Some of these books were patently meant for the enjoyment of learned readers and were structured to produce meaning from the reader's engagement with pictures; primary among these were books of arcane and carefully engraved spiritual emblems, and discourses on coins or medals that provided indispensable iconography.[1] The engravers, and the printers who

OPPOSITE PAGE   Etienne Dupérac, title page, *Speculum Romanae Magnificentiae*, Rome, Antonio Lafreri (detail of fig. 4).

1   Title page, *Benedicti Pererii Valentini e Societate Iesu Commentariorum in Danielem prophetam libri sexdecim*, Rome: in aedibus Populi Romani; apud Georgium Ferrarium, 1587. F★SC5.P4143.587c, Houghton Library, Harvard University.

printed their work on presses made for that purpose, as well as the woodcutters whose images were more affordably in demand for the book market, were an integral part of the networks – simultaneously intellectual and economic – formed between readers, authors, publishers, booksellers, typographers, and printers of type. These printmakers were associated, on the one hand, with anyone working in the arts of *disegno* in any medium. On the other hand, they were associated with the literary people who frequented print shops and bookstalls and knew what opportunities authorship could promise in the practical world. From these associations across families and neighborhoods streamed diverse books that used pictures ostentatiously, sometimes seemingly gratuitously, but which in fact were highly self-conscious publishing acts that employed images in ways that allow us to see how pictures were used in speech and text in everyday life.

Documentation in Roman archives of the complex financial arrangements that recorded the contracts between authors, printers, publishers, booksellers, and engravers of copper or wood made with the goal of producing a printed book rarely mention the particular project that was afoot. Instead a temporary partnership, or *societas*, was formed with the intention of accumulating funds, and sometimes only the recognizable names of the people involved indicated that printing was to be the outcome. Two examples involving the Roman *libraio* and publisher Giacomo Tornieri, in which the projects were actually contracted in detail, show the importance of the booksellers in the look of the printed books, how many people could become involved in the production of a single volume, and how easily a lack of communication could cause the enterprise to fall apart. In 1586 Tornieri contracted with a priest representing the Spanish Carmelites in Rome to print a missal for them. He promised in front of a notary that all 1,500 of the little volumes would be printed on good white Fabriano paper, that he would engrave fifty small "figurette" in wood and two beautiful friezes for the adornment of the missal, that he would have made, and would print, four larger images engraved on copper plates and put them in every missal according to the example brought by the Reverend Father, who would provide him with copies of the work approved by the Inquisitors, that printing would begin the first of June, and that he would print at the rate of one *foglio* per day until the edition was completed. An illustrated Carmelite missal with Tornieri's imprint appeared dated 1587, and that year Tornieri contracted with Thomaso de Victoria of the Carmelites to sell missals for them. Shortly later, however, we find Tornieri registering a legal protest because Thomaso de Victoria, on examining the finished results, did not find the job completed properly and did not want to pay for them.[2]

The same notary, a little later, records a similar contract between an association of the booksellers (called *librarij*) Domenico Basa, Bartolomeo Grassi, and Tornieri, and their client Cardinal Antonio Carafa, at whose palace near Piazza S. Eustachio they were gathered, and who wanted printed copies of a commentary on the prophet Daniel, written by a Jesuit priest.[3] The printer (*impressor*) Francesco Zanetti signed the document in the role of witness. The Stamperia del Popolo Romano was begun by the papacy during the reign of Paul IV and then given to the city under the directorship of Paolo Manuzio. By the 1570s it was being run by an association of booksellers including Giorgio Ferrari and Domenico Basa. When the book that Carafa was brokering with the booksellers, *Commentariorum in Danielem prophetam libri sexdecim*, appeared in 1587 it bore the imprint of the Stamperia del Popolo Romano and the address of Ferrari, who had been a business partner of Grassi until they fell out in court in 1583 (fig. 1). Included with the contract was a note written in a different hand on thin, crisp white letter paper, tucked into the notary's legal files, specifying the terms of the negotiation between the client and the publication team:

> Things to negotiate with Basa and with the others who will print the work about Daniel: the good quality of the paper, and of the ink, the quality of the printing, and to whom to give the corrected copy, and then the first impression, and that they will proceed to print in this way. If some books are on bad paper or incorrectly printed, and the cardinal judges that they have to be reprinted they will promise to do so. They will promise to make one *foglio* per day, and to send it for correcting to the Collegio del Gesù at a convenient time that they will arrange between them. They promise to give the author of the work sixty volumes gratis, and, for sending to various colleges of the religion, to sell the author up to fifty or sixty volumes at the price of three or four *giulii* less than what the book would sell for to the public here in Rome. The fonts they should use in printing the *Daniel* should be these: for the text, *sopra-*

silvio; for the commentary, *silvio*; for the citations in the commentary, *cursive*; and for the marginal annotations, a beautiful *lettera tonda*.[4]

The cardinal, more sophisticated in his patronage than the reasonably well-organized Carmelite father, not only included a highly detailed list of specifications about what he wanted the book to look like, but he also installed regular checkpoints throughout the printing process. Thinking about the look of the book and the quality of its manufacture as well as its faithfulness to the original manuscript, Carafa was also punctilious about its content, knowing full well that just because one printed book from an edition was without errors, this did not mean that the rest would follow suit. Each individual volume was therefore checked by a trusted corrector working for the authorial group. The idea of a single author, and faith in the printed text as an exactly repeatable, reproducible object indexical to a pristine original, were concepts that were unusual in the practice of producing books in early modern Roman print shops. As elsewhere, shops printed more than one book at a time, and especially when illustrations were involved, books could be printed simultaneously in different shops, complicated proceedings that were coordinated by booksellers-in-charge who acted as publishers.[5] The next year the evidently satisfied Carafa worked with Basa again, contracting with him to bring a new Bible through the press.[6]

The process of producing a printed book was, and continued for some time to be, regulated primarily by the church for its content and by no one in particular in terms of quality. Competition seemed to maintain acceptable standards, and customers of printing or of printed books were liable to the same protections as consumers of any service or product. The owners of Venetian print shops and bookshops had organized themselves into a non-compulsory corporation by 1549 with officers drawn from the most prestigious print shops in the city.[7] While the guild offered the usual ritual and spiritual assurance for each member and met in a chapel at the church of SS. Giovanni e Paolo, it also at its origins dedicated itself to ensuring that printers ignorant of the art did not bring disgrace to the other Venetian printers by issuing badly printed books. At some point, an examination for entry into the guild was established, designed to demonstrate familiarity with the techniques of printing and binding different kinds of books. By the end of the sixteenth century the number of guild regulations were inflated with the addition of procedures for dealing with the Inquisition, offices that Roman printers accessed more directly on their own. Roman printers and booksellers remained rather unregulated – except by the demands of the church – until the beginning of the seventeenth century, at which point they responded to encouragement to reform their confraternity in the wave of general lay reform.[8] They were given the church of Santa Barbara in a piazza off the Via de' Giubbonari, in the center of the district where most of the shops relating to book and picture printing and sales were located, but they were never so much a crafts guild as they were, like most of the *arte* in late sixteenth-century Rome, a trade-related religious confraternity.

This made Roman printed books, somewhat like paintings, often the product of a complex patronage system and dynastic workshops. While the printing of modern books is likely to be outsourced to a country that the author has never visited, early modern Roman books were often printed by people personally known to the entire authorial team of the book in question, whose print shops were close by those of the booksellers, and who tended to engage artists and engravers from the same area.

In spite of the many legal and financial risks, this period also created new genres and opportunities for enterprising publishers: helpful tracts on the newly revised Gregorian calendar, spiritual aids in the vernacular, the exemplary lives of the recently beatified, funerary orations, regularly updated confraternity statutes, and, above all, pilgrim's guidebooks were safe standard fare for Roman printers that found a good market throughout the sixteenth and seventeenth centuries. Illustrations like the woodcut *figurette* in the Carmelite missal could be seen as adornment, like the printed friezes that signaled the beginning and end of chapters, or the historiated initials that were a standard part of a printer's stocks. But they also created opportunities for conversations across different genres of printed media.[9]

Pictures in early modern printed books, or alongside text in broadsides and on the printed proclamations called *bandi*, enabled and relied on conceptions of authorship that were particular to the period. Pictures in books accompanied the universal practice of reading out loud that, as we read in the *Dialogues* of Magino Gabrielli (Chapter 5), might take place in a farmhouse, an urban piazza or streetcorner, or a fair where charlatans and mountebanks read out the miraculous attributes of their wares, performed their efficacy in theatrical skits, and sometimes published broadside or pamphlet versions of what they had to say. While some of these charlatan tracts are rough broadsides, others are ornate advertisements for products and methods among the printed ephemera of this period.[10] But Magino also makes clear that the pictures both elucidate text and rely on text in order to be deciphered. This is not likely to be a problem for anyone in his audience who might look at pictures without being able to read because "we have to suppose that at least one person per house will know how to read, and if not, as sometimes happens in some strange villa, a neighbor will understand, and one teaches it to many."[11] Earlier in the book his interlocutor, engaged in examining the illustrations that demonstrate the techniques the author is teaching, repeatedly assures him that she is paying attention with both senses: "I keep my ears out for what you are saying, even if sometimes I turn my eyes to the picture."[12]

Speaking printed text out loud was an act of both publishing and reading, in the sense that certain texts (public pronouncements in the form of the printed and pronounced *bandi*, for example, but also new inventions of all kinds as well as popular songs) were routinely made known through the act of declaiming them in public spaces.[13] *Trombetti* signed the backs of archived copies of early modern documents attesting to the fact that they had read their contents aloud at previously agreed locations, at least originally also "trumpeting" their arrival to announce the broadcasting of their information. They also had to affix printed copies of the texts to a wall in such previously agreed upon public places so they could be seen and read by all. As late as the last quarter of the eighteenth century *trombetti* swore after each act of declamation that they had not only affixed their documents but also declaimed their contents clearly in a loud voice.[14]

According to Michele de Certeau, the mediation of the reading voice itself made early texts (both printed and written) comprehensible, and he called attention to the loss of the rhythms and cadences of reading when the process became silent and solitary: "To read without uttering the words aloud . . . is a modern experience. . . . In earlier times, the reader interiorized the text; he made his voice the body of the other; he was its actor. Today, the text no longer imposes its own rhythm on the subject, it no longer manifests itself through the reader's voice. This withdrawal of the body, which is the condition of autonomy, is a distancing of the text."[15]

De Certeau saw the emancipation of the reader from the tyranny of a printed text as a characteristic of modernity, a newfound freedom from determinate readings or what we would recognize as an authorial voice. Roger Chartier, who dedicated a book on reading, writing, and publishing to de Certeau, pointed out one of the major difficulties in de Certeau's call for writing a history of the reader. To describe the history of a practice as intimate, unbounded, and particular as the reception of printed information read silently or out loud would seem to require nothing short of masses of individual microhistories. Chartier wrote: "The paradox underlying any history of reading . . . is that it must postulate the liberty of a practice that it can only grasp, massively, in its determinations."[16] The history of the book, which has come to rest on a history of reading, has in fact been well served by the development of the genre of microhistory, but the history of the reception of pictures has developed along art historical lines. When a reader is also a viewer of pictures, both in and outside of the book, how much does the literary context change the tone of the picture, or the picture the text?

Early modern authors of illustrated books wrote with deliberate reference to the reader's share in interpreting their printed works. Sometimes the illustrations they included in their books were attempts to call the reader back to particular meanings. Just as often, they provided material for readers' active memories, concerned with recall of real events and objects, and also their active imaginations, which synthesized ideas along an entire spectrum of personalized mental structures.[17]

The creative team responsible for the appearance of a printed book, who all shared aspects of its authorship, counted on the reader's ability to parse images according to varying familiarity with Renaissance pictorial idioms.[18] Repurposed or cannibalized images from other sources brought with them the sense of their meaning in other contexts, as they were often supposed to do. By the same token, once a book entered the market, it could appear in ways that might have surprised the original authorial group. Old text could be updated with new plates freely copied from more modern books on similar subjects, as we see in later editions of Juan de Valverde's famous illustrated anatomy textbook updated with copies of Vesalius' even more famous anatomy, or in Pietro Paolo Magni's treatise on bloodletting (Chapter 4). Old plates could equally well find new life in the reiteration and retranslation of saints' lives and other religious books, as in the case of engravings originally made for an illustrated life of St. Benedict (Chapter 2). Texts that seemed authoritative once were retooled by publishers whose idea of intellectual property was very different from our own, and the legal protections of patents and privileges that were established to protect the authors and publishers of written texts were mostly ineffective in keeping images from being copied, if not right away, then certainly after the period of their protection had expired. Most amazingly, scores of images that had currency when they were made in the first half of the sixteenth century continued to provide good information to artists and delight viewers into the next century through the reprinting of old plates or, more often, the readaptation of figures and motifs in new contexts. In these ways pictures and even pictorial styles and their visual references made connections across time and space through printed volumes and pictorial media that appealed to certain social strata and also participated in forming reading communities through the use of common visual vocabularies.

In researching this material I realized, as have historians of the book before me, that there would be no such thing as a useful survey of the early modern illustrated book, just as it is difficult to say much useful in a survey-like way about early modern books in general.[19] There is no over-arching history of book illustration that could synthesize the reasons authors and publishers invested in the considerable trouble and expense of adding pictures to their books between the middle of the sixteenth and the middle of the seventeenth centuries, a span of time that defies most periodization in either historical or art historical studies, but which does have a certain purchase in the study of literature. Mikhail Bakhtin saw it as the last century of a carnivalesque conception of the world, coinciding with an enthusiasm for publishing that had not yet settled into the forms and genres that are most familiar today.[20] The lack of crystallized conventions allowed a range of expectations for publishing a book, many of them improbable to readers like us.

This century also saw the formation of the Index of Prohibited Books, the establishment of the Inquisition in Rome, and the Decrees of the Council of Trent, all of which shaped and were intended to shape the way printed books would be made and consumed. As Gigliola Fragnito and others have shown, the terms of church censorship were inefficiently being worked out at the time, and rather than conceiving of its effects as unified and paralyzing, sixteenth-century censorship is best understood as a regime under which people lived, read, wrote, and thought about their relationship to the accumulation of other people's knowledge and the modes of dispersal of their own.[21] John Heilbron characterized the Inquisition in the seventeenth century as a "fact of . . . many people's lives, a sort of low-level background terrorism" that people learned to live with according to their circumstances and even their temperaments.[22] Turning attention to the effects of regulation on the publishing industry rather than the terms of the regulations, Fragnito points to the development of particular styles of publication in the face of increased requirements and restrictions, and the capacity of authors and publishers to adapt. Censorship joined the many other social, legal, and economic circumstances exerting pressure on early modern publishing associations in which the character of these printed books was forged.

~

The Formation of Printing Associations

The temporary associations between booksellers and printers of text and pictures that formed to realize specific printing projects, like the ones involving Tornieri above, were labile and not always harmonious; the most dramatic of these include murder and stolen books as well as the usual complaints about inaccuracies, miscommunication, and lazy workers. Many of these famously came together in the paratext of Principio Fabrizi's exhaustive treatise on the Boncampagni family's dragon impresa illustrated with 256 copper engravings. There were supposed to be 302. Over the course of four pages the author ventilates his frustrations with the printing process. Forced to bring the book to press still short of the illustrations to introduce six chapters, Fabrizi liberally meted out blame, writing that in the first edition important images had to be left out

> through the defects of the engravers, who took so much more time to work the copper than before that, if I ever wanted to see an end to it, I was forced to leave some parts imperfect, as I did. Then, Bartolomeo Grassi, too, the printer of this work, after taking every consideration with the engravers for a space of three years, was forced to send it out without the six big images that I mentioned and without the medals of the Virtues, and the other Titles, that are brought up in the margins, as you can see in the space left on the copper plate. . . . The printers, no less than the engravers, take so long with the work that they stop doing it, and time always produces some accident.

He then admits that the printers were facing a difficult undertaking:

> It is very true that if in the other things they do not deserve to be excused, they deserve much praise if they bring the work to an end because of the great expense they had, and the nuisance in printing the pages twice – that is, the letters and the copper plates separately and at different times and sometimes in different places, because of the multitude of plates, and for the great amount of work. [Grassi] said that he had never engraved more than are in this book, nor seen others engrave even half the number of prints that there are in this one, which comes to 256 figures counting both large and small.

Finally collapsing into self-pity at the end, Fabrizi condemns the resources available to a foreign researcher in Rome that seemed continually to damn his enterprise: "To the above-mentioned difficulties are added the lack of books about so many arts and professions, which are in keeping with the other great inconveniences that Rome presents to foreigners, not being a great friend to Mercury."[23]

The engravings for his book were designed by or under the direction of Natale Bonifacio, while Fabrizi's statement implies many engravers and at least two sets of printers, perhaps more. The engravings had to be printed on a press with a roller that could accommodate the engravers' thin copper plates, the text stamped in a letterpress from characters set into frames. Some print shops seem to have had both kinds of presses; other associations were probably formed expressly to print such hybrid projects. The most straightforward way to print pictures in books was to carve them into woodblocks, which were the same thickness as the frames that carried the type, and could be printed together with it. Not surprisingly, it was unusual for books to contain engraved illustrations printed on the same page as text printed from type, and the most famous early example of such an effort, Cristoforo Landino's commentary on Dante illustrated with engravings designed by Botticelli and engraved by Baccio Baldini, was never completed.[24] When the increased delicacy and richness of engraved illustrations was demanded, most publishers tended to print the pictures on separate sheets, inserting them into the pages of text in the binding process and often foregoing the chance to use the verso of the image for printing text. This was how the publisher Bartolomeo Bonfadino managed the printing of Pier Paolo Magni's treatise for barber-surgeons illustrated with engravings (Chapter 4), while Antonio Blado took the same arduous road as Grassi when he issued Camillo Agrippa's fencing manual with half-page engraved illustrations placed above their letterpress explanations (Chapter 3). The publishers of the life of St. Benedict discussed here in Chapter 2 avoided the problem altogether by limiting the verbal narration to a miniscule amount of text engraved on the same plates and in the same technique as the large pictures. When they included translations, as in the Spanish edition printed later, the publisher simply added narrow oblong plates with the engraved Spanish text set under the original Latin. As Fabrizi lamented, any version of this accessorizing added considerable expense and a good deal of time to the printing process, making each page more difficult and exacting to produce and increasing the chance of error.

In these books publishers, authors, and printers address their readers directly to brag about or decry their adventures in publication. Authors' anxieties and opinions about publication come through clearly and with astonishing directness. As odd as it may seem to modern readers, complaining about the printing process, and even the specific experience of printing the book in the reader's hands at that very moment, was a subject that many authors of vernacular illustrated books took the opportunity to address, to the point where it could even become a secondary subject matter. Sometimes this information is found in the paratexts – the letters and dedications in which the authorial team of any book addresses particular patrons and dedicatees and also, through Letters to the Reader, a general readership. This is how the publisher Bartolomeo Grassi chose to call attention to what was good and notable in Fabrizi's book, jumping in with a peremptory letter to the reader at the beginning, many pages before the reader would encounter Fabrizi's complaints about his Roman publishing experience at the end of the volume. In his prefatory letter the publisher calls attention to what Fabrizi mentions at the end: that the copious number of figures engraved in copper (256), which he had printed at his own great expense and with much care, is a remarkable and laudable fact. He also takes credit for another contribution: the addition of numerous pages of tables and indices at the back that attempt to organize the rambling *materia* of the book.

Illustrations were a feature of these books that could increase readership, but they also significantly increased production costs and added another layer of expense, logistics, and personalities to what was already an intricate and uncertain process. Illustrations in books like Fabrizi's were more than an improvement, a helpful and desirable addition, a clarification to the text, or a marketing point. The figures were often the point of the book, a way of speaking parallel to text, but more often the cause of, or prompt for, textual explanation.[25]

As historians of the book have pointed out since Roger Chartier's work beginning in the 1970s, reading communities also included authors, patrons, and publishers. The presence of these many alternate authors, too, looms large in the history of a targeted readership that seems, in these early modern texts, to be local and, by today's standards, intimate and knowable.[26] The large proportion of sixteenth-century books in the dialogue format published in Italy in this period, many of which contain dedicatory letters at the beginning, attests not only to the primacy of conversation in Italian literature, but also to the importance of the speaking voice in the mediation of learning and in didactic writing. Illustrated books written in the vernacular are full of the sorts of asides, excuses, addresses, and directions to the reader that we associate more readily with speaking or with informal letter writing than we do with printed treatises, even when they announce themselves as such. Reading out loud in a group, or imagining printed text as spoken voice, seems today like a particularly intimate way of absorbing its meaning; it was a literal reiteration of the written words that was a vital part of sixteenth-century learning.

Printing and Picture Books in Renaissance and Early Modern Rome

Pictures appeared in printed books almost immediately after the first press started up in Italy in 1464, under the German printers Sweynheim and Pannartz, slightly northeast of the center of Rome in a Benedictine monastery in Subiaco.[27] However, they were not a necessary feature of the didactic books that abounded in the following century. Many of the first herbals, geographies, guidebooks, and medical books, even emblem books such as Cesare Ripa's *Iconologia*, were first issued without pictures, offering readers verbal descriptions of visual objects or ideas to appeal to the imagination.[28] Illustration in the earliest printed books, versions of classical or patristic works in Latin, worked within the rubric of memory aids developed for manuscripts and were designed to make the books richer textual objects.

The authors whose works are discussed in this book were perfectly aware of the most usual uses of illustration and design conventions in both printed and manuscript treatises. Pietro Paolo Magni's treatise on bloodletting was written for barbers who were already familiar with the phlebotomy charts that, though they changed in their appearance, were in use from antiquity through the early modern period. Camillo Agrippa knew what fencing manuals looked like, but he also knew the effect on the reader of showing his fighting men in the timeless undress of nude athletes, and that his innovations in conceiving of the practice of dueling gained credibility through depictions of precise measure and motion in the visual language of printed geometry tracts. The practical value of illustrations in mediating between what a viewer might be seeing and what an author was discussing was responsible for many of the naturalistic illustrations in early modern books, whereas schematic images such as charts, wheels, and other diagrams helped to teach the organizational knowledge necessary to absorb the broader theories of any topic.[29] As such, this kind of pictorial image continued to help readers from the fifteenth through the seventeenth centuries in the identification of plants or veins, and also participated in learning by providing ready-made images for the art of *memoria*, as necessary to novice barber-surgeons who wished to avoid killing their patients as it was to Benedictine monks learning the proper methods of prayer. The style of these illustrations changed to keep up with modern imagery, nowhere more important than in books of natural knowledge or explanations of technology.

The character and proliferation of press-related activity in Rome, where printing arrived early, is inseparable from the city's status as the center of the Western church, the seat of the ancient empire and its would-be successor, the Papal States, and the home of the diverse and chancery-oriented papal court.[30] The astonishing array of printed self-portraits of the city of Rome that issued from Roman presses between 1500 and 1800 suggests that Rome was most famous for transforming itself from a city of picturesque rubble, as in the series of city views by Etienne Dupérac, to the city of impressive marble that dominates the later *vedute* of Giuseppe Vasi and Giovanni Battista Piranesi. The medieval city had been a conglomeration of the decaying monuments of antiquity, dangerous to passersby and their inhabitants alike; the once-magnificent forum was a silted up cow pasture, broken aqueducts had left sections of the city without drinkable water, and the cyclical flooding of the Tiber regularly inundated the areas along its banks with its infamous muddy overflow. The city had no strong central governing body: even the papacy had left for France. But from the beginning of the fifteenth century the papal court was back, and old St. Peter's was turned over to the series of expert remodelers who worked the building into its current form over the next two centuries. For a long time, pictorial printing in Rome was to remain tied to the urban concerns of the papacy and of the pilgrims and tourists who came to the city to witness its transformation and to take part in the benefits offered by the presence of many sacred relics and important rituals that occurred from day to day, and especially in Jubilee years.

By the end of the sixteenth century, resurrecting the obelisks that the ancients had brought to the city as victory monuments, and repurposing them by setting them up to provide triumphant nodal points for sites of papal importance, were only the most dramatic of the engineering feats that came to represent literary and engineering careers. Wonders such as the unsupported dome of the Pantheon stood as architectural exemplars for aspiring builders, and the cow pasture in the forum provided artists with as much material as did the collections of sculpture preserved at the Vatican and on the Campidoglio. Urban features created to attract attention to the pagan monuments that had acquired Christian meaning, such as the Castel Sant'Angelo, and the necessary problems of access to accommodate church ceremony with its many outdoor processions, were uniquely Roman projects. Semi-clandestine exploration of the catacombs for their peculiar combination of mysterious relics and information about antiquity was a kind of spiritual spelunking

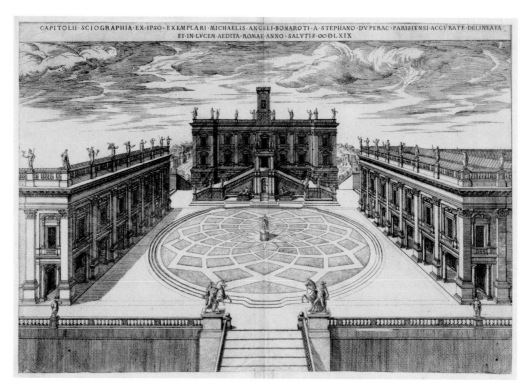

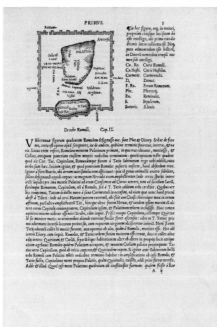

ABOVE 2   Etienne Dupérac, *Capitolii sciographia ex ipso exemplari Michaelis Angeli Bonaroti a Stephano Duperac Parisiensi accurate delineata*, Rome, 1569, etching with engraving, 371 × 540 mm, British Museum.

LEFT 3   The Capitoline Hill and Palatine Hills at the time of Romulus, woodcut, p. 3, Bartolomeo Marliani, *Urbis Romae topographia nuper ab ipso autore nonnullis erroribus syblatis emendate . . .* Rome: Valerio and Luigi Dorico, ca. 1549. fIC5.M3438.534acaa, Houghton Library, Harvard University.

that only the bravest or most obsessed cared to try, but it was particularly Roman in character. To these were added dredging the Tiber and attending to its bridges. Indeed any hydraulic work was seen as a reprise of the native Roman ability to move running water around the city or celebrate its emergence architecturally at meaningful points, which provided an opportunity for the popes and even for smaller aristocratic landlords to act in imitation of the emperors who had first built the aqueducts.[31] French gardeners and Netherlandish hydraulics experts, along with a good portion of more native talent, arrived in Rome to lay the pipes and install the plants that would establish formal gardens for the resident population of princes and clergy whose building projects rivaled the villas of antiquity. For the pilgrims who were able to return to the city to visit the important, restored religious sites and view the many new or renovated churches, for the antiquarians who resurrected the ancient city on paper through study of the remains that were daily unearthed in the ongoing construction projects, for the engineers, gardeners, sculptors, architects, and tourists drawn through its massive ancient gates, Rome increasingly became a focal point for technological practices in every field, as well as for the effective staging of Christian ritual and pageantry in an antique style.

Some of those attracted to the city were literary men and scholars of various levels of education who gained employment as secretaries to the Roman courts dispersed throughout the city, established by high-ranking clergy as well as old Roman nobility. The Massimo family were Roman nobles who traced their name back to the imperial Maximus family, became fervent supporters of the Oratorian founder Filippo Neri, and housed Sweynheim and Pannartz (and then their successors) in their own *palazzo* after they left Subiaco, establishing there the earliest print shop in Rome. Many of the cardinals and other clergy in palaces around the city came from courts in other parts of Italy and Europe, and although many of them struggled to preserve the level of ostentation their status demanded, their presence promised patronage opportunity for artisans and men of letters who could claim to share their homeland or known interests.[32]

The urban improvements of the eternal city, however, were completed in print more fully and much sooner than they were completed in stone, at every level of expense and authorial aspiration. Luxurious volumes of engraved maps, architectural motifs, Renaissance fantasies of antique vases, portraits of emperors, popes, and saints joined pocket-size woodcut broadsides of important disasters and single-sheet images of the religious processions that marked the calendar year. Artists were busy making plans and views of the ever more rational layout of the city as it was in the present, as it was hoped to become (such as the many engravings of Michaelangelo's Campidoglio before it was constructed), and as it was understood to have looked in antiquity (figs. 2–3). The physical construction of modern Rome proceeded alongside the intellectual reconstruction of ancient Rome in books and prints that paid careful attention to the archaeology of the ancient urban structure, and recorded the many works of pious patronage that transformed the chaotic medieval city into a theater of papal ambition.[33]

By the third quarter of the sixteenth century the publisher Antonio Lafreri had begun to issue a title page for print-buying visitors to the city so that they could put together their own

LEFT 4   Etienne Dupérac, title page, *Speculum Romanae Magnificentiae*, Rome, Antonio Lafreri, *c.*1574–7, engraving, 440 × 275 mm. NYDA.1951.001.00001, Avery Library, Columbia University.

OPPOSITE PAGE 5 AND 6   Pasquino in the Jubilee year, 2000. "In memoria dei motociclisti caduti sulle buche di Roma."

collection of the many representative images of Rome he offered for sale in the form of single-sheet engravings (fig. 4).[34] Advertising its prints as accurate depictions of ancient and modern Rome, the title page promised the reader to reveal and clarify in print what could be observed only in the rubble: "Rome holds many monuments immersed in her own ruins. . . . Therefore most benevolent reader, examine this book closely, it will show how great was the former grandeur of the city." In different hands the *Speculum Romanae Magnificentiae* could be a volume of Christian churches augmented with images of saints and popes, or an obsessively complete collection of antique temples, rituals, sculpture, coins, and architectural motifs that might also include a map of the ancient city.[35] It could also, of course, represent both ancient pagan and modern Christian Rome, which would be how most of the print-buying public, whether pilgrims, clergy, merchants, or humanists, would experience the city streets. Along with these images, and in conversation with them, providing ready-made scenography for myriad Roman narratives, were the many guidebooks available at an array of possible prices, with or without illustrations. By the end of the eighteenth century and the full launch of the Grand Tour, the genres of printable Rome had been mined for every variety of pictorial project, and gave Roman printing its own character.

This is not to say that initiatives by the papacy and individuals to civilize the city were immediately successful, and as Fabrizi complained, Rome held out more promise than it delivered for

ambitious people with skills that might seem to be in order. Projects benefiting papal reputation that were consonant with papal interests inevitably received more attention than those that made the city commodious for its more humble residents. Even during the Jubilee year of 2000 the loquacious statue Pasquino was moved to write a poem mourning the passing of native Romans whose *motorini* had been swallowed up in the terrible potholes that had gone unfilled while a new bus line was added to bring the faithful more expeditiously to the pilgrimage churches (figs. 5–6). Foreigners' accounts of the city alternated between wondering at its marvels and complaining about its filth, corruption, and inconveniences. St. Benedict of Nursia began his life of monastic withdrawal and contemplation after being repelled by the city's squalor and many vices at the end of the fifth century, while the more reluctant Grand Tourists of the eighteenth century could find Rome disappointing and even disgusting, like the Count of Bonneval who found that the mythical Tiber was nothing more than "an ordinary river, at most half as wide as the Seine. Its waters, always reddish and troubled, are good for nothing. Due to its frequent flooding and the havoc it causes, it can be regarded as the plague of this capital."[36]

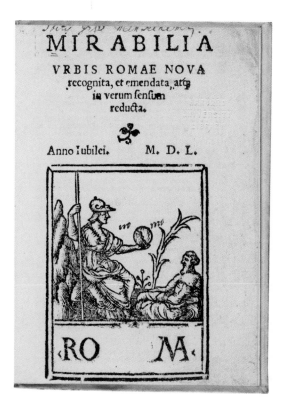
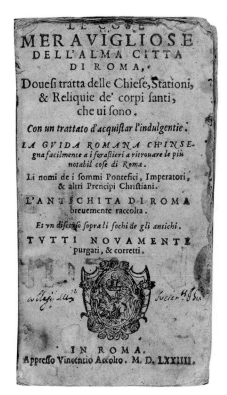

ABOVE AND OPPOSITE PAGE 7   Some Roman editions of the *Mirabilia urbis Romae*: a. Title page, *Mirabilia urbis Romae*, Rome: Antonio Blado, 1550, engraving. Arc 767.62*, Houghton Library, Harvard University. b. Title page, *Le cose meravigliose dell'alma città di Roma*, Rome: Vincentio Accolto, 1574. AA1115 C82, Avery Library, Columbia University. c. Title page, *Las cosas maravillosas de la sancta ciudad de Roma*, Rome: Por Tito y Pablo Dianos, 1589. AA1115 C822, Avery Library, Columbia University. d. Title page, Pietro Martire Felini, *Trattato nuovo delle cose meravigliose dell'alma città di Roma*, Rome: Bartolomeo Zannetti; ad instanza di Gio. Antonio Franzini & heredi di Girolamo Franzini, 1610. Typ 625 10.387, Houghton Library, Harvard University.

Printers' Networks

And yet printing began as an immigrant enterprise in Italy, with the first printers at Subiaco, as almost everywhere, being foreigners from Germany. The very busy trade in and practice of pictorial engraving and the making and selling of single-sheet prints was rather dominated by Frenchmen for most of the sixteenth century: in particular, Etienne Dupérac, Antoine Lafreri, Claude Duchet, Nicolas Beatrizet and later, when the field was much more diverse, Andrea and Lorenzo Vaccari (de la Vacherie), Philippe Thomassin, and Paul Maupin. A printer like the Lebanese Giacomo Luna, who took an Italian name, could find work in Rome because of the Vatican interest in printing

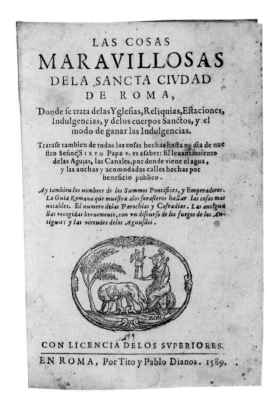

Oriental texts, and he branched out to other papal-inspired printing projects. Even immigrants from Friuli or Naples could be seen as benevolent outsiders with their own allegiances and customs. Still today the city is marked by "national" division into neighborhoods associated with the French, the Germans, and the Spanish, but also Lombards, Florentines, and Venetians, each with their own church. The streets around Campo di Fiori, particularly the Via dei Pellegrini and the Via Giubbonari, were the famous hubs of most of the print and book shops.

Along with the flow of visitors and pilgrims, job seekers, members of the Curia, ambassadors, and other travelers, there was a steady stream of people in the book trade who set up shop in Rome and lived there, raising families who went into the business and who married into other local printing businesses. In this way the character of certain shops, like that of Giovanni Osmarino Gigliotti, or the Zannetti shop, or Antonio Blado's, grew, branched out, specialized, contracted, were continued by their heirs, and eventually ceased to exist by the middle of the seventeenth century. The authors, printers, artists, and publishers dealt with in this book knew each other, married into each other's families, saw each other on the street, in churches, and at meetings of the artisans' confraternity (the Compagnia di San Giuseppe at the Pantheon) to which many of them belonged, or in publishers' shops, bookstores, and print shops. They witnessed each other's legal procedures, valued each other's work for contractual purposes, they hired each other to help when jobs became too overwhelming, and provided financial backing when it was profitable for both sides.[37]

Antonio Blado, Giovanni Gigliotti, and Bartolomeo Bonfadino, who probably owned presses capable of printing copper plates as well as text, all figure strongly as publishers and *librai* in the realm of Roman illustrated books.[38] Antonio Blado (1490–1567) began printing in Rome in 1516, and at least by 1534 his shop was located "in Campo Florae in aedibus D. Io. Baptistae de Maximis." Books bearing the imprint of Sweynheim and Pannartz from 1467 to 1475, after they left the Benedictine monastery at Subiaco, give their address as "Romae iuxta Campum Flore in domo Petri et Francisci de Maximis."[39] It is not known if this refers to a building on the land where Baldassare Peruzzi's Palazzo Massimo alla Colonna now stands or another property owned by the Massimo family on the Via Mercatoria that was even closer to the busy Campo de' Fiori, but either way it is evident that there was a predisposition on the part of the Massimo family to rent to printers, and so Blado continued in the tradition of the first Roman printers.[40]

Blado's first book, published in 1516, was a *Mirabilia urbis Romae*, maybe the most Roman of all books. Although versions of this medieval guide to the city were regularly issued by printers in Venice as well as in Naples and Milan, and in Latin, French, and Spanish as well as the increasingly more usual Italian, most were printed, quite regularly, in Rome, with notable increases in volume and publishers during the Jubilee years. The *Mirabilia* as a literary compendium of historical information on the city and its antiquities had a history that began long before the age of printing, and took off in print in ever expanding formats. It became a pilgrimage guide that eventually included information about the renovations of Christian Rome and its churches, a *lunario*, instructions for how to obtain indulgences, explanations of church ritual, and other relevant material for pilgrims including tables of all the emperors and popes with their *imprese* (fig. 7).[41] Over the course of his sixty-year career Blado is credited with having printed almost 1,500 items, alone and in association with other printers. In 1535 he obtained the position of Stampatore Camerale; as such it was his responsibility to print whatever was needed for the papal bureaucracy. This would include the latest results of the Council of Trent, or whatever was wanted by the associated clergy. Besides the usual medical, juridical, and spiritual texts that any sixteenth-century publisher might produce, and the *avvisi*, books on local architecture, monuments, and ritual that any Roman printer might produce, Blado's press issued papal bulls, and statutes of the guilds and confraternities. He also printed the official *bandi* posted around the city announcing chancery decisions ranging from indulgences to laws affecting dress, restaurant regulations, and use of the Tiber, to name just a few examples. The last publication bearing his imprint was a book of Roman statutes.

The publishing list of Blado's son-in-law Giovanni Gigliotto (active *c.*1570–1586) represented most of the genres of Roman printing, both with and without pictures: there were books of exotic travel describing the famous visits by the Japanese ambassadors, local confraternity statutes, law books and medical texts, histories of ancient Rome, books describing important weddings, entries into the city, and funerary orations, and many examples of those particularly Roman publications, the *avvisi*, and books on the rituals of the Jubilee years.[42] Gigliotti also published ever-popular guidebooks like *L'antichità di Roma* and *Le cose maravigliose dell'alma città di Roma* (fig. 7), which he issued almost annually between 1571 and 1585.[43] He also published the elegantly illustrated *Ludovici Demontiosii Gallus Romae hospes* (1585), a Latin treatise on ancient art and architecture in Rome with illustrations of the Vatican obelisk, the Pantheon, and monuments in the forum.[44] Equally useful and popular books, like guides to the Gregorian calendar and the uses of Agnus Dei – the wax discs blessed by new popes and useful in healing – grammar books, popular hagiographies and prayer books, a book of embroidery patterns, and the occasional bawdy or carnival song, insured a steady income.[45]

The Brescian printer Bartolomeo Bonfadino (active 1583–*c*.1607), who also located his shop in the Via Pellegrini, printed over a hundred books on his own and another forty in partnership with Tito Diani in his more than twenty-year long career.[46] He was the brother-in-law of the printer Giacomo Bericchia, who was a frequent publishing partner of Giacomo Tornieri. His list looks much like that of Giovanni Gigliotti, except that a little bit more than a quarter of his production, thirty books, contained illustrations beyond a simple frontispiece or title page. Most of his books bear only his own imprint, and one of the most beautiful and whimsical printer's marks of any Roman printer: a porcupine rattling its quills (figs 42 and 71).

Bearing in mind the ties between Roman print shops over generations, the illustrated books I discuss in the following chapters were bound by more than a common geographic center of production, or shared cultural affinities in a general sense. This is not clear from a glance at their bibliographic information, but in mapping their production at the end of the sixteenth century surprising connections between print shops and authors emerge. In 1584 Bonfadino was printing Camillo Agrippa's fourth book, his *Dialogo . . . sopra la generatione de venti, baleni, tuoni, fulgori, fiumi, laghi, valli, & montagne* (Chapter 3) as well as the first edition of Magni's *Discorsi* on medicinal bleeding illustrated with tipped-in engravings (Chapter 4). Agrippa, a Milanese engineer who had come to Rome with the idea of moving the Vatican obelisk, an adventure he discussed in a book he published with Francesco Zanetti in 1583, had published his first book, a fencing manual, many years earlier at the shop of Antonio Blado. He published his second book, a sequel to the first one, with Blado's heirs twenty-two years later, in 1575. In 1585 Bonfadino printed Agrippa's *Dialogo . . . del modo di mettere in battaglia presto & con facilità il popolo di qual si voglia luogo con ordinanze & battaglie diverse*, illustrated with several engravings. Ten years later Agrippa brought out his next book, on the art of navigation, with Domenico Gigliotti, the son of Giovanni and his wife, Agnese Blado, the daughter of Antonio Blado. After Giovanni's death in 1586, and before Domenico struck out to publish under his own name in about 1594, Giovanni's widow, Agnese, along with Domenico and his two sisters, Elisabetta and Tarquinia, published under the imprint Eredi di Giovanni Gigliotti, just as Giovanni had published for a couple of years (1570–72), at her father's shop with her mother and brothers as Haeredes Antonij Bladij & Ioannes Osmarinus Liliotus.[47] After Giovanni Gigliotti set up his own shop in 1570, Blado's widow and children continued the business as Heredi di Antonio Blado; they were operating under a version of this name when Agrippa turned to them for producing his second book, published in 1575. Gigliotti's heirs, under the name Eredi di Giovanni Gigliotti, produced the *Dialoghi di Magino Gabrielli* on how to raise silkworms, discussed in Chapter 5.

A prolific author as well as a busy engineer, Agrippa moved over to one of the publishers involved with the Medici Oriental Press, Stefano Paolini, for his last, unillustrated, deeply philosophical combination memoir and spiritual text about his life's work, *La virtù, dialogo di Camillo Agrippa Milanese sopra la dichiarazione de la causa de' Moti* (1598). In 1585 Bonfadino partnered with Bartolomeo Grassi as publisher to print Giulio Roscio's book of the works of charity with very ambitious full-page engravings by Mario Cartaro.[48] That same year, with Tito Diani and the publishing team of Bericchia and Tornieri, he printed an illustrated herbal by the papal physician Castor Durante, with woodcut illustrations by Leonardo Parasole.[49] In 1586 he printed the second edition of Magni's bleeding book, with the frontispiece by Cherubino Alberti. In 1587 he was occupied with Angelo Sangrino's quarto version of the luxurious reading manual for the Benedictines with pictures by an unknown engraver based on those by Passeri for the *Vita et miracula sanctissimi Patris Benedicti* (discussed in Chapter 2).[50] He was also at that time printing Pietro Angeli's illustrated commentary on obelisks in association with Bartolomeo Grassi.[51] In 1588 he published Magni's second treatise, a tract on prophylactic cauterization, with a portrait of Magni and woodcut illustrations of instruments, and in 1595 he published Giovanni Battista Cavalieri's picture book of portraits of all the popes, a perennially lucrative subject by a popular engraver of religious images and book illustration.[52] The last publication bearing his name was an elegantly illustrated book on navigation by Bartolomeo Crescenzio, *Nautica Mediterranea*.[53] He was still alive in August 1609 when he acted as a witness to a document in which the woodblock printer and publisher Leonardo Parasole rented property to the head of the Medici Press, Giovanni Battista Raimondi.[54]

From the case of Agrippa we can see how an author who seems to be doing business with different publishers is in fact working within a coterie of related companies, sometimes even the same print shop or printers in their next incarnation. The printing families knew each other's businesses better than anybody else did, and acted as expert estimators of each other's goods when such valuation was necessary. For example, in 1587, when he was deeply into Fabrizi's endless publication project, the *libraro* Bartolomeo Grassi provided values for the printed books, and the publisher Pietro Paolo Palumbo for the engraved copper plates made for Magni's book, left at the death of her husband, for the widow of the Mantuan engraver Adamo Scultori (Chapter 4). Often one expert would oversee the evaluation of goods for the seller and one for the buyer; it was important that both come to an agreement in order for the valuation to go forward. Therefore, in 1589 when the printer-engraver Giovanni Battista Cavalieri of Trent calculated the value of the prints and plates owned by Pietro de' Nobili, who had died in 1587, he was speaking to the interest of the dealer's infant son, Pietro Paulo de' Nobili. The engraver Aliprando Capriolo, also of Trent and responsible for carving the illustrations for the *Vita* of St.

Benedict in Chapter 2, acted on behalf of de' Nobili's business partners, Marcello Clodio of Teramo and Girolamo Arbotti, of Rome, participating in the same calculation. At the end of the exhaustive inventory, both men signed the document, affirming that they agreed.[55] In 1607 Bartolomeo Bonfadino estimated the value of the goods belonging to the business of his deceased colleague, the printer Luigi Zannetti of Venice, on behalf of the man who was buying Zannetti's Roman business, in this case Zannetti's brother Bartolomeo. The Vatican printer Curzio Lorenzini was called on to estimate the value of the business for Zannetti's heirs, his wife, Francesca de' Orlandi, and daughters, Faustina and Sabina.[56] Bonfadino also appears in 1593 as an expert estimator when the painter Rosato Parasole disputed the value of some of his work decorating the walls of a room at a *vigna*, and in 1609 as the witness to a document in which the Zannetti print shop was divided among the family after the death of Luigi.[57] The families and goods, wealth and life events as we might call them today, of people in the Roman book and print business were intimately intertwined.

For these reasons, opening a Roman illustrated book from the period between about 1575 and 1625 is a dizzying experience, a freefall into a dense set of circumstances and skills woven together by a set of intricate business propositions, the tangible result of which is the book. The book itself then has a life of its own, entering a world of reading, copying, commenting, and reissuing in which the original authorship and circumstances are often lost and the physical artifact changed to the extent that the work takes on a very different meaning and tone. This was the case with Agrippa's fencing treatise, discussed in Chapter 3, that was reissued in 1568 with much reduced illustrations; the life of St. Benedict discussed in Chapter 2, whose images provided prototypes for many different versions of the saint's life over the course of the seventeenth century, but never for the same reason as the first book was made; and the phlebotomy treatise of Pietro Paolo Magni discussed in Chapter 4, which was reprinted in different formats and with updated illustrations for over 200 years. Only Magino di Gabrielli's treatise on silkworms (Chapter 5) was destined to appear in a single printing.

In the last quarter of the sixteenth century the formats of illustrated books, while already recognizably modern, still belonged in unexpected ways to antiquated systems that were on the verge of changing; the books therefore are deceptively modern. The cultures in which they were read were still powered by premodern systems, the origins of which are not immediately obvious to us since visible vestiges of these systems have been retained in the books we read today. We still dedicate our books, illustrate them, index them, footnote them, number the pages, and try to commit parts of them to memory. But the important cultural institutions from which all of these practices evolved were very much in flux at the end of the sixteenth century. They include practices of authorship, censorship, or dedication; practices of writing in a standardized vernacular or else in Latin. How to conceive of the relationship between image and text, as well as of the relationship between reading and memorization, or between reading and the retention of what has been read were issues that were open to different manners of resolution. Practices like alphabetizing, page numbering, footnoting – apparatuses that organize the information in a book and help us to access it without actually reading it – seem so natural that we have become newly aware of the establishment of the conventions of printed books through the need to re-examine the way we read and handle information in the organization and development of the Internet.[58]

It helps to look at the evolution of the meanings of the Latin word *illustrare* to understand the work of printed images in late sixteenth-century books, and their relationship to texts that may or may not appear with them. Coming from the Latin *illustro*, *illustrare* means to illuminate in the sense of "to shine a light on," to make something more visible.[59] It was related to the idea of clarification in terms of the reception – or better, the fabrication of a mental image – of a thing. This seems to be a sense of "illustrate" that belongs to the sixteenth century. The idea that the image in a book might be didactic in the sense of providing instructive examples, or might be decorative, appears to dominate in the seventeenth century.

In the late sixteenth century illustrated books often identified themselves as such in their titles or subtitles. But the word *illustrata*, even when used in the title of a book that happened to contain illustrations, usually referred to some other aspect of the book that made it illustrious: "accuratissima opera deuotissimaque expositione illustrata."[60] Leandro Alberti's illustrated description of all of Italy, printed for the first time with no illustrations in 1550 and reissued many times with the additions of tables and indices, finally appeared in 1568 in a version that was illustrated with maps: *Descrittione di tutta Italia di f. Leandro Alberti Bolognese, nella quale si contiene il sito di essa . . .; la qualita delle partisue; l'origine delle attà, de castelli, signorie loro . . . imonti, i laghi, i fiume, le fontane, i bagni, le miniere, gli uomi famosi, che di tempo in tempo l'hanno illustrata. Aggiuntaui la descrittione di tutte l'isole, all' Italia appartenenti, con i suoi disegni . . . Con le sue tauole copiosissime.* In Venetia; appresso Lodouico degli Avanzi, 1568. In this case the title announces an illustrated book that deals with those famous men who made the country more illustrious, and to which illustrations, called drawings (*disegni*), were not added to the description until thirty-eight years after it first appeared.

Giovanni Panteo's 1550 book on the transmutation of metals does announce that it is made *illustrata* by the addition of the pictures of things: "iconibus rei accommodis illustrata."[61] But many books, such as editions of Aristotle that were also illustrated, were *illustrata* with *scholijs*, or great learning; or an unillustrated book could be *illustrata* because of its many learned annotations, or a particularly good index, or interesting digressions, or just the quality of the authors whose words were included within.[62] A book wishing to proclaim in its title that it included pictures was likely to specify that the things making it *adornata* or *illustrata* were *figure* (or *figuris*).[63] More often than not, the figures were added in a new edition, and – like the addition of tables, indices, or commentary – were announced as an improvement and selling point in the title, such as the famous *Iconologia* by Cesare Ripa, reissued *adornata* with woodcut illustrations for the first time in 1603.[64]

This book, then, is an attempt to add to histories of the book a discussion of the work of images. It is written very much in the grain of the specificity of local production – concentrating on the ambitions of uncanonical authors who felt that the press was available for their use as well as for the noble elite – that has already been so productive for the history of reading as formulated over the last fifty years by Roger Chartier, Robert Darnton, Carlo Ginzburg, and Adrian Johns. Constructions of authorship by Chartier and by historians of science such as Mario Biagioli have shown how texts are shaped in the molds of specific forms of patronage, making it more reasonable than ever that a book about books could aim deep without leaving one neighborhood for very long. The work of Jon R. Snyder, Virginia Cox, and of course Mikhail Bakhtin, who have written with such great historical sensitivity of the spoken voice in literary texts, have made me wish to adorn the discussions of early modern dialogues with the addition of the dimension of printed pictures, already there to welcome the reader into a fully populated conversation, ushering us into a room where a brilliant party is in full swing.

Carnis resurrectionem ui
tam eternam amen.
✠Rux christi salua me.
✠crux christi deffen
de me. ✠crux christi pro
teoe me. ✠in nomine patris.
✠et filii. ✠et spiritus san
cti Amen.

Questa fu facta per noi ragazi
E anchor perquei che son bestial epazi

2

Pictures for the Ear of the Heart: The Life and Miracles of St. Benedict

The large, richly detailed engravings of the *Vita et miracula sanctissimi Patris Benedicti* were carefully crafted to cue a conversation, mediated by young Benedictine readers, with the select volumes of holy books purposefully stashed in their well-stocked minds (fig. 8).[1] The folio-sized collection of pictures that narrate the life of St. Benedict for initiates into the monastic life was based on the second book of the late sixth-century *Dialogues* of St. Gregory the Great.[2] It was first published in 1579 through the corporate authorship of the congregation of Spanish Benedictines in Rome under Gregory's name, and the images they requested were designed to attract the attention of the young readers to whom these large, densely figured pictures are primarily addressed. A hybrid mode of reading that is part monastic and part modern, which this book promotes, and the relationship of this book to others like it that encouraged a particular kind of reading with pictures, are the subject of this chapter.

Of the books treated here, this is the only one that is advertised, on its title page, as having been written collectively. However, this is not the only collaborative literary enterprise among the group, which otherwise all bear the name of a single author whether or not this proved to be the case. Noting their own work as a selection of stories from Gregory's dialogues, the Benedictines produced an edition of Gregory's foundational narration of the life of this important saint. As such, the author of this book is usually given as Pope Gregory I (*c.*560–604), who died nearly a thousand years before the Spanish Benedictines made this version of their founder's life into a luxuriously engraved picture book, and whose dialogues were usually issued in textual, not visual form.

The scenes selected to narrate Benedict's life and teaching would easily have confirmed Protestant belief that the Benedictines were worst offenders when it came to sensational religious narrative. Richly packed with attack dragons and miracle-working household objects, demented young ladies cavorting with pinwheels, and naked, sex-crazed young hermits rolling in thorn

8   Frontispiece, *Vita et miracula sanctissimi Patris Benedicti*, Rome, 1579, engraving,. Typ 525
79.674, Houghton Library, Harvard University.

bushes, the book also contained images from Benedictine and Roman history that privileged violent scenes of war and natural disasters. As such, it was a powerful tool for holding the attention of initiates into the order, and in teaching the monastic reading that was an important part of Benedictine practice. The use of specially crafted images to make a connection to the intellect through the sense of vision was a medieval technique of monastic study, and images associated with this technique usually broke down topics in need of memorization into easily digestible, discrete parts.[3] In charging the artist Bernardino Passeri (*c*.1540–1596) and the engraver Aliprando Capriolo (fl. 1575–1599) to create the important images, the members of the congregation of Spanish Benedictines sought out illustrations that used a unified field, single point perspective, and the other visual properties of High Renaissance and Counter-Reformation narrative painting. The book taught traditional processes of monastic reading, updated with contemporary imagery to reach a modern audience raised on pictorial styles not usually associated with meditative reading and the carefully cultivated techniques of *memoria*.

St. Benedict of Nursia (*c*.480–547) founded the Benedictine order, and over the course of fifty numbered, folio-sized pages the engravings tell the story of how he took leave of his noble family, turned his back on the university training expected of people of his social rank, and transcended the temptations of this world to pursue a life of withdrawal into silence and prayer. Although he began as a solitary hermit in a cave in Subiaco, his sanctity soon attracted followers eager to learn from him. Overcoming many obstacles with the help of other holy men, loyal animals, and the strength of his faith, he founded monasteries where he instituted a way of life that became codified in the *Regula Benedicti*, the monastic rule based on work and prayer to which Benedictine monasteries, among others, adhered. Although there were other religious founders and monastic rules in Benedict's time, the model of monasticism instituted at Montecassino, the abbey he is said to have founded in about 529 and where he lived until his death, is commonly considered to have shaped Western monastic life in the Middle Ages. Dante seated Benedict in heaven just below John the Baptist with the other famous medieval founders of religious orders, Francis and Augustine, and he is still often called the founder of Western monasticism.[4] It is not surprising that versions of the *Vita et miracula* appeared in several vernacular languages and Latin, and was issued in various editions and formats.

The title page credits the congregation with choosing – or as they put it, "collecting" – those notable events of Benedict's life that should be illustrated. The frontispiece identifies the order's protector as Cardinal Giacomo Savelli, but there is no indication of a publisher for the first edition of the book other than the congregation itself.[5] A line at the bottom of the title page, "Bern[ardi]nus Passarus inventor," identifies the artist of the fifty very painterly images as the Roman painter and print designer Bernardino Passeri, and rightly gives him authorial billing along with the Benedictines and St. Gregory.[6] His drawings were skillfully engraved by Aliprando Capriolo of Trent, who worked in the style of the popular Netherlandish engraver Cornelis Cort. Capriolo had arrived in Rome a few years earlier, probably under the protection of powerful fellow Tridentines Cardinal Cristoforo Madruzzo and his nephews, Cardinal Giovanni Ludovico and

Giovanni Federico Madruzzo.[7] Both cardinals were active in hosting the Council of Trent, and Giovanni Federico was known as a book collector and supporter of the arts. Passeri's signature and Capriolo's artisanal monogram appear at the bottom of several of the plates throughout the book.[8] Today the richly engraved series is most conspicuously an example of a highly developed Northern engraving style used in the service of some very Italian subject matter.

The *Dialogues* of Gregory the Great was the closest thing to a primary source that there was for a biography of St. Benedict. Gregory, the revered pope saint and father of the Roman church, claimed to have received his information directly from the abbots who succeeded Benedict at Montecassino and Subiaco, who were therefore eyewitnesses to the events he recounted and to the character of the man.[9] Whereas Book One of his *Dialogues* was made up of twelve chapters each devoted to a single holy man, Book Two was composed of thirty-eight chapters devoted entirely to the life of Benedict. It was the task of the Spanish congregation to break down the stories into discrete moments, much as they would any sacred text they wanted to commit to memory for further spiritual use. The duty of the artist they chose was to reassemble them in order to teach moralizing moments in the story of the life of their founder at the same time as it instructed novices in the best ways to read them.[10]

Although the monks considered Gregory the most reliable source for their project, the authorship and subject matter of the *Dialogues* of Gregory the Great had been called into question since the mid-sixteenth century. An early Protestant humanist, Huldreich Coccius (Koechlin) of Basel, was the first to publish doubts about Gregory's authorship of the *Dialogues* in 1564, on the basis of literary style.[11] The idea was taken further by other Protestants eager for examples of the Roman church concocting its origins and sacred history. They supported the belief that the sloppily written *Dialogues* promoting pagan adherence to miracles and magic was among those documents that best demonstrated the need for church reform. They found it effective to attribute the credulous work to Gregory anyway, since the collection of what they considered superstitious miracle tales provided proof, as Melancthon put it, that even a sainted early pope and father of the Roman church could be dismissed as the "dance-leader and torch-bearer of a theology going to ruin."[12] When the *Vita et miracula* appeared in 1579 this debate was still being carried on publicly in print.[13]

In spite of the doubts cast upon Gregory's authorship, the *Vita et miracula* links its pictorial engravings of Benedict's triumphs over disasters, dragons, and demons as closely as possible to Gregory's stories. Each plate bears a number citing the chapter of the *Dialogues* in which the story appears, while a sequential plate number at the foot of each illustration secures a consistent narrative order.

At the beginning of the *Dialogues* a young deacon named Peter comes upon the morose Gregory and, asking what the matter is, finds the older man regretting that in his own life he has failed to attain the perfection of the soul that characterized past men of God who were notable for their retirement from the world to a life of contemplation. Gregory asks the younger man to draw him out on the subject: "But that which I have nowe said will be far more plaine, and the better perceived, yf the residue of my speche be dialogue wise distinguished, by setting downe eache of our names, you askinge what you shall thincke convenient, & I by answere, giving satisfaction to such questio[n]s as you shall demande at my handes."[14] Gregory is more than willing to expound on the miracles attributed to the church fathers, and uses Peter's prompts as encouragement to proceed with further explanation.

The highly detailed, richly worked engravings of the *Vita et miracula* relate each of the stories in settings that were familiar to the Roman Benedictines, showing their particular brand of monastic life mandated by the Rule of St. Benedict that governed every aspect of the monks' daily existence. The Rule prescribed the two activities that made up the order's motto: "ora et labora," fervent inward prayer accompanied by hard physical labor. Images of hard work – most of them related to the physical construction of the many churches Benedict was said to have founded – are presented in the context of the moral lessons which make that work efficacious. Those lessons, hidden in the images and meant to be discovered by the trained reader, are the goal and true meaning of the deceptively simple narratives. The engraving on Plate 19 of the book is a good example of the willingness of the young monks to work and the imperfect state of their awareness of the true nature of the many-layered world they occupied. It shows a group of monks who, in the process of building a monastery, strain to move an intransigent rock until Benedict perceives and evicts the tiny but potent devil who had taken up residence on top of it (fig. 9). Benedict could perceive the devil, while the diligent monks, good at hard labor but not yet attuned to the evil lurking in the world around them, could not.

Benedict's special gifts in perceiving good and evil are a persistent theme throughout the book, extending to the more mundane and recognizably schoolmasterish tasks of disciplining the imperfect monks in his care, who are given to the recognizable transgressions of all young people. Plate 22 (fig. 10) shows Benedict catching two monks in a lie about their whereabouts that had involved disobedient eating and associating with a woman outside the monastery.[15] The structure of the image is typical of that of the rest of the pictures in the book. A large central scene shows the boys kneeling before Benedict in the role of Master, who lectures them while standing at the head of a closely packed line of more obedient monks. He holds the Rule in one hand and points with the other out through a window where the reader plainly sees, as the all-knowing Benedict could, the two monks enjoying a forbidden meal at a table outside a house in another city. Set behind the room in which the vision appears, in a space that would be architecturally difficult to produce but is pictorially convincing, the two boys appear setting the tables of their own plain monastic refectory, engaged in good work towards the preparation of a sanctioned communal supper. A guardian raven that had saved Benedict from eating poisoned bread in a previous scene pecks safely at his own licit meal on the refectory floor.

In the primary scene, the mendacious monks, kneeling in penitence at the foot of the saint, are astonished that he knew they had eaten forbidden food away from the monastery when he was not present. The image combines the important lesson (Benedict is always with the monks in spirit, and will ferret out lies) with a framed and vivid scene of a past transgression (framed

Aedificant cellas Monachi; Sathanas grauat uni
Saxo, ne tolli poßet. adauxit onus .
Agnouit tacito seßoris pondere preßum .   19

Sanctus, mox format sacra sigilla Crucis.
Mobile fit subitò, depulso Dœmone, Saxum
Fortiter et cœptum continuatur opus.   CII

9   Aliprando Capriolo after
Bernardino Passeri, Plate 19,
*Vita et miracula sanctissimi Patris
Benedicti*, Rome, 1579,
engraving. Typ 525 79.674,
Houghton Library, Harvard
University.

as a window) and the future penance and decorous result of the Master's discipline (setting the tables), while seeming at first glance to be a continuous image of the benefits of communal life in any well-run Benedictine monastery. In a rare scene of monks talking, an activity that was particularly discouraged to make room for hearing God and to avoid just such mistakes as the telling of lies, two of the tonsured monks in line speak among themselves, choric figures discussing the lesson the reader is learning while reading and viewing.

Besides praying and working, reading, or *lectio divina*, was a third important activity for which hours were set aside each day.[16] *Lectio divina* was a difficult and active form of reading specific to monastic orders. It emphasized memorizing important texts, and involved a four-step process of

Ingreſsi Monachi matronæ tecta pudicæ
Sumunt, clauſtrali lege uacante dapes .
Ut rediere foris se pransos esse negarunt . 22

Mendaces blando corripit ore Pater .
Qui locus, atq; cibi fuerint, quot pocula, dicit,
Reprenſi ueniam supplice uoce rogant . C. 14

Bernardinus jusserit mit

10   Aliprando Capriolo after
Bernardino Passeri, Plate 22,
*Vita et miracula sanctissimi Patris
Benedicti*, Rome, 1579,
engraving. Typ 525 79.674,
Houghton Library, Harvard
University.

reading, meditating, heartfelt prayer, and interpretive contemplation.[17] These carefully constructed images, in which books appear everywhere, were charged with teaching the methods of *lectio divina* to Benedictine initiates in the form of the highly imitable life of the Master. They provided an opportunity for the young monks to exercise their memories and scour their own souls for traces of the transgressions to which they were especially vulnerable.

In the role of corporate author and pictorial mouthpiece for Gregory the Great, the Spanish Benedictines took every opportunity to bring the reader to an active awareness of, and therefore into control of, an inner life experienced simultaneously with the highly regulated exterior life communal to all the monks. The images conjoined the known exterior world and the developing interior world that monastic life was supposed to inculcate in each of the cenobites. They accomplished this through detailed pictorial description, legible and meaningful use of chiaroscuro, and tightly constructed backdrops, often architectural, on which the narratives played out. Aside from the terse, six-line encapsulation of the relevant part of the *Dialogues* separated from the image by two thinly ruled lines at the bottom of each page, the printed book itself provides almost no text. There is a tiny footnote to the chapter number from the *Dialogues* at the bottom right, and aside from this there are the numbers that keep track of the plate's order in the volume. There is no introduction or dedication in the first edition of the book; in fact, there is no further text visible at all. In part, this is because the interior life cultivated in the combination of reading and prayer revolved around books without always necessitating a book in hand.

Though the use of rigorous perspective and the meaningful play of shadow and light make each picture seem to show a complete, unified narrative scene, the educated reader would encounter each page already equipped with his own interior library with which to roam purposefully in the images and discover the fuller meaning of each exemplar. The monks heard scripture read out loud each day as they ate their meals. Il Sodoma (1477–1549) showed this in his frescoes of the life of St. Benedict in the cloister of Monte Oliveto in Siena, painted about seventy-five years earlier but famous among cycles of Benedictine imagery (figs. 11 and 12).[18] In this way, eating and reading both became acts of digestion, one physical, the other spiritual. The monks were also encouraged to read and pray out loud, sonorization of the texts converting them to more easily memorizable form. This activity was called *meditatio*, and was the primary tool for accumulating a usable library of texts in mental reserve.[19]

Lectio divina is modeled for the reader in an image of the noble Benedict after he first fled the corrupt life at the university in Rome to withdraw into a cave in Subiaco, an adolescent youth bent on teaching himself a more devout form of study by way of seclusion with good books. Plate 5 of the *Vita* (fig. 13) visualizes this first retreat with an image of Benedict as a wavy-haired teenager in the company of several books and a simple crucifix, emphasized through its doubling in the projection of its shadow on the cave wall. A devil mocks the youth's studiousness and, since he cannot distract him, tries to starve him by throwing a rock to break the bell that calls him away from prayer to accept food brought to him once a day: the *Dialogues* says that the devil "envied the charity of one and the reflection of the other."[20] The point of the story is the saint's rapt absorption in his books as he wrestles with steps one and two of *lectio divina*: reading with a state of focused alertness and tense attention. Mary Carruthers compares this with that of a fisherman watching out for a fish: "Monastic reading is reading undertaken with 'intention' and *sollicitudo* – directed, engaged feeling that is regarded as the foundation of all good meditation."[21] Ideally, this engagement with a text should be accompanied by heartfelt prayer, and even by tears. Withdrawal from distraction, and the ability to read with attention and full corporal identification, would be fundamental to all further meditation. It was step one.

The emphasis on reading and prayer was not essential to the most straightforward version of the narrative of Benedict's *Life*, but was instead an anomaly of this printed book, made to

11   Luca Signorelli and Il Sodoma (Giovanni Antonio Bazzi), *Scenes from the Life of St. Benedict, c.*1505, fresco, Chiostro Grande arcade, Monte Oliveto Maggiore.

teach monastic reading as much as to convey a particular text. The same scene frescoed in Monte Oliveto abbey (fig. 14), meant to be encountered as the monk walks past it in meditative circumambulation of the cloister, shows the saint as a young man instead of as a teenager, praying with no visible books. The emphasis in the *Vita* on the role of absorptive reading in the construction of a model monk is absent in the fresco, and the image emphasizes withdrawal instead.

LEFT 12   Il Sodoma, *Benedict Served Poisoned Bread*, *c.*1505, fresco, Chiostro Grande arcade, Monte Oliveto Maggiore.

OPPOSITE PAGE   13   Aliprando Capriolo after Bernardino Passeri, Plate 5, *Vita et miracula sanctissimi Patris Benedicti*, Rome, 1579, engraving. Typ 525 79.674, Houghton Library, Harvard University.

Working through the *Vita*, the reader develops – and is then given the opportunity to exercise – his ability to move around in the engraved images and through the other texts that he has memorized, particularly the fascicules of scripture mentioned in the *Regula*, the *Regula* itself, the *Dialogues* of Gregory the Great, and stories of desert hermits and holy men from sources like the medieval *Historia monachorum*. The most usual method for guiding the reader of the *Vita* is the use of a picture within a picture that extends the narrative across time and place, as we saw in Plate 22 (fig. 10).[22]

Some of the fictive windows provide a screen for projections of Benedict's prophetic visions. Plate 24 (fig. 15) explicitly demonstrates this prophetic power in military terms, as the Ostrogoth king Totila sends his squire, dressed in his clothes, to meet the saint and see if Benedict can divine that the man is an imposter. He passes this test instantly, and then wins over the king in the next plate. Plate 26 (fig. 16) shows Benedict describing to the bishop of Canusium how it is not actually the barbarian Totila who will destroy Rome, but the forces of nature in the form of tempests, earthquakes, and fire that will bring down the "walls, baths, temples, theaters, and homes" of the

Obscuro sedem figit Benedictus in antro,   Allati poßent ne dare signa cibi.
 Romano modicos suppeditante cibos.   Nec tamen hoc pietatis opus perfringitur, aptis
Hoc grauiter Sathanas fert, tintinabula frangit,   Panes hic horis porrigit, ille capit.

city.[23] A turbaned groom paired with the bishop's mule in the shadowy background is echoed by a well-lit monk in the foreground, who, in placing a copy of the Rule next to a string of rosary beads on a table in front of the saint, also places it squarely before the reader. Benedict, who has been interrupted in his reading, holds the place in his book with his index finger as he turns in his chair to face the bishop and point to his vision of the destruction of Rome. The convincing eloquence of his prophecy is fully realized outside the window as a torrential deluge issuing from black clouds onto the city, depicted in detail with the Colosseum, Pantheon, and two tall victory columns.

Other images use the device of a picture within a picture to encourage the reader to cast back and to the future in the same scene, much as we saw in the image of the monks who lied about their meal. After attaining a reputation as a holy man at his cave, Benedict is asked to preside at a monastery that had no abbot (Plate 9, fig. 17). Reluctantly, he leaves the wilderness to take up that position, warning the monks of a future event: they will not be happy with his reforms. The primary scene in the foreground proves him right, showing how at dinner one

OPPOSITE PAGE 14　Il Sodoma, *The Devil Intercepts Benedict Receiving his Meal*, *c.*1505, fresco, Chiostro Grande arcade, Monte Oliveto Maggiore.

RIGHT 15　Aliprando Capriolo after Bernardino Passeri, Plate 24, *Vita et miracula sanctissimi Patris Benedicti*, Rome, 1579, engraving. Typ 525 79.674, Houghton Library, Harvard University.

night a jealous monk served the saint a glass of poisoned wine, but due to the power of the saint's salvific blessing the glass miraculously shattered before he drank from it. In the *Dialogues*, Benedict returns to the wilderness without rancor, gathers disciples around him, and goes out to found his own order. Pictorially, that scene functions seamlessly as a framed painting in the dining room above the saint's head. In terms of narrative, the smaller image does double duty: it portrays Benedict's first foray from the cave to the bad monastery, and also, for those who know the story and linger with the picture a bit longer, it extends the story to the second departure from the cave, the one that will result in the formation of the very monastic experience that the reader himself is now experiencing, governed by what became the Rule of St. Benedict.

The reader, calling on memory to fill in the parts of the story that the picture elides, oscillates between the framed image as past and future. Both concepts pivot on the largest, foreground scene of the proof of the saint's holiness in the miraculous effect of his blessing. The attentive reader looks for these pictorial signs as cues to active engagement with the pictures based on

A duentu Totilæ populadam dum gemit Vrbem
Præsul flens canus, maxima Roma, tuos.
Sanctus ait Non barbarico delebitur ense;

Tempore sed multo fulminibusq, cadet.
Quæ cecinit uates, quis non ita diruta cernit,
Mœnia, cū thermis, templa, theatra, domos;

26

C-18

16   Aliprando Capriolo after Bernardino Passeri, Plate 26, *Vita et miracula sanctissimi Patris Benedicti*, Rome, 1579, engraving. Typ 525 79.674, Houghton Library, Harvard University.

recollection of a familiar story suggested in the caption at the bottom. Because Benedict's life imitates the life of Christ, aspects of the story will also refer to scripture. The image calls on the viewer to work through his memorized texts as he moves from picture to picture within each scene before indulging in the pleasure of experiencing the image as a unified narrative, enriched by the meaning that binds the scenes together.

This mode of illustration is used throughout the book, and was also used by Cosimo Rosselli (1439–1507) in the Sistine Chapel cycle that would have been the most well-known source for this sort of pictorial forecasting in Rome. In the last fresco of the wall series, the *Last Supper* is shown as the moment of the institution of the Eucharist, and also as the moment when Christ foretells his betrayal and death to the horrified apostles while a tiny black demon whispers his instructions in Judas's ear. Above the apostles' heads the present and the future are brought

17   Aliprando Capriolo after
Bernardino Passeri, Plate 9,
*Vita et miracula sanctissimi Patris
Benedicti*, Rome, 1579,
engraving. Typ 525 79.674,
Houghton Library,
Harvard University.

together in a series of three scenes showing the temporally consecutive events that were the
results of the devil's ability to influence Judas and the strength of Christ's faith (fig. 18). We see
the prayer in the garden at Gethsemane, the betrayal of Christ, and the Crucifixion in a continuous landscape outside the window above Christ's head, events that he prophesizes but have
yet to take place.

Rosselli's use of a picture within a picture was a very different visual enterprise from that
undertaken a little later by the Spanish Jesuit Jerome Nadal when he worked up a pictorial
meditation on the Sunday gospel stories to teach the Jesuit technique of devotion minutely
described in Ignatius Loyola's Spiritual Exercises.[24] The Spiritual Exercises, known to Ignatius's
followers by 1533, were composed in final form by 1541. Charges that the Spiritual Exercises were
actually plagiarized from Benedictine sources were active in the seventeenth century, and addressed

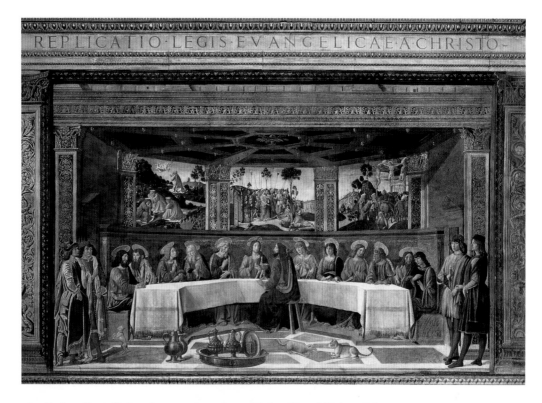

18   Cosimo Rosselli, *Last Supper*, 1481–2, fresco, Sistine Chapel, Vatican Palace.

and rejected then by the Benedictines at a general assembly of the congregation of Monte Cassino. The main difference that the Jesuits cite between the Spiritual Exercises and other forms of mental prayer is that their form leaves nothing at all to chance, and is practiced according to the degree of each individual's commitment and the state of his soul. The images for Nadal's book, made by Passeri and others after the *Vita et miracula* was printed, make that clear. They bristled with annotations throughout the pictures and included an identifying key in the captions that corresponded with letters posted at points in the image. Sometimes the images contained little roundels, as in the reading from John 16:5–14 for the fourth Sunday after Easter, when Christ tells the apostles that he is going away and the Holy Ghost will come and reprove the world (fig. 19). The roundels in the images of the Spiritual Exercises that make little emblems of Christ's words to the apostles are visually disassociated from the narrative structure of the scene, appearing as self-contained, secondary emanations from Christ above his resplendent halo. The images form a trio of precious and arcane medallions containing shorthand versions of Jesus's vision of salvation, ready-made memory images. Like charms or amulets, they evoke the presence of the Holy Spirit that the legend below helps us to understand will stave off ever-present evil in the form of a devil, visibly lurking outside the window in a roiling black cloud. These illustrations were followed by a detailed textual commentary that structured the reader's recall of scripture down to the last imagined pebble and bush.

The main novelty of Nadal's book lay in the fact that in its engagement with the reader it indeed left nothing to chance. This made it useful for teaching within the community of Jesuits, as well as for spreading the proper practice of the Spiritual Exercises into the New World and the Indies in a way that kept as closely as possible to the letter of Ignatius Loyola's method of prayerful meditation. The Spiritual Exercises are considered to have originated during the beginning of Ignatius's holy life in Catalonia where he stayed in 1522, praying and meditating under the direction of a Benedictine confessor, and where he was probably introduced to the *Imitatio Christi* and Benedictine spiritual exercises published in 1500 as the *Exercitatorio de la vida espiritual.*[25] Pictorial control through the use of annotations and captions was aimed at maintaining maximum attentiveness on the part of the reader, who moved in a linear – or at least alphabetized – way from one point to the next throughout the picture with the aid of a spiritual advisor, himself aided by the directions in the annotations. This not only kept the mind from wandering away from the pictures – which had to be highly detailed to aid the reader in imagining himself inhabiting the precise space in which the biblical stories played out – but also restrained the mind from wandering too much into aspects of the pictures that might make them more enjoyable as art than as a means to the proper conduct of prayer. To that end, "It was crucial," wrote Diego Jiménez in the *proemio* (preface) to the book, "to recruit the very best artists."[26] The drawings made by Passeri and Martin de Vos were not finished before 1587, and were then engraved by the Flemish Wierix brothers, who must have been chosen for their notable fineness of line and clarity of detail. To the Spanish Jesuits in Rome, Passeri (who made over a hundred of the designs) fitted the bill as one of the best artists, yet – as the Jesuits had reason to point out when the originality of the Spiritual Exercises was called into question – Benedictines used imagery differently in training young cenobites in their own monasteries in the practice of attentive reading, or *lectio divina*, from how the Jesuits did in missionary practice.

The active reading advocated by the Benedictines, then, was a matter not of rote memorization, but of memorization in the service of knowledgeable interpretation. The monks learned how to negotiate the pages of the *Vita* in a way that required reading, meditation on text and image, and the application of judgment to aid in weighing each aspect of the printed page. By the time a reader was ready to turn the page, the pictures had ceased to function as simple narratives, and instead had become cues for discourses about prayer, faith, obedience, and the soul. In the *Regula Benedicti* the term *lectio divina* is used for the reading of holy texts whether alone or out loud in a group. Although the Benedictine Rule frowns on unnecessary speaking, it is difficult to imagine that reading words and pictures did not sometimes take place communally and informally along with focused conversation. It recommends that the monks gather as a group (especially at night) and that "someone should read the *Conferences* or the *Lives of the Fathers*, or at least something else that will edify the hearers."[27]

The reader of the *Vita* was kept busy enough navigating the many ingenious pictorial devices by which Passeri and Capriolo depicted important actions in the text, such as dreams, visions, and intentions that would seem difficult to illustrate in an unambiguous manner. While these aspects of the tales seem less dramatic than the appearance of monsters, women, and demons, their clear depiction in black and white helps an active reader, especially one learning the methods of monastic reading, to perceive the powers of the wise abbot in the realms of imagination and memory. Plate 10 of the *Vita* (fig. 20) shows Benedict receiving the children of Roman nobles outside his monastery and at that moment envisioning the establishment of eleven more monasteries.

The strength of the saint's intention and clarity of his vision are revealed to us in the depiction of a hilly landscape dotted with exactly eleven monasteries visible through an architecturally improbable but pictorially convincing lunette above the doorway in which he stands.[28] Though reaching out to welcome the boys Maurus and Placidus, Benedict focuses inward with an upturned gaze that directs us to the image of the monasteries above, which we simultaneously understand as Benedict's idea or plan, and proleptically as a result of his past fame and virtue that

Ber. *nus paſſerus inuen*

Ad teſqua interea plures trahit æmula virtus,   Præficiens paucos aggregat ipſe ſibi;
Bis ſex clauſtra quib⁹ conſtruit ampla Pater. Eutitius Maurum, Placidum Tertullus eisdem
Biſſenos illis vitæ morumꝗ magiſtros   IO   Addunt, Romana nobilitate ſatos   C.4.

will materialize his idea in future time. Once again, the reader is presented with a single image that can be visually negotiated to show the saint's will, faith, and character within an otherwise canonical-seeming pictorial narration of his life that rarely veers from the linear and factual, staying very close to the textual source.[29]

The central figure of the saint, columnar in his white habit and towering over the bowing children and their eager parents, is framed by the stone doorway flanked by shadowy cenobites who peer over his shoulders with pious approval at the new arrivals. In the orderly, unified image of simple monks welcoming fussily dressed nobles before a meticulously rendered stone building, we are nudged towards recognition of the full significance of an event (Benedict's fame spreads and worthy citizens consign him their well-bred children) through a past accomplishment (the building of a magnificent monastery and the installation of cenobitic life) and a future result (the multiplication of monasteries and spread of the order). Images such as this one, in which the past is used to extrapolate a plausible future through attention to the present, are evidence for the muscular use of *memoria* "for thinking, for inventing, for making a composition in the present that is directed towards our future."[30] Although Carruthers here describes a medieval system of reading, we see it cued in late sixteenth-century imagery, evidence that at least in the monasteries printing had not changed the methods of meditational reading advocated by medieval monastic readers.

The caption beneath the image relates the event from the *Dialogues* as straightforwardly as possible, allowing a teacher reading the Latin to help a less literate viewer expound on the image, and to appreciate Benedict's capacity for understanding and remembering the future in the present: "Famous, Benedict gathers numerous followers. He founds twelve monasteries and in each places twelve monks and one abbot to teach them. The Roman nobility entrust their children to him: Eustichius brings Mauro, Tertullus brings Placidus."[31] While the six-line verses of the *Vita et miracula* rarely moralize, keeping to a condensed, event-driven declaration of the stories, the structure of the images demands that the reader search each of them until sense is made of the different scenes, and they are related to the large central scene with the help of the text. At this point, focused wandering may stop, and the page that at first appeared unified, then fragmented, is comprehended as a complex structure of separate parts that also includes the text. Here, interpretive or moralizing work is done in the process of the reader's discovery rather than through authorial narration or pictorial revelation. What is said in text in a seemingly direct way and what is pictured in an indirect and at first mysterious way are constantly in tension, and can be interpreted as an example of the "other speaking" (*alieniloquium*), in which what is said is other than what is signified, which constituted the basis for Hugh of St. Victor's definition of allegory.[32]

The authorial team of the printed volume – Passeri, Capriolo, and the Spanish Benedictines – reinforce the reader's progress through this and other texts guided by pictorially well-integrated symbols and easily recognizable situations in which the reader is repeatedly stimulated to recognize familiar signs in practical circumstances. A good example is Plate 33 (fig. 21), in which Benedict appears to sleeping monks in a collective dream and gives them minute instructions for building another monastery. The monks, who had been idly waiting for him to come and tell them what to do, wake up the next morning, tell each other their dream while waiting for the saint to come, and then return home thinking they have been stood up. Benedict admonishes them for not having faith; they did not understand that their dream was in fact an effect of his own agency on their imaginations. In the *Dialogues,* young Peter tells Gregory the Great that he needs a better explanation. Gregory explains, with some exasperation, that obviously the spirit has a more mobile nature than the body, and if scripture tells us that the prophet Habakuk, in Israel, could be transported to Babylon along with an entire meal to serve to Daniel in the lion's den over miles and miles of terrain, then why would anyone doubt that Benedict's spirit could appear a few cities away to a group of sleeping monks?

The sieve leaning against a little bucket, bathed in light emanating from the lamp in front of the crucifix on the wall above, is provided in part as testimony to the very strength of Benedict's spirit that made it able to be broadcast to the sleeping monks over a great distance. A sieve was the vector of the saint's first miracle: having been borrowed by his beloved nursemaid, it accidentally broke, causing her incredible anguish. It was made whole again after Benedict fervently prayed for it to be fixed, illustrated in Plate 3 of the book. The hand tools decoratively piled on the floor in the foreground are the instruments of the monks' own good works with humble objects, and along with the sieve and the crucifix are emblematic of their motto, *ora et labora.* The tools, the sieve figured here along with the familiar cots, and the shoes and robes that were the monks' only allowable possessions, animate the humble scene for the book's audience. They render the saint's blueprint for construction, pictured through the window above the collectively dreaming heads of the sleeping laborers, as tangible to the reading monks at home as were the robes, cots, and tools.

The engravings, then, aimed to engage the reading monks in a process of recognition and interpretation of multiple scenes, objects, and texts, as well as clarifying their roles in the physical and spiritual construction of monastic life. Evocative depiction enabled a reader to perceive and emulate the powers of the wise Benedict in the realms of both imagination and memory. This was a primary goal of the readers of this picture book, who were, in reading it, being trained in the methods of *lectio divina* through the use of images that contained as little text as possible. In this way, engagement with the pictures activated use of the texts they had already committed to memory. Less interested in evangelizing foreign lands than in training young minds at home, the Benedictines' book traded on the fact of a homogenous reading audience that approached the pictures with similar preparation, and probably under guidance of a presiding abbot. Key to this training was instilling in each boy an acute awareness of the operations of the reading self. This lesson, too, was given an image and verse.

The *Dialogues* narrates an event in which Benedict is informed of a monk who persisted in wandering during prayers, even after he was warned not to by the abbot of the monastery where he lived. In the *Dialogues* Benedict then visited the monastery, and after the reading of the Psalms ended and the monks "betook themselves to prayer, he perceived a little black boy who pulled

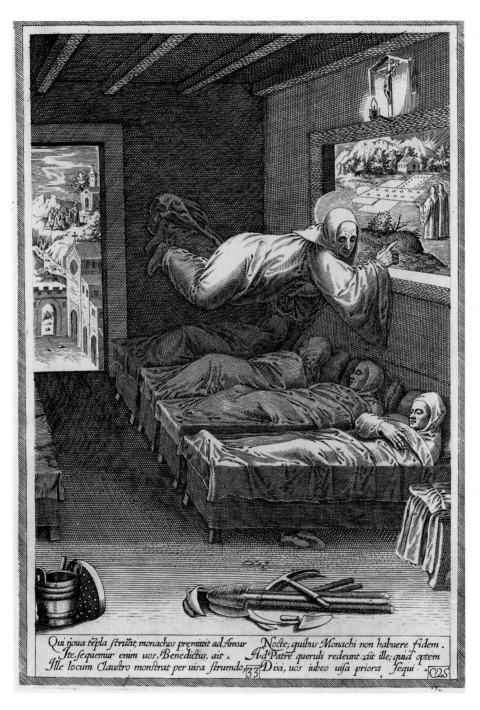

Qui ṇoua tēpla ſtruāt, monachos preṃittit ad Arar   Nocte; quibus Monachi non habuere fidem.
Ite, ſequemur enim uos, Benedictus. ait .   Ad Patrē queruli redeunt ait ille; quid optem
Ille locum Clauſtro monſtrat per uiſa ſtruendo.   Dixi, uos iubeo uiſa priora ſequi .

21   Aliprando Capriolo after Bernardino Passeri, Plate 33, *Vita et miracula sanctissimi Patris Benedicti*, Rome, 1579, engraving. Typ 525 79.674, Houghton Library, Harvard University.

this Monke out by the hemme of his garment."[33] He mentioned the apparition to the abbot, and to his special disciple Maurus. After two days of intense prayer, Maurus was able to see the black demon, while the abbot still could not.

The *Dialogues* continues: Benedict then struck the boy "with a wane for his obstinacy and boulddenes of heart, and from that time ever after the Monke was free from the wicked sugges-tion of the black boy, and remained constant at his prayers. For the wicked fiend as if himself had bin beaten, durst no more tempt him to the like offence." The caption to Plate 11 (fig. 22) explains the scene this way: "A monk warned by the father's voice that he should not leave the temple at the hour of prayer / Nonetheless left it; / The father saw him being deluded and dragged out by the Ethiopian specter, / Which he also let Maurus see, / He lashed the monk with a scourge / Who was never afterwards bored with praying as he was before."[34] It tells the reader that Benedict allowed Maurus, who had mastered the art of heartfelt prayer, the privilege of seeing the Ethiopian specter, and Passeri and Capriolo allow us (or rather, the monk-reader) to see him, too.

This beating administered by the father, which cured the boy of wandering from prayers, is the primary subject of the event as it is illustrated in the *Vita*. The attentive reader would remem-ber that the *Dialogues* describes the devil leading the monk away by the hem of his robe, a detail not mentioned in the caption but shown in the picture. Here the devil, tugging at the monk's robe, dislodges its hem into the caption. The tiny corner of hem, trespassing into the space reserved for text, points like an arrow at the words that narrate the act of wandering from the oratory during prayers. Although the piece of cloth overlaps the boundary of the image by just a fraction of an inch, the engraver emphasizes it with a shadow that makes it the agent of a sharp visual pun on the monk's, and the mind's, propensities to wander from their proper place. The reading monk, perhaps leaving the realm of attentive *memoria* for a moment in order to relish the idea of one of his mates being caught out and punished, will be snapped back to attention as the reading self and the punished monk are momentarily brought into alignment.

Benedict dedicates a whole chapter of the *Regula* to a description of the monk's habit, which is his primary article of clothing and a symbolic garment marking his separation from the world and immersion within a smaller community. Plate 4 (fig. 23) shows young Benedict receiving the habit, illustrating his decision to abandon the world and assume the life of a penitential hermit in the cave at Subiaco.[35] In Benedictine life, the robe of anonymity symbolizes the soul of the monk in a religious community.[36] As the lowest part of the monk's garment, the hem represents the basest, or most dishonorable, part of him. Also, because all holy lives are lived in imitation of the life of Christ, the reader will remember that the hem of Christ's robe held the power to heal, as described in the Gospels (Matthew 9:20): "And behold a woman who was troubled with an issue of blood twelve years, came behind him, and touched the hem of his garment." The cenobite would recall this bit of scripture repeated often through his own vocalized reading as well as echoed in the voices of his fellow monks. In Gregory's narrative, it is the base hem of the monk's robe that becomes available to the nature of the Ethiopian specter, Ethiopians being understood,

Orandi, monachus, ne templa relinqueret hora,
Voce Patris monitus, deserit illa tamen.
Vidit ab æthiopis spectro ludiq; trahiq; **11**

Hunc Pater, et Mauro posse uidere dedit.
Quem simul ut virgæ percussit verbere, nunqua
Orandi passus tædia prima fuit. **C·S·**

Ber.^{nus} passerus inuen.

22   Aliprando Capriolo after Bernardino Passeri, Plate 11, *Vita et miracula sanctissimi Patris Benedicti*, Rome, 1579, engraving. Typ 525 79.674, Houghton Library, Harvard University.

23 Aliprando Capriolo after Bernardino Passeri, Plate 4, *Vita et miracula sanctissimi Patris Benedicti*, Rome, 1579, engraving. Typ 525 79.674, Houghton Library, Harvard University.

24 Aliprando Capriolo after Bernardino Passeri, Plate 8, *Vita et miracula sanctissimi Patris Benedicti*, Rome, 1579, engraving. Typ 525 79.674, Houghton Library, Harvard University.

in the medieval legends of saint's lives, as potent tempters into sins of a sexual nature and also of laziness, since they were known to tempt improperly prepared ascetics to give up hard work of the type that was so fundamental to the Benedictines.[37]

Benedict had overcome the propensity to allow his own imagination to dwell with the devil by practicing mortification of the flesh early in his own religious formation, while still a young man at the cave in Subiaco. In a startling image of a suddenly mature and god-like Benedict that appears early on in the book, we see how a demon in the form of a blackbird tormented him and distracted him from his prayers with persistent arousing visions of a woman's body (Plate 8, fig. 24). He was able to cure himself permanently by removing his habit and rolling naked in a thorn bush until his skin was painfully lacerated. This image is accompanied by a rare moment of textual moralizing in the caption: the wounds he received cured his soul while piercing his flesh.

External exercises such as the one pictured here were efficacious for several reasons. The Aristotelian body, in which soul and biological organism were one and the same, was particularly vulnerable to affecting the heart through the arousal of the senses of the flesh. Through the voluntary regulation of the senses' perceptions, the caption implies, the body's appetites are rendered ineffective. According to Aristotle, the Sensitive Soul is made of motive and perceptual faculties. Perceptual faculties are divided between internal senses (the perception of absent sense objects, as achieved through cogitation, memory, fantasy, imagination, and common sense) and external senses (the perception of present sense objects through the senses of vision, hearing, smell, taste, and touch). Motive faculties are appetitive, which produce emotions, and progressive, which produce physical movements. Mortification of the flesh substitutes disciplined perceptual faculties for appetitive ones.[38] This image substitutes the agitation of the persistent visions, in the form of the fluttering blackbirds acting on the perceptual systems embodied in the heroic image of the nude Benedict, with the pleasures afforded by the merciful thorn bush. Chronologically, Benedict appears first as a tiny, clothed ascetic praying outside his cave in the background, tormented by the blackbirds. In the foreground we see an athletic, naked Benedict decoratively caressed rather than gruesomely wounded by sinuous branches. In this image Benedict appears more carnal and embodied than anywhere else in the book. The lower third of the plate, in which the saint submits to the curative powers of the thorn bush, shows him in an unmistakably sexualized pose. This seems difficult to reconcile with the work of an artist in control of his graphic rhetoric to the degree that Passeri was, and it is hard to imagine how an oversight committee as involved as the Spanish Benedictines would tolerate such an obvious visual surplus of meaning.

Chapter 2 of the *Dialogues* recounts Benedict's suffering from "the remembrance of a woman which sometime he had seene" that was "so lively represented to his fancy by the wicked spirit, and so vehemently did her image inflame his breast with lustful desires, that almost overcome with pleasure, he determined to leave the wilderness." Luckily, he was assisted by divine grace, as the *Dialogues* continue:

> Seeing near him a thicket full of nettles and briars, he threw off his garments and cast himself naked in to the midst of them, there wallowing and rooling himself in those sharpest thorns and nettles, so that when he rose up, his body was all pitifully rent and torn. Thus, by the wounds of his flesh, he cured those of his soul by turning pleasure into pain, and by the vehemence of outward torments, he extinguished the unlawful flame which burnt within him, overcoming sinne by changing the fire.[39]

The passage recounts an important formative episode in the life of the founder, and also recalls a seminal scriptural text, God's angry words to Adam and Eve in Genesis (3:17–18) after they became aware of themselves as naked beings: "cursed is the earth in thy work; with labour and toil shalt thou eat thereof all the days of they life. Thorns and thistles shall it bring forth to thee, and thou eat the herbs of the earth." The connection with carnal sin is made pictorially, and through the matching of texts that the reader expects to do, the landscape around the cave in Subiaco is elided with that of the early desert fathers who became hermits (those whose meditative strength Gregory the Great explicitly says he envies at the beginning of the *Dialogues*) and the Garden of Eden at the moment of the Fall.

In not showing the lacerations of the flesh, the image instead presents the reader with a cured and whole body, framed by the delicately arching branches of the accommodating thorn bush. Rather than rolling in pain, the saint seems to embrace his punishment as he might have the woman, whose troubling memory seems to linger still in the twining brambles. The surplus of visual information in the body of the saint, who never in any picture before or after looks as virile and god-like, connects the image with the young reader, as does the straying hem. Benedict's Herculean powers of prophecy and imagination are depicted in the ideal body of a Greek statue, associated now with self-restraint and control. The rocky grotto with its thorns and nettles opens invitingly to the reader just beyond the caption, offered to Benedict's followers as a healing promise to cure troubling emotions from lovesickness to hunger to homesickness, torments associated with wandering attention and all the subject of illustrations in the book.

The work of the thorns is explained in one of the few instances of interpretive commentary to appear in the captions. The story of Benedict's coming of age in the scene with the thorn bush was meant to be instructive and provided the reader with several important teachable moments in the *Dialogues*. After that episode, having proved his ability to withstand temptations to vice, Benedict was now fully prepared for the rigors of a life of solitude, and also qualified to lead others in that practice. At this point in the *Dialogues* the young Peter says that he has understood a great deal, but "I pray make it more plain unto me." Gregory spells it out: "In youth, the temptations of the flesh are greate, but after fifty natural heat waxeth cold. Now the souls of good men are the holy vessels, and therefore while the elect are in temptation it is necessary that they live under obedience and be wearied with labors, but when by reason of their age the fervor of temptation is assuaged, they . . . may become instructors of souls."[40]

The lives of the early Christian hermits who had retired to caves in the Egyptian desert, seeking a spiritually pure and healthy existence, were recounted in a fourth-century collection by Rufinus of Aquileia known as the *Historia monachorum*.[41] These exemplary lives were among the texts that the monks would hear read to them. Benedict also advised cenobites to read the *Conferences* of John Cassian, an ascetic who lived for years in the Egyptian desert. Cassian's stories included those about monks tortured by lustful thoughts inflicted on them by demons. The goal of the monk's self-discipline, learning to negotiate the warring influences of his demonic and angelic natures, was to feel nothing in the face of these thoughts, and to avoid all desire and therefore pleasure as well.[42] The tales of desert hermits recount examples of those who were improperly prepared for advanced solitary meditation, whose wandering attention at prayer allowed a vacancy for the Ethiopian demon to enter. The "Life of Antony" is the earliest of these *Vitae* in which the devil is presented as a black Ethiopian boy, and associated with temptations of the flesh and memories of the solace of home and family.[43]

Aspiring holy men of the desert were not allowed to retire alone to meditation and prayer until they were sufficiently prepared to discipline their minds and bodies, and to reject the temptations that were an unavoidable part of a life of solitude and inward focus. In both the monastic tradition of the desert fathers and in the *Regula* the body at leisure was always available to sin, which took it over sometimes as a visitation from outside, and other times as a natural state that had to be kept at bay by constant occupation such as manual labor or *lectio*.[44] Such preparation is difficult to teach, because even in confessing to such temptations or discussing them they are made present and real, as is the case in this image. Evoking sexual transgression is itself both a temptation and a pleasure, but speaking of temptation can also allow the self to overcome its effects.[45] The *Vita's* image of the cure of the soul seems to make this clear in its triple manifestation of the far-off robed and tormented saint, an image that is visually rather underplayed; the powerful saint in the middle-ground in the process of disrobing, evoking both the end of the story in the thorn bush and also the transitional moment when giving in to temptation and giving up his calling was still a consideration; and, in the foreground, at the foot of the picture and showing a prone, half-obscured body directly above the reader's caption, the cured saint. The picture emphasizes the moment of multiple possibilities, and the moment of choice.

In Plate 11 (fig. 22) the wandering monk, led away from prayer in the circular oratory by the little black boy, is cured of wandering thoughts, and is called back to self-discipline through discipline administered by the father. For Benedict, the monastic oratory, like his cave, was a place of withdrawal accompanied by heartfelt, attentive prayer that should be inward-looking and marked by "purity of heart and tears of compunction"; this required trained, focused attention.[46] Unlike Benedict in Subiaco, the wandering monk was still in need of the communal education the monastery provided, with its structured living that included prayers and reading in common and, importantly, the oversight of the wise master, pictured here as the fatherly Benedict. The father with the wand is the external agent of the cure that a holy man such as Benedict could conjure from within.

Sodoma's version of this scene also confronts the viewers viscerally with the sight, probably familiar to the monks of Monte Oliveto, of the bare and bloodied back of one of their own, as Benedict inflicts the discipline that the monks would also have been encouraged to undergo themselves, as voluntary penance, in their cells (fig. 25). In the context of a book rather than a fresco on a cloister wall, the engraving also relates the monk's punishment closely to more quotidian methods of reading and learning and focuses attention on the lessons at hand, in that the whip was also an attribute of the secular schoolmaster. We can see this demonstrated in an illustration from a book made for the little prince Massimiliano Sforza, which threatens boys who might ignore their reading lessons and therefore lose sight of the path of virtue. In another book made for the same little boy we see him choosing that path correctly, in rightful emulation of the young Hercules (fig. 26 and 27).[47] Like a more childish version of Benedict's temptation, the Sforza prince is shown in a moment of indecision with grave consequences for the rest of his life as he considers the lures of temporal pleasure and the duties and decorum becoming to a ruler.

25   Il Sodoma, *Benedict Frees a Monk from Possession by the Devil*, c.1505, fresco, Chiostro Grande arcade, Monte Oliveto Maggiore.

While calling to mind the rewards of the upward path in the coda to the story seen at the far left, and presenting the consequences of wandering from a proper mental or physical place, Passeri is also careful to construct his image so that the crime and the punishment are visibly connected (fig. 22). Benedict's shadow in declarative perspective clearly links the beating to the site of the transgression: the oratory is shown as a small, domed central plan structure inside a walled court-yard. The building is a fairly accurate portrait of Bramante's (1444–1514) Tempietto on the Janicu-lum Hill (fig. 28). Funded by Spanish monarchs and built in the courtyard next to the church of San Pietro in Montorio, on property ceded by a Spanish pope to his Spanish confessor, it has

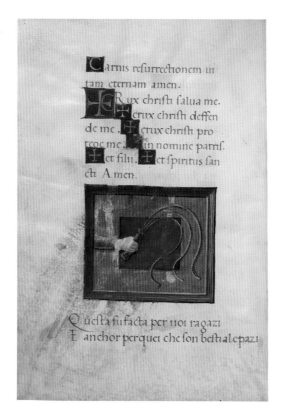

been identified with Spain ever since.[48] As an invitation to *compositio loco*, the imagination of a familiar and specific place for vivid meditation recommended in monastic reading practice, it would have resonated strongly for the congregation of Spanish monks in Rome who commissioned the engravings.

In the *Vita* the exemplary life of St. Benedict was carefully presented as a picture book guided, rather loosely, by captions, and very suggestively by the collection of images that composed each illustrated page. It took part in the rituals of reading that were integral to life in Benedictine monasteries. Books are present in many of the *Vita*'s images: the saint often holds a book that may be understood as the *Regula* or scripture, and at the cave in Subiaco he is surrounded by them as he studies to become a holier person through the consumption of texts. The texts alone are not enough to produce right knowledge; John Chrysostom claimed that reading could only produce a holy life with other knowledge added to it, citing the case of the "Ethiopian eunuch of Acts 8, who could 'read' but did not 'understand'" until he had received further teaching.[49]

The captions in Passeri's picture book were translated into Spanish, and also reprinted in Latin in 1596 by Paolo Arnolfini, a bookseller and publisher from Lucca who was active in the short-

OPPOSITE PAGE LEFT 26 Attributed to Giovanni Ambrogio de' Predis, *Liber Iesus.* Archivio Storico Civico e Biblioteca Trivulziana, MS 2163, fol. 4v.

OPPOSITE PAGE RIGHT 27 Attributed to Giovanni Ambrogio de' Predis, Prince Massimiliano Sforza Choosing Between Vice and Virtue ("Qui tutto alla virtute el conte e dato / E la donna di vitii ha refutato"), Donatus, *Ars grammatica* of Massimiliano Sforza. Archivio Storico Civico e Biblioteca Trivulziana, MS 2167, fol. 42v.

RIGHT 28 Donato Bramante, Tempietto, Rome, *c.*1502–8.

lived printers' and booksellers' confraternity at the church of Santa Barbara dei Librai.[50] In the Spanish edition, there is a dedicatory letter to Philip II and a new image of the saint above an engraved sheet with a short life of Benedict in Spanish and a history of the order in Rome, by Fr. Alonso Chacon (fig. 29). The Spanish translations are engraved on small strips of copper as secondary captions, printed underneath the original Latin captions and exhibiting their own platemarks.

Passeri's images, almost instantly popular, were copied and printed in smaller versions by different publishers for different markets. Rather soon the connection between the pictures and *lectio divina,* as practiced by the Spanish Benedictines, was lost. For Angelo Sangrino (Angelo Faggio), an Italian abbot of Montecassino in this period, neither the connective shadow, nor the Janiculum Hill, nor the subtleties of the intrusive hem provided the necessary pictorial context for instructing children in the life of the order's founder. In 1586 he had the images of the *Vita* copied half-size with the addition of pages of moralizing prose, and printed the work in Florence and, a year later, in Rome. The illustrated Roman edition, printed at the shop of Bartolomeo Bonfadino, was dedicated to the thirteen-year-old prince Odoardo Farnese, who eventually did become

29   Frontispiece, *Vita et miracula sanctissimi Patris Benedicti ex libro ii Dialogorum Beati Gregorii Papa, et Monachi collecta per Thomam Triterum, et e Latina in Hispanicum lingua conversa per D. Franciscum Cabrera / Adiuncta vita, et effigie eiusdem. / S. Benedicti / Ad Philippum Hispanioarum et Indiarum principem*, Rome: Paullini Arnolfini Lucen[sis], 1597, engraving. Biblioteca Casanatense, CCC K.II 21.

30   Plate 13, Angelo Sangrino, *Speculum et exemplar Christicolarum. vita beatissimi patris Benedicti, monachorum patriarcae sanctissimi. Per r.p.d. Angelum Sangrinum abbatem Congregationis Casinensis carmine conscripta,* Rome: Bartolomeo Bonfadini, 1587. MLs 144.30, Houghton Library, Harvard University.

a cardinal of the church and whose great-uncle, Cardinal Alessandro, was the Cardinal Protector of the Benedictine Order (fig. 30).[51] This didactic version of the book contained more words than pictures, presenting Benedict's life by way of written arguments and discourses as an exemplum of moral virtue for lay people who practiced a very different form of reading than did the monks-in-training.[52] The effect of adding a squared pavement inside the courtyard in the foreground of the Sangrino engraving of the story of the wandering monk emphasizes the separation of the parts of the narrative, creating an overtly Renaissance perspective box with a map-like surface on which to dispose them. The Passeri image (fig. 22), by continuing the rock-strewn, rough dirt path leading from the church into the courtyard and piazza in front of the oratory, connects the two spaces visually and metaphorically. The church and the healing blessing that the reformed sinner receives are connected, through the stony, uphill path, to the places of prayer and penance below. Both images show Benedict casting a deep shadow behind him, a device that emphasizes the figure, allowing it to occupy more space without distortion, but the Benedict of the Sangrino engravings wears the more majestic buttoned jacket of a cardinal of the church, enforcing his authority with terrestrial markers of ecclesiastic rank. The patient expression of the patriarchal figure in the Passeri print is replaced by gaunt determination in the Sangrino version, making him a sterner disciplinary figure. But most importantly, in the Sangrino we miss the fragment of overlapping hem that makes the crucial connection between the sinner in the story and the reader who might take a certain amount of self-congratulatory pride in the fact that it is not he who has been caught giving in to his inner devil.

Religious humanists in Benedictine circles after the Council of Trent worked to dispel the cloud that hung over them because of the evidence that another popular religious publication, *Il Beneficio di Cristo* of about 1543, which had recently been denounced and listed in the Venetian Index of Prohibited Books, had been written by a Benedictine monk.[53] Although the order had a history of strict reform, it was not in good odor in the years immediately following the last Council of Trent, and the further suspicions thrown on their origins through the Protestant denials of the authorship or credibility of Gregory the Great added to their uneasiness. Issuing a persuasively rendered visualization of the *Dialogues* while another Gregory, Pope Gregory XIII, occupied the throne of St. Peter was a move to revivify what must have seemed like the rapidly dissolving public image of the most important historical ecclesiastic figures associated with the order. Other monastic orders, like the Augustinians who had no particular regard for the importance of labor, emphasized reading and withdrawal from the world even more than did the Benedictines. However, it was important here to emphasize those aspects of reformed Benedictine spirituality that most differentiated it from the newer mendicant orders, or versions of the monasticism that preceded it: fervent, heartfelt prayer proceeding from profound inward searching, filial obedience to God, and, most in question, the primary importance of an unwavering faith evidenced in good works and penance in order to attain salvation.[54]

In the end, the 1579 *Vita* becomes a rather practical manual for reimagining (or reinventing) a post-Tridentine Benedictine spirituality. As inventor of the images, Bernardino Passeri was responsible for a good deal of how the pictures eventually looked. However, the program that divided the Gregorian story into primary and secondary parts, the collapsing of different moments of the stories into single, representative scenes that could be read backwards and forwards, the inclusion of important symbolic elements such as Spanish-Roman architecture, could not have been his alone. Such an intricate program required the collaboration of a monastic reader who could envision a late sixteenth-century format for the practice of *lectio divina*: that focused, alert, allegorical manner of reading in which the Benedictine monks – who, we can guess, were the primary intended audience for this book – were trained. This training was in fact at the heart of the whole enterprise of living in the monastic community; it was a way simultaneously to strengthen the self and control its imaginative capacity, and to teach the important precepts of renunciation and asceticism as modeled in scripture and in the narrative of Benedict's life.[55]

"Listen, O my son," begins the Rule of St. Benedict, "to the teachings of your master, and incline the ear of your heart, and cheerfully receive and faithfully execute the admonitions of your loving father, that by the toil of obedience you may return to him from whom, by the sloth of disobedience, you have gone away."[56] We think of printed books today as made on commercial presses, published with the idea of financial gain and fame for the author, and written for the widest possible audience. The *Vita* has no single named author, but unusually, as we saw, it is fairly easy to identify its readers: a specific group of youths whose mandate was to work, to pray, to read, and to keep silence. They were to avoid bad words entirely, and good ones as much as possible. For them, pictures in a book were a machine for exercising their minds, strengthening resistance to lustful wandering, and increasing their ability to understand scripture through the life of their founder by the discipline of *memoria*.

3

Camillo Agrippa's Cosmology
of Knowledge

Camillo Agrippa's *Trattato di scientia d'arme* was a fencing treatise, published in Rome in 1553 by Antonio Blado, who had at that point been the papal printer for just under twenty years.[1] A half-length portrait of the Milanese author at the beginning shows a tense-looking bearded man with athletically close-cropped, curly hair and a military bearing, staring out from a stony oval niche bearing his name in lapidary letters (fig. 31).[2] Aside from the portrait, the book, ambitiously dedicated to Cosimo de' Medici, contains two full-page narrative scenes, a diagram of an armillary sphere, and fifty-five further elegant engravings of sinewy nude fighting figures that give a step-by-step demonstration of Agrippa's method and theory of fencing. Although the images are fundamental to the author's explanatory text and are referred to constantly, neither the artist nor the engraver is named in the book. Thirty-four years later, the Milanese art theorist Giovanni Paolo Lomazzo praised the skill of the Lombard painter Carlo Urbino in displaying Camillo Agrippa's art of fencing.[3]

The *Trattato* was the first of seven books that Agrippa produced in Rome over the course of his lifetime, most of them relying on the use of engraved illustrations. The pictures he included in books of natural knowledge were designed to characterize his profound familiarity with mathematical principles in ways that were meant to secure his reputation as a courtly purveyor of natural laws with practical applications. This first effort is the most elaborately illustrated of the books Agrippa carefully crafted for Roman presses, creating a cycle of publications that framed his working life. Mostly through engagement with pictures and diagrams in dialogues staged between himself and Roman nobles or men of letters, Agrippa portrayed himself as embodying through practice a profound knowledge of the natural laws that governed the theories and practice of civil and military engineering.[4]

Agrippa came to Rome from Milan, bringing with him ambitions fueled by a familiarity with Aristotelian natural philosophy, Christian Neoplatonism, Leonardo da Vinci's ideas about motion

OPPOSITE PAGE   *Academic Disputation*, Camillo Agrippa, *Trattato di scientia d'arme, con un dialogo di filosofia di Camillo Agrippa Milanese* (detail of fig. 55).

31   Author portrait, frontispiece, engraving, Camillo Agrippa, *Trattato di scientia d'arme, con un dialogo di filosofia di Camillo Agrippa Milanese*, Rome: Antonio Blado, 1553. Folger Shakespeare Library.

and measure, engineering skills, and an engagement with the world of letters that encompassed an interest in literature, scientific writing, and a sharp sense of the possible benefits of being a published author.[5] Much of his work and the fulfillment of his ambitions played out in the realm of letters and publishing. Through his connections to the publishing world, he fostered relationships with potential aristocratic patrons, which was to be expected, but he also made literary overtures to a deep strata of men much like himself: cardinalitial secretaries, sculptors, artists, antiquities collectors, minor aristocracy, engravers, musicians, and cartographers. These men appear in his dialogues as interlocutors, in his treatises as dedicatees, and in the fencing manual as the learned gentlemen who stand up for his honestly earned right to claim expertise in a practical world of applied natural principles. As for his own job description, Agrippa is probably most accurately termed an engineer, although our more specialized conception of such an occupation does not begin to cover either his interests or activities.[6]

The Cosmology of Knowledge

Only a few of the projects known from archival records or printed books resulted in documented practical outcomes apart from publication. Although Agrippa writes that he came to Rome in 1535, at about the age of fifteen, the earliest record of his presence in the city is a notice in 1543 that he was already famous for the work he had done on repairing the ancient aqueduct system, which much benefited the Roman people.[7] He was most praised for having engineered the hydraulic structure that he made to bring water from the ancient *aqua vergine* that emerged by the Trevi fountain up the steep Pincio Hill to irrigate the gardens of Cardinal Ricci da Montep- ulciano. Aside from pleasing an art-loving cardinal, Agrippa's hydraulic invention – which involved powering a pump by means of a waterwheel – was described as new and never before seen in Rome, and first made a garden possible at one of the city's highest points.[8] The villa was sold in 1576 to Ferdinando de' Medici, who built an artificial mountain called Parnassus on top of the highest point in the garden (fig. 32). This necessitated calling Agrippa back to rework his previous triumphant invention, which he did by 1588, the year that the cardinal gave up his religious career and moved back to Florence to become Duke Ferdinando de' Medici. The guidebook *Mercurio Errante* (1776), reciting the splendors of the villa and noting its connections to antiquity, called attention to "a tall round hill surrounded by Cypress plants: here in ancient times was the Temple of the Sun, as many would like. In modern times the grand dukes had made a huge fountain, conducting the water by mathematical instruments, it being a place too high for water to come naturally, even if today the aqueducts are all destroyed; to get to the top one climbs a stair of about sixty *scalini*."[9] The second project was recorded in a patent that Agrippa took out for an invention for coral fishing in 1570, of which nothing more is known.[10] He never wrote about either of these works of engineering.

Two more projects took shape as printed books. One, the invention of a homemade navigational instrument related to a nocturnal astrolabe, was engraved and printed on paper in such a way that it could be cut out, backed with wood or heavy paper, and attached to lead weights by a piece of string (fig. 33). It appears at the beginning of a treatise on navigation printed in 1595.[11] In it he tells his readers almost incidentally about the scope of his publishing ambitions.

The first thirty-five pages of the book are a narrative tract on the use of the instrument he invented. At the end Agrippa adds a commentary using his preferred convention of a dialogue between himself and a well-born interlocutor, here named Alfonso, an inquisitive and somewhat lazy gentleman who has been given the role of asking leading questions designed to draw the author out about how his invention should be used. Having read the first part of the book, he languidly comments that it is particularly tiresome that in order to use Agrippa's instrument he has to remember the names of all the winds and their directions. He immediately receives from the author a simple way to remember the names of the four most important winds that will not overtax his mind. Armed with this new technique, he and Agrippa turn to a minute, step-by-step

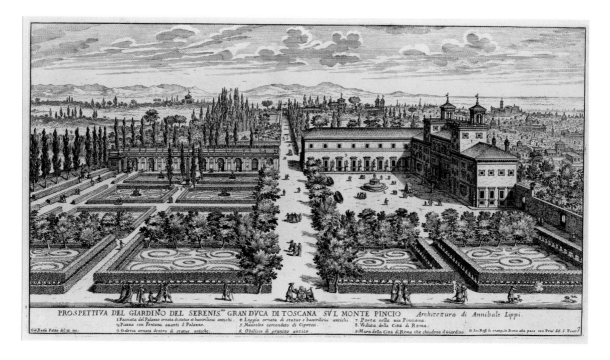

PROSPETTIVA DEL GIARDINO DEL SERENIS.^{mo} GRAN DVCA DI TOSCANA SVL MONTE PINCIO   *Architettura di Annibale Lippi.*

1.Facciata del Palazzo ornata di statue et bassirilievi antichi .   4. Loggia ornata di statue e bassirilievi antichi .   7. Porta nella via Pinciana .
2.Piazza con Fontana auanti il Palazzo .   5.Mausoleo cercondato di Cipressi .   8. Veduta della Città di Roma .
Gio:Batta Falda del et inc.   3.Galeria ornata dentro di statue antiche .   6. Obelisco di granito antico .   9.Mura della Città di Roma che chiudono il Giardino .   Gio: Iac.Rossi le stampa in Roma alla pace con Priu: del S. Pont.

32   Giovanni Battista Falda, *The Gardens of the Villa Medici* (with detail of Parnassus), etching, Plate 8, *Li giardini di Roma con le loro piante alzate e vedute in prospettiva disegnate ed intagliate da Gio. Battista Falda nuovamente dati alle stampe con direttione e a cura di Gio. Giacomo de Rossi*, Rome: Giovanni Giacomo de Rossi, 1680, British Museum.

practical explanation of how to assemble and use the instrument. The reader inhabits the role of the restless Alfonso, learning from Agrippa as a pampered teenager would learn from a patient teacher.

The lesson proceeds in an orderly manner as Alfonso continues to demand a concrete example to back up each of the theories behind Agrippa's ideas about navigation. But then the student muddies the lesson by suddenly making a free-form association having to do with the ocean: he asks if Camillo knows how to get fresh water from the bottom of the sea. "Don't start," warns the author, suddenly stern, "that discourse here, because you will ruin the book, which is about something else, and besides, you will ruin the order that we have begun." "What order is that?" asks Alfonso helpfully. Camillo responds:

> You know that I have composed twelve books, each one of them titled with the name of the months, intending that the whole work together be called *The Year of Agrippa*, so that the first will be called January, the second February, the third March, and so forth, one by one according to the order of the months. Now, of these twelve I have only published three, that is, January, February, and March, and this one will be the fourth, April.[12]

The Year of Agrippa was conceived as a cosmology of texts strategically released from the author's storehouse of skills and information, each labeled at the end with the name of a month of the year. The author at the center of the world, in charge of a virtual zodiac of knowledge, was popularly pictured in images such as an engraving showing a perpetual calendar for prognostication (fig. 34). The calculating astrologer Angelo Gallo, shown with his mathematical instruments presiding at the center of the moving band of the heavens, personifies the Aristotelian preference for representing man as the unifying principle in a well-ordered, intelligible, and (perhaps more importantly) accessible scheme. It also demonstrates the association from the medieval period of cosmological and theological principles, with man as the microcosm uniting all human experience. In Agrippa's case the well-known astrological idea bound the older association to a more modern one: the expectation that mathematicians in universities up through the time of Galileo would also undertake teaching how to cast horoscopes for the physicians in training who would need this information in their practice.[13]

Agrippa's image of himself at the controls of a universe of knowledge also correlated to relatively new conceptions of authorship in regard to sixteenth-century publishing. Book publishers made a point of characterizing their readers as avid collectors and organizers of knowledge. At least since the middle of the century publishers had been issuing books and even images in series, united by a title page like the single-sheet prints in Antonio Lafreri's *Speculum Romanae Magnificentiae*, or simply by a common title. *Collana* was a word that evoked the image of a circle, meaning simultaneously a necklace and a series of connected publications.[14] Agrippa's vision of publication meant that eventually his small, easy to read, illustrated works would forever lead from one to another, in monthly doses measured by natural time, each book concise and informative while also drawing the reader into further understanding of natural phenomena that accorded with

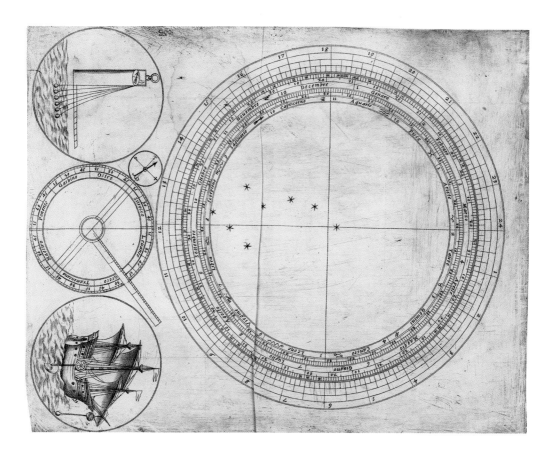

church teachings. Bound together, as Renaissance readers might bind them once they were acquired, the books would form a *libro unitario*, a single body of work, a corpus of one author's knowledge and experience.[15]

The material available to an author for the illustration of such books comprised any printed image that appeared on the visual horizon of late sixteenth-century Romans. The fencing manual was illustrated with images newly made for that text alone; the book was structured and conceived around the images, which also made up about half of its actual content. The book on navigation used images differently. While the printed instrument itself, being Agrippa's invention, was designed specifically for the book, the frontispiece was a thoughtful bit of visual engineering illustrating Agrippa's literary dedication to the reforming Milanese theologian, Cardinal Borromeo. Agrippa used a copy, made by the Friulian engraver Camillo Graffico, of a detail of an engraving by Cornelis Cort (1533–*c*.1578) of an image by the academic painter Girolamo Muziano. It shows Peter floundering in the waves as he makes his way towards Christ, who, skimming the surface of the water himself to approach the apostle's fishing boat, reaches out to grab his wrist and steady

OPPOSITE PAGE 33 Navigation tool, engraving, Camillo Agrippa, *Nuove invenzioni di Camillo Agrippa Milanese sopra il modo di navigare*, Rome: Domenico Gigliotti, 1595, John Carter Brown Library.

RIGHT 34 Angelo Gallo, *Rota perpetua per conoscere l'anni fertile e sterile*, 1578, engraving. C.2.1/17, Biblioteca Angelica.

him (figs. 35 and 36).[16] The image shows an example of navigation without mechanical instruments in the company of the best guide, and it illustrates Agrippa's assurance to the cardinal in his dedication that he understands there are two kinds of navigation: navigation of the sea, which this book will treat, and navigation of the tempests of the soul on earth, for which God is the best compass.[17] Printed in a caption below Graffico's engraving is Christ's exhortation to Peter from Matthew 14: "Modicae fidei quare dubitasti" (You of little faith, why did you doubt?), which in the role of a frontispiece does double duty by exhorting the reader to believe in the author and his instrument as well.

Altogether the treatise contains four elaborate engravings of very different sorts as part of the *proemio* or preface, thereby setting the tone of the work and directing it to a desired audience. The borrowed image by Cort via Graffico is followed by the dedication to Carlo Borromeo, then a letter to the readers, and an unusual engraving, seemingly custom-made, of a complicated allegory featuring Agrippa addressing Nature herself in the form of Diana of Ephesus (fig. 37), to which we will return. The allegory is followed by the carefully composed printed navigational

DOMINE SALVVM ME FAC. MODICÆ FIDEI, QVARE DVBITASTI. *Matth. 14.*

instrument (fig. 33) and then a medallion-like author portrait showing Agrippa looking appreciably older than he did in the fencing manual (fig. 38).

In these simple juxtapositions Agrippa refined an image of himself as an expert tutor with an understanding of nature so broad and deep that he was able to simplify it into the exposition of essential principles. The students he instructs informally in his books were not planning on becoming natural philosophers or even professional engineers. They were a select group of gentlemen amateurs, consumers of knowledge about the natural world, eager to hear their own understanding of it confirmed, enlarged, and explained in comprehensible and practical terms.[18] In a study of natural knowledge and practical need in Elizabethan England, Eric H. Ash proposed the rise of a particularly sixteenth-century figure he termed the "expert mediator."[19] Ash argues that expertise became an important commodity when noblemen needed a knowledgeable figure to coordinate the execution of elaborate military or capitalistic ventures involving navigation, mining, or similar pursuits. The desirable properties of such a mediator were a sound theoretical knowledge of the technical issues involved and skills in communication to be able to explain the complex undertaking to his employer. Such an entrepreneurial figure stood as a go-between uniting the rough world of artisanal craft and the refined world of patronage, and in that way

OPPOSITE PAGE 35   (Jan I?) Sadeler after Cornelis Cort after Girolamo Muziano, *Calling of Peter*, engraving, 180 × 272 mm. New Hollstein 52, copy d. British Museum.

RIGHT 36   Camillo Graffico after Cornelis Cort, *Calling of Peter*, engraving, Camillo Agrippa, *Nuove invenzioni di Camillo Agrippa Milanese sopra il modo di navigare*, Rome: Domenico Gigliotti, 1595. Houghton Library, Harvard University.

was analogous to the figure of the court artist, perhaps an extension of such a figure into a capitalist market. In this way and under such circumstances, the value of an expert figure was shifted from practice to theoretical discourse about the practice. Mario Biagioli, in a study of the diversity of the practices of early modern mathematicians in Italy, proposed that this variety in practice was an indication of high social and professional mobility, particularly among the mathematicians who did not teach at or attend universities. He attributes a good deal of the improvement in the status of non-university trained mathematicians, from the culture of the abacus schools, to changes in artillery, primarily the use of cannons, and the resulting new fortifications that this demanded. With the formation of a group of mathematically literate aristocrats, whose military engineers were sometimes ennobled for their work, mathematics came to be associated with knights as well as with accountants, and surveyors and mathematicians lobbied for salaried roles in the service of a prince or a duke.[20] Agrippa was not quite any of these figures, and yet the emerging idea of a resident expert, and the opportunities for patronage for a man with theoretical and practical knowledge of mathematics, account for much of the role of books in Agrippa's life and career. He did not have the university education in which Ash's mediators founded their claim to theoretical knowledge tempered by practical dabbling. He did fit the bill as a working inventor and

Intendi nobil donna le querele delle Muse, e delle regolate scientie, e le differentie che sono tra noi, e secondo
la sapientia tua giudica: le nostre tante contrarieta, acciò uiuiamo in pace.

37 *Allegory of Agrippa with Nature and the Muses*, engraving, Camillo Agrippa, *Nuove invenzioni di
Camillo Agrippa Milanese sopra il modo di navigare*, Rome: Domenico Gigliotti, 1595. Houghton Library,
Harvard University.

38   Author portrait, engraving, Camillo Agrippa, *Nuove invenzioni di Camillo Agrippa Milanese sopra il modo di navigare*, Rome: Domenico Gigliotti, 1595. Houghton Library, Harvard University.

an explicator for a certain class of gentlemen interested in mathematics, astronomy, engineering, and natural philosophy. His status as a practitioner and a gentlemen perhaps lent him a certain immunity to charges from both sides of trespasses arising from ignorance of the culture of the other.[21] Rather than launching new scientific ideas, he offered explanations for natural phenomena as they were already understood, in terms that his audience was predisposed to accept. Like the astrologer operating from the center of the spinning world, he offered himself and his cosmology of texts as the unifying basis for beliefs commonly held about natural knowledge in the sixteenth century.

Elio Nenci has characterized Agrippa's explanations as based on mechanical models derived from experience, theorized in terms of Aristotelian precepts remade into a Christian Neoplatonism that also seamlessly and unselfconsciously included astrology, magic, and cabalistic explanations that reflected popular (but by no means 'low') learning in the sixteenth century.[22] Nenci follows Agrippa's philosophical arguments with sensitivity to the strains of knowledge and belief that Agrippa references. Although he often cites human sources in his work in the form of characters present as interlocutors, or names resorted to for further testimony about his successes, there are no citations in Agrippa's writing. Nenci bases his characterization of Agrippa, as Agrippa himself did, in terms of the men whose names appear in his writing, an assortment of artisans who also appear in the membership list of the Compagnia di San Giuseppe, people who made their living practicing the mechanical arts. To this group are added papal secretaries, high-ranking clerics, collectors of antiquities, and men of letters.

Agrippa's coterie of interested parties appears several times in his works. The dedication at the beginning directs each book to its most highly placed desired audience, and certainly to a desired patron. Certain of his books contain signed sonnets at the beginning in praise of the author and his inventions. All the dialogues take place between Agrippa and a locally recognizable gentleman of more minor status than the dedicatee, but someone who fills the role of both an equal and a different kind of patron — a figure whose good opinion seems to count towards the author's credibility, and also opens doors for him that might lead to increased patronage. Alfonso, for example, is identified by his full name only in the last line of the dialogue, which comes as a kind of punchline, and is a technique Agrippa uses in other dialogues as well so that only at the end do you find out that the indolent Alfonso is Alfonso Soderini, or that Mutio is Muzio Frangipane.[23] The reader enjoys the company of men he might consider equals and is pleasantly surprised at the end to emerge among the elect. Soderini was a minor baron from an old Florentine family who married into Roman nobility. His interests in architecture, artillery, and engineering played out in the massive modernizing of the fortifications of a medieval *rocca* that he had acquired from the Savelli family, making him the first baron of Colle Sabino, a Lazio hilltown near Rome. He and another of Agrippa's interlocutors, Tiberio Astalla, appear as subscribers in raising funds for the decoration of the Oratorio del Santissimo Crocifisso in the church of San Marcello al Corso, where the Frangipane family owned a chapel that today contains a posthumous sculpted bust by Alessandro Algardi of Muzio Frangipane as an ideal military leader.[24] Other interlocutors, each one named and therefore pre-endowed with a personality, reputation, and position that plausibly interested him in the knowledge Agrippa had to offer, came from similar backgrounds that made them obvious lay consumers of practical engineering knowledge, often military in application, and its theoretical principles.

At the time of his death in 1600 Agrippa had written his way through the month of May, not quite a half year, with his cycle of published vernacular treatises. The published components of *The Year of Agrippa* (see table) give a good idea of the aspects of natural knowledge that were considered most important in Roman literary and artisanal circles in the last quarter of the century. Agrippa's larger publication project brought together such varied characters as printers, artists, cardinals of the church, poets, gentlemen, publishers, scholars of Oriental languages, calligraphers, cartographers, inquisitors, mathematicians, and architects, among others. Each of these embodied a different set of skills and interests, they came from all over Italy, and for a time they all inhabited an area of a few square miles in the center of Rome. Most of them were already personally known to Agrippa. They were brought together in the production of printed illustrated books, most of which — more than 400 years later — we can still hold in our hands and read if we like, although with the exception of the fencing manual, they no longer explain or confirm much of anything that provokes modern curiosity about mathematics or science.[25]

Table of Agrippa's Publications.

Year	Title and publisher	Notes
1553 (Julius III) [fencing]	Trattato di scientia d'arme Rome: Antonio Blado	Dedicated to Cosimo de'Medici Dialogue with Annibale Caro Agrippa age 33
1575 (Gregory XIII) [cosmology]	Modo di comporre il moto nella sfera . . . Rome: Heredes Antonio Blado	Agrippa age 55
1583 **Genaro** (Gregory XIII) [mechanics or physics]	Trattato di trasportar la guglia Rome: Francesco Zanetti	Dedicated to Giacomo Boncompagno Dialogue between Camillo, Fabritio and Agapito Fossani Agrippa age 63
1584 **Febraro** (Gregory XIII) [meteorology]	Dialogo . . . sopra la Generatione de Venti . . . Rome: Bartolomeo Bonfadino and Tito Diani	Dedicated to Cardinal Luigi d'Este Dialogue with Tiberio Astalla Agrippa age 64
1585 **Marzo** (Gregory XIII) [military]	Dialogo . . . del modo di mettere in battaglia . . . Rome: Bartolomeo Bonfadino	Dedicated to Henry III of France Dialogue with Mutio Frangipane Agrippa age 65
1595 **Aprile** (Clement XIII) [navigation]	Nuove invenzioni sopra il modo di navigare Rome: Domenico Gigliotti	Dedicated to Cardinal Borromeo Dialogue with Alfonso Soderini Agrippa age 75
1598 **Maggio** (Clement VIII) [causes of motion, &c.]	La virtù, dialogo . . . sopra la dichiarazione de la causa de' moti Rome: Stefano Paolini	Dedication to Clement VIII (No illustrations) Dialogue with the spirit of Luigi d'Este Agrippa age 78

January, published in 1583, was Agrippa's original scheme for how to move the Vatican obelisk, one of his few historically documented projects apart from a published book.[26] It was Agrippa's third book, although the first to be written after he had conceived the notion of a series for the Roman presses. Like the 1553 fencing manual and the navigational treatise of 1595, it contained a narrative exposition followed by a dialogue expanding on the preceding material. In it, Agrippa notes, "At the time I came to Rome, which was on the 26 October 1535, I thought about how to move the obelisk securely to the Piazza San Pietro, and at that time the worthy gentleman Antonio Sangallo, and the great Michelangelo Buonarotti, and an infinite number of others were grappling with the problem of that undertaking."[27] This was entirely reasonable; moving the *guglia*, as it was called, became a kind of Excalibur to men of mathematical or engineering ambitions already in the city. It was a beacon for foreigners who came to Rome hoping for papal commissions and to make their names in related fields such as engineering or architecture.

The obelisk had originally stood at the center of the Circus of Nero, and St. Peter's Basilica was eventually built alongside that site of pagan games. When the church was rebuilt in the sixteenth century, the obelisk, marking nothing at all, was an increasingly irritating focal point. Its removal to a more meaningful location had been thought about already during the pontificate of Paul III, which was when Agrippa had arrived in Rome, and then became an urbanistic priority under Sixtus V, when the obelisk was finally moved.[28] The Vatican obelisk was the only one of the many needle-shaped trophies of Eastern conquest in Rome that was still standing in the sixteenth century, and the broken pieces of the fallen obelisks lying around or recovered in other parts of the city were eloquent testimony to the risks involved in challenging the engineering skills of the pagan ancients. Agrippa was indeed a serious entrant in the unofficial contest. In 1581 he went to the Vatican to present his plan for moving the obelisk in an upright position, winning an audience with the pope and his personal physician, Michele Mercati, who also wrote a book about the obelisks.[29] Agrippa's book is dedicated to Giacomo Boncompagno, the highly placed son of the then sitting pope, Gregory XIII (1572–85), who most encouraged the study of mathematics and cartography, and instituted the reform of the calendar. Agrippa wrote four of his seven publications during his papacy.

Gregory XIII died two years after Agrippa's treatise appeared with an accomplished fold-out engraving of his scaffolding for moving the obelisk (fig. 39), a full explanation of his model, and an informative dialogue with Fabritio and Agapito Fossani. These two gentleman, probably Milanese, had both served as administrative heads of their *rione* in the 1580s. At the beginning of his treatise they had been his confidants about the project for over thirty years.[30] When Sixtus V ascended to the papacy St. Peter's was almost completed and the matter of the obelisk had become urgent. Although competitive bids for its removal to the center of the piazza in front of the church were still entertained, it became clear that the winner would be Domenico Fontana, the man who was already acting in the capacity of papal architect and engineer. The year after Sixtus took office the obelisk was moved according to Fontana's plan, which he later published in a detailed record that depended heavily on minutely rendered large-scale technical illustrations (fig. 40).[31] At the beginning of his treatise Fontana included a rather demeaning engraving showing all the losing entries in the ongoing competition; Agrippa's is labeled "B," and shows how he planned to move the obelisk in a vertical position with an emphasis on the awkward shape of the elaborate scaffolding (fig. 41). Fontana's winning scaffolding hovers above, triumphantly supported by winged putti.

February, published the next year, was a meteorological work titled *Dialogue on the Generation of Winds, Lightning, Thunder, Rivers, Lakes, Valleys and Mountains* (fig. 42).[32] Like many who dedicated themselves to interpreting the book of nature, Agrippa began by reminding his readers that to obey the laws of nature is to obey their creator. This was not a platitude; it had practical application as it was one reason why nobles, rulers, and theologians did attend to this subject matter, professing or truly having a curiosity about natural philosophy that should rightly be harnessed to curiosity about the creator whose work nature was.[33] It was also true that Aristotelian science

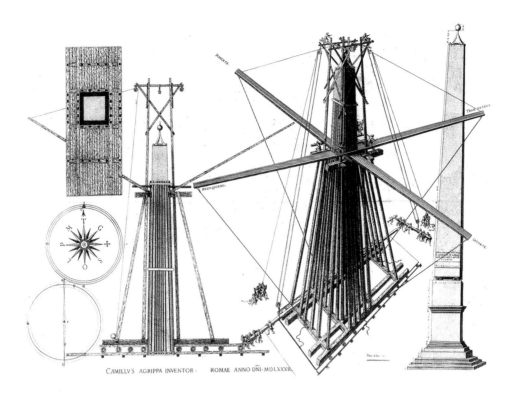

39 Agrippa's model for how to move the Vatican obelisk, engraving, *Trattato di Camillo Agrippa Milanese di trasportar la guglia in su la piazza di San Pietro*, fol. 28r, Rome: Francesco Zanetti, 1583. Stamp. Arch. Cap. S.Pietro B120 p.28, Biblioteca Apostolica Vaticana.

had raised religious objections from the early Middle Ages, and was most recently the object of a mandate in the Fifth Lateran Council (1512–17) for natural philosophers to remain alert to how their work supported the doctrine of the immortality of the soul.[34] Therefore, Agrippa's introductory note was protective, taking advantage of the flexibility of the conventions of the *proemio* as a platform from which this kind of disclaimer could be launched. This flexibility and use of introductory material was one of the aspects of early modern publishing that was shaped by the prevailing atmosphere of the church's scrutiny of printed material.

The book was dedicated to Cardinal Luigi d'Este of Ferrara, who had inherited the nearby villa at Tivoli with the fantastic gardens and fountains built by his uncle, Cardinal Ippolito d'Este (fig. 43). Educated in the classics as a philologist, Cardinal Luigi was also introduced early on to the normal chivalric activities of the Ferrara court, such as hunting, music, and fencing. He evolved into an active supporter of poetry, theater, and literature in Italian. He rose rapidly

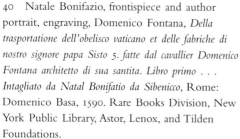

40   Natale Bonifazio, frontispiece and author portrait, engraving, Domenico Fontana, *Della trasportatione dell'obelisco vaticano et delle fabriche di nostro signore papa Sisto 5. fatte dal cavallier Domenico Fontana architetto di sua santita. Libro primo . . . Intagliato da Natal Bonifatio da Sibenicco*, Rome: Domenico Basa, 1590. Rare Books Division, New York Public Library, Astor, Lenox, and Tilden Foundations.

41   Unsuccessful models for moving the obelisk, engraving, Domenico Fontana, *Della trasportatione dell'obelisco vaticano et delle fabriche di nostro signore papa Sisto 5. fatte dal cavallier Domenico Fontana architetto di sua santita. Libro primo . . . Intagliato da Natal Bonifatio da Sibenicco*, Rome: Domenico Basa, 1590. Rare Books Division, New York Public Library, Astor, Lenox, and Tilden Foundations.

through the ranks of the church, becoming archbishop of Ferrara at age fifteen and cardinal at twenty-five. He also became the protector of the French crown in Italy, appointed by King Henry III, and he was particularly involved at the time this book was written in renovating and adding to the miraculous fountains at the Villa d'Este, the kind of work at which Agrippa had already proved himself in cardinalitial circles with his famous extension of running water to the Pincio Hill. Luigi d'Este's court included engravers, astrologers, philosophers, musicians, and alchemists.[35]

DIALOGO
DI CAMILLO
AGRIPPA
MILANESE
SOPRA LA GENERATIONE
de Venti, Baleni, Tuoni, Fulgori, Fiumi,
Laghi, Valli, & Montagne.

CON LICENTIA DE'SVPERIORI.

IN ROMA,
Appreſſo Bartholomeo Bonfadino, & Tito Diani.
M. D. L XXXIIII.

42   Camillo Agrippa, *Dialogo di Camillo*
Agrippa Milanese sopra la generatione de venti,
baleni, tuoni, fulgori, fiumi, laghi, valli, &
montagne, Rome: Bartolomeo Bonfadino
and Tito Diani, 1584. Houghton ★IC5
Ag854 584d, Houghton Library,
Harvard University.

In 1580, he became the patron of Giovanni Battista della Porta, who came to live at the villa and who dedicated his work on physiognomy, the *De humana physiognomonia* (1586), to the cardinal.

The cardinal hired decorators for those parts of the Villa d'Este that his uncle had not completed. Among the painters who frescoed maps and landscape scenes onto the walls of the ground floor rooms was a professional cosmographer from Pisa named Matteo Nerone, who had made a name for himself as a printer and painter of maps and globes (fig. 44).[36] Nerone was not only a painter of landscape frescoes, but also of *mappa mundi*. He had worked in Gregory XIII's famous Hall of Maps in the Cortile del Belvedere at the Vatican, and painted globes for members of the Jesuit order, in which Cardinal Luigi was particularly active. Nerone also supported himself as a merchant of scientific instruments, measures, and mathematical texts. At the beginning of Gregory XIII's pontificate he had regular work as the *proto*, or overseer, of the Tipografia Orientale Medicea, the print shop founded by Cardinal Ferdinando de' Medici after 1584 with a mandate to publish and disseminate the writings of the evangelists in Arabic-speaking countries. Along

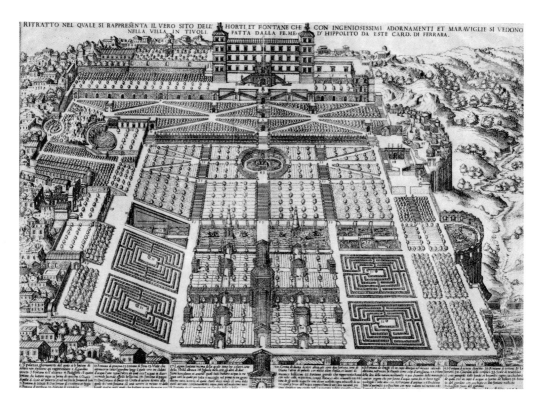

43   Mario Cartaro, *Ritratto nel quale si rappresenta il vero sito dell' horti et fontane che con ingeniosissimi adornamenti et maraviglie si vedono nella villa in Tivoli fatta dalla fe. me. d'Hippolito da Este Card. di Ferrara*, 1575, engraving, 370 × 523 mm, British Museum.

with supplying the Christian message to the infidel in their own languages, the press produced the medical, mathematical, philosophical, and geographic works of Euclid and Avicenna, as well as grammar books in Arabic and Latin. Nerone himself was only responsible for the works in Arabic and Chaldean, although there is no evidence that he read either language.[37]

The director of the Medici press, Giovanni Battista Raimondi, was a mathematician and Arabist from Cremona, a northerner like Agrippa. He was one of the best Arabists in Italy, and he worked tirelessly to keep the press producing, searching out reliable Arabic language manuscripts for material to print.[38] During 1586–7 Bernardino Baldi, the polymath abbott of Guastalla who was in the process of writing a *Lives of the Mathematicians* based in part on Vasari's *Lives of the Artists*, was in Rome studying Arabic with Raimondi, and living at the palace of the literary print collector, Cardinal Scipione Gonzaga.[39]

Nerone was also Agrippa's brother-in-law, and in 1593, when he was brought in front of Roman law courts and jailed on charges of illegally selling some of the products of the Medici press at prices that ruinously undercut those of the print shop, Agrippa and his cousin Giacomo were

44   Landscape paintings in the Villa d'Este, *c.*1565, fresco.

present to support Nerone and to identify the books in question. Among them were an Avicenna, some Arabic grammars, and (possibly the most relevant both to Agrippa and to Nerone) an anonymous book later attributed to the geographer Muhammad ibn Muhammad al–Idrisi, *De geografia universali.*[40] In the depositions from the case of the theft, one of his colleagues described Nerone as someone who took pleasure in drawing and also in painting globes.[41] Agrippa, then, was personally connected to an artist cosmographer who worked as a decorator for Cardinal Luigi d'Este, with a professional interest in the sale, manufacture, and utility of scientific instruments and books. He was also familiar enough with the products of the Medici press to be called as an expert witness in their identification at Nerone's trial. It has been proposed that the hydraulic system Agrippa used in the gardens of the Villa Medici in the late 1570s, a system that was said to have been entirely new and never seen before in Rome, might have been inspired by Arabic technology, which used a hydraulic wheel to activate a pump.[42]

1585 saw the publication of March, a work of military engineering dedicated to Henry III of France, who was probably known to Agrippa as the patron of his patron, Cardinal d'Este. The book was titled *Dialogue of Camillo Agrippa Milanese on a way to put into battle quickly and easily people from anyplace you like, with various ordinances and battles.*[43] Written in the form of a dialogue with a member of the noble Frangipane family, the book explained how to get Romans from every neighborhood into an orderly battle formation on the Piazza San Pietro in a timely fashion, with whatever weapons they had available. Agrippa provided diagrams of armies in a variety of marching formations shaped like triangles and columns as a practical aid in diminishing the chaotic nature of such an endeavor (fig. 45). The book also included designs for weapons Agrippa had invented himself, such as the "Hexagon for Protecting Foot Soldiers" (fig. 46), which he dates to 1557.

Gregory XIII died the same year that the book appeared with the dedication to the king of France, and Sixtus V was elected to power. This pope almost succeeded in completely breaking off relations with France, a rift that was eventually mended through the ambassadorial work of Cardinal d'Este. Whatever Agrippa might have meant by dedicating to the king of France a book on how to organize Rome into an armed camp in that year, as was the case with the timing of the obelisk book, the advent of a new pope seemed to be unfavorable to Agrippa's chances of winning patronage in those quarters.

In 1595 the dialogue on navigation was published. In 1598 came Agrippa's final, unillustrated book labeled "May," and titled *La virtù: Dialogue on the declaration of the causes of motions, taken from the words written in the Dialogue of the Winds.*[44] The book was published by Stefano Paolini, who also worked for the Medici press and was renowned for his elegant typography.[45] It began with a memorial notice about Cardinal d'Este, who had since died, which Agrippa embedded in a short dedication to the pope, Clement VIII. The dialogue proceeds with the help of Cardinal Luigi himself in the form of "La virtù dell'illustrissimo Sig. D. Aloigi Cardinal D'Este," who is Agrippa's phantom interlocutor in this retrospective look at the 1584 treatise on the motion of winds, a book that had originally been dedicated to the cardinal. His presence is also felt at the end of the book, where Agrippa posted a very brief tract on a method for dredging the river Po in the cardinal's native Ferrara, another hydraulic enterprise and another project that may have come to nothing through the death of a patron.[46]

Like most of the practical schemes Agrippa wrote about, the Po project was not realized in any other form than that of publication. The books made Agrippa an author and purveyor of knowledge independently from the realization of his ideas in the physical world. Armies, for example, did not have to put into battle in Rome in order for Agrippa to claim credit for his method of organizing the city into battalions; the navigational instrument and its treatise and dialogue answered many questions for the reader about how to use navigational instruments and how to remember the winds, even without cutting out and pinning together the shapes on the first page, or taking to the sea. Until the stone shaft was actually moved and Fontana's treatise appeared with its authoritative method for obelisk moving, an event fully documented and described in print, the Vatican obelisk proposal – which in the end did have something to contribute to Fontana's solution – would have been of considerable interest to any of the apparently many readers with a stake in the obelisk problem.[47]

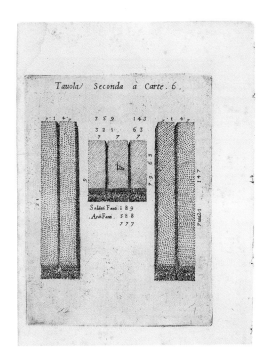

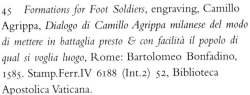

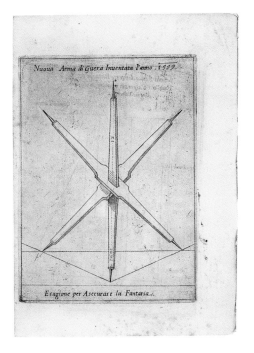

45   *Formations for Foot Soldiers*, engraving, Camillo Agrippa, *Dialogo di Camillo Agrippa milanese del modo di mettere in battaglia presto & con facilità il popolo di qual si voglia luogo*, Rome: Bartolomeo Bonfadino, 1585. Stamp.Ferr.IV 6188 (Int.2) 52, Biblioteca Apostolica Vaticana.

46   *Hexagon for Protecting Foot Soldiers*, engraving, Camillo Agrippa, *Dialogo di Camillo Agrippa milanese del modo di mettere in battaglia presto & con facilità il popolo di qual si voglia luogo*, Rome: Bartolomeo Bonfadino, 1585. Stamp.Ferr.IV 6188 (Int.2) 54, Biblioteca Apostolica Vaticana.

The sixteenth-century press created a library of practical ideas, methods, and techniques for readers with an appetite for minute descriptions of engineering or warfare, who enjoyed steeping themselves in the detailed instruction, lively arguments, and useful images in the pages of a book. Hybrid projects like Agrippa's would continue to be produced, though rarely, by publishing polymaths like Cornelis Meyer in the next century.[48] But Fontana's documentary play-by-play account of his own engineering successes, and other illustrated treatises in the vernacular that aimed to display the variety of an author's knowledge, sometimes in the form of published theaters of machines, would in the future most likely celebrate and record inventions that had already been proven in practice.[49]

When Alfonso Soderini asked Agrippa, in the *Dialogue on Navigation* of 1595, if there were other books besides these, Agrippa answered that there were indeed: some written and some that existed only in his mind. "And about this I do not need to say more," he adds, "because in time they will speak."[50] Agrippa's two earliest treatises, one on the motion of celestial orbs from 1575, and the first book, the work on fencing of 1553, were written before he began to think of his publications systematically as a corpus of knowledge.

The Treatise on Fencing: the Trattato di scientia d'arme

Although Agrippa had already made a name for himself as a hydraulic engineer in Rome by 1543, it was the fencing treatise published a decade later with which he first introduced himself as an author and dispenser of practical knowledge through the Roman press. The first half of the book was written as a technical manual proposing a method of fencing, carefully building from preliminary explanations of guards, steps, and the theories behind them to more advanced movements that incorporated the principal axioms and lessons, and put them into action. It gives rules for defensive and offensive movement based on a peculiar conjoining of practical geometry, optics for painters, and the ritualistic movements traditional to fencing. His contribution, besides some guards that were new to the activity, was to require the practitioner to act in terms of using the point rather than the side of the sword blade, and to think in terms of sight lines modified by perspective distortions. Perhaps more importantly, he used both mathematics and the language of geometrical and mathematical treatises to teach the reader to visualize himself and his enemy as a series of points in a revolving perfect shape rather than as a particular and psychologically complex individual of dubious history and frightening and capricious movements. Also unusual at that time for a treatise about fencing is the almost complete lack of discussion of honor, vengeance, and retribution – the slippery pretexts for any duel. Instead, Agrippa explains, he is interested in the "Intelligence of Arms", not of the issue of justice.[51] The text is closely keyed to the images that usually appear at the end of the textual explanations. The fencing treatise is followed by a dialogue (advertised in the book's subtitle) between Agrippa and Annibale Caro, Farnese courtier and man of letters, in Caro's apartments. It serves as a pretext for Agrippa to present himself as an active but abstract thinker skilled in mathematics and astronomy with wide interest and application, and to obtain the unofficial, but evidently necessary, imprimatur of the man he is presenting to the reading public as his literary friend.

Read in connection with the dialogue, the fencing treatise becomes a showcase for the marketable skills Agrippa claimed to possess, and the dialogue substantiated his claim. Besides changing the look of the genre of fencing books, Agrippa's treatise laid out the social and professional terms for the conduct of his early years in Rome; it shows how he continued to use his Milanese connections and education, and how he managed the Roman relationships he nurtured afterward. In addition, we learn about his conception of who possessed the right to be an author. He did not, we find out in the fencing book, have a university education, although he had specific ideas about the sources and goals of knowledge. He populated his book with accomplished illustrations of the naked human body in motion, mobilizing lithe figures to perform his theories of fencing through a scrim of geometric shapes that allow us to see their minds at work as they calculate

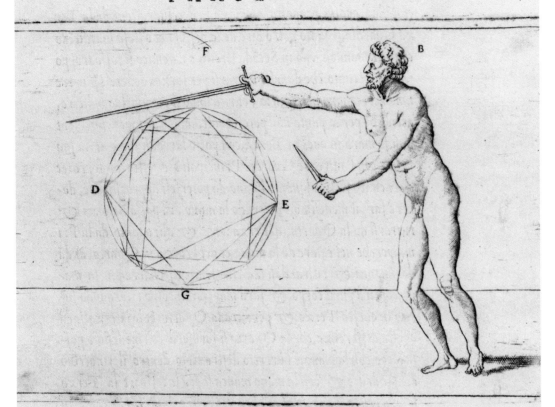

DE LA TERZA GVARDIA
Signata per C. Cap. VI.

 Eſtaria di ragionare anchora di queſta Seconda
Guardia Stretta, come de la Prima pur'aſſai, non
dimeno douendoſi dir' in altri lochi de li altri effetti ſuoi, & ſeparatamente, & tutte inſieme, Se-

47   *Second Guard*, engraving, Camillo Agrippa, *Trattato di scientia d'arme, con un dialogo di filosofia di Camillo Agrippa Milanese*, Rome: Antonio Blado, 1553. Typ 525 53.126, Houghton Library, Harvard University.

their moves in time and space (fig. 47). His books are also crowded with references to real men whose association could bring him patronage, or vouch for the viability and utility of his ideas; the names of the gentlemen and nobles who turn up as interlocutors and living testimonials to his credibility in Agrippa's treatises and dialogues constituted both his literary and character references at the same time.[52] Agrippa is the medium through which practical experience is filtered and expressed as illustrated scientific literature.

Agrippa himself does not seem to have been a noble of any sort, although he makes it clear that he was welcome in aristocratic company. In Italy the term *nobile* was less used to denote nobility than gentility of character and learning, but Agrippa invoked it when calling on his friends for support of his command of natural knowledge, and the right to reveal it. The duel was itself an act of violence in a civil society that ostensibly took place only between gentlemen, as only a gentleman could be said to have honor to defend.[53] While other fencing manuals and the far more common diatribes against the duel outlined in detail all the reasons that a duel could be legitimately or illegitimately provoked, Agrippa gives notice in his preface that he is not going to touch on that sort of information, saying that each person must decide for himself whether or not to take up arms when all other reason has failed, and that those who do so in prideful anger and animosity bring themselves infamy rather than fame. References to loss of life or to the duel as a violent act are surprisingly few in the treatise, which sticks closely to its purpose of describing the theory and practice of guards and steps. The pictures in which an enemy is shown run through with a sword are completely bloodless and seemingly free of pain (fig. 48).

Although one of Agrippa's immediate relatives, Giorgio, was actually employed as a bombardier in Florence, and Agrippa would eventually write a treatise on weaponry and how to raise an army, in the dedication of his *Treatise on the Knowledge of Arms* to Cosimo de' Medici he defines his subject by drawing a clear distinction between fencing and artillery fighting:[54]

> Due to the diabolical invention of artillery, it seems that there is nothing left of the beautiful ancient order of knights than the duel. . . . Therefore, I contrived . . . by what little skill I have been conceded either by nature or by God, to help in what measure I could, which is, to show how a man can, with his wits, with art, and with his valor, defend himself as much in fencing, as in unexpected assault with arms. . . . It seemed appropriate to me to consecrate . . . my work to . . . your Excellency, because everyone knows that if your worthy halls of learning were the true restorers of literature and of the beautiful study of the sciences, as well as of precious languages, you, accompanying arms with letters, are the true supporter of letters and of arms.[55]

Agrippa's dedication, more elaborate and personal in this first book than in the books that followed, flatters those Platonic concepts of good leadership that make the book an attractive bid for patronage to a drawing-literate, intellectually sophisticated humanist duke. He also acknowledges Cosimo's emphasis on state-building through his patronage of art and letters, and his encouragement of the Tuscan vernacular though his establishment of a ducal press and the patronage of the universities at Florence and especially the study of medicine at Pisa.[56] The dedication

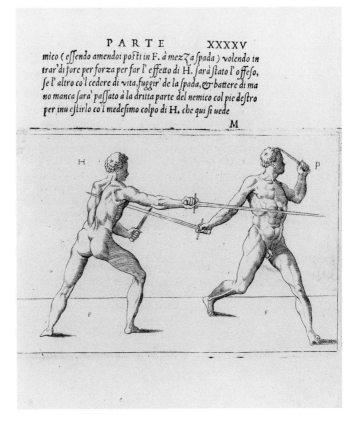

48   *Fencing Figures*, engraving, Camillo Agrippa, *Trattato di scientia d'arme, con un dialogo di filosofia di Camillo Agrippa Milanese*, Rome: Antonio Blado, 1553. Typ 525 53.126, Houghton Library, Harvard University.

therefore sets the vernacular illustrated fencing manual squarely into halls that foster the study of letters, the preservation and encouragement of the Tuscan tongue, and beautiful studies of science. In terms of Agrippa's relative Giorgio the bombardier, and taking into account the success of military experts in obtaining patronage at the Sforza court at Milan, those studies most useful to a powerful court outside a university setting would be related to military strategies and technologies. Leonardo da Vinci's letter of self-introduction to Francesco Sforza described his expertise in military planning and invention in detail while noting as a postscript that he could also paint and sculpt, a sequence of priorities that would have been familiar to the Milanese engineer.[57]

Agrippa's fencing lessons involved discussions of acute and obtuse angles, bodies bisected by points and lines, and carefully choreographed steps divided into unvarying and infinitely repeatable lengths and tempos, explained at every possible point in terms of geometry, optics, physics, or mathematics (fig. 49). The illustrations of the Spartan, athletic human body attacking looming Euclidean shapes were annotated with letters and numbers that keyed them to explanations in the text. Figures in motion are overlaid with impressive schematic diagrams, and there is an ostentatious use of perspective in many of the illustrations. The visual application of precise

DE LA PRIMA GVARDIA

ſignata per A. Cap. IIII.

Sſendoſi moſtrato diſopra in figure le Quattro
Guardie Principali inſieme, ciaſcuna ſignata per
la ſua littera, in ordine del' Alfabetto: et dicchiara
to la cauſa de li nomi loro, tolta , ragioneuolmente
da l' origine de la prima: Et dettoſi anchora perche ſiano le Prin

49 *Another Geometrical Figure*, engraving, Camillo Agrippa, *Trattato di scientia d'arme, con un dialogo di filosofia di Camillo Agrippa Milanese,* Rome: Antonio Blado, 1553. Typ 525 53.126, Houghton Library, Harvard University.

50 Title page and author portrait, Camillo
Agrippa, *Trattato di scienza d'arme. Et un
dialogo in detta material*, Venice: Antonio
Pinargenti, 1568, Folger Shakespeare Library.

geometric principles to the practice of fencing, Agrippa's own invention, was new and very
modern.

Also original, and expensive, was the decision to print the humanist, italic script of the text
onto the same sheet as the half-page copper engravings, necessitating the same problems that
Bartolomeo Grassi was to encounter later in Fabrizi's *Delle allusioni . . .* , requiring different kinds
of presses for the thin copper plates and the letters, carried along with woodblocks and historiated
initials in typographer's frames. Each illustrated page of Agrippa's treatise had to be printed twice
for each side of the paper that was printed, once for the image and once for the text, on two
separate presses. The Venetian publisher who issued a version of Agrippa's book in 1568, after the
original privilege had expired, made things easier for himself by copying all the images onto
copperplates printed two or four per page, and adding the pictorial pages in between the text
pages (figs. 50 and 51).[58] This was a cheaper, more manageable way to produce a book, even
though it forced a reader to move back and forth between the pictures and the text, and necessitated a less direct address to the reader from the author, a submersion of the "I" that we have
come to think of as responsibly authorial.

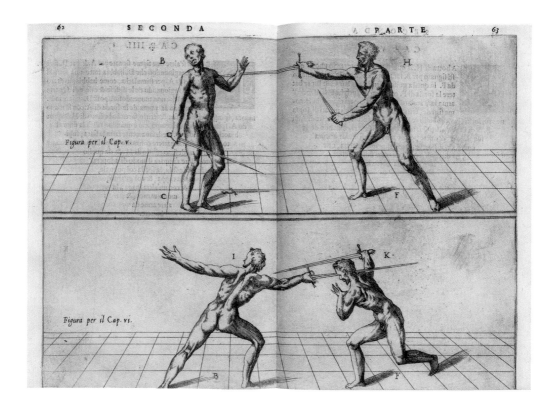

The most famous illustrated fencing manual printed up to then, and the one with which Agrippa is usually contrasted, was the 1536 *Opera nuova* of Achille Marozzo.[59] This book had used finely cut woodblock depictions of Renaissance dandies in slashed finery to model the motions of gentlemanly violence (fig. 52). Unlike Agrippa, the Bolognese Marozzo was a fencing master who called himself "gladiatore" in the title of his work. His book explains the ritual as well as the motions of defending honor, and although he adds some theoretical discussion of the duel, his theories are distinctly traditional. His diagram of *Fencing Man*, for example, which appears close to the beginning of his book, is a straightforward rendering of the medieval memory images modeled after pre-Vesalian medical diagrams for bleeding or healing wounds (fig. 53).[60] Similar figures commonly appeared at the outset of treatises having to do with the human body and were included in memory handbooks as guides for the loci of invention, meaning that they also appeared as guides for finding the wounds of Christ in frescoes on church walls, and in every other place where a human body was the template for affective or emotional interaction with ideas. Marozzo's leading image declares his book useful in the tradition of illustrated manuals for literate practitioners, and he updates the schematic flatness of his models by setting his fencing men on a squared pavement rendered in perspective.

OPPOSITE PAGE 51   Giulio Fontana, two etched pages, Camillo Agrippa, *Trattato di scienza d'arme. Et un dialogo in detta material*, Venice: Antonio Pinargenti, 1568, Folger Shakespeare Library.

RIGHT 52   Achille Marozzo, *Calling of a Duel*, frontispiece, woodcut, *Opera nouva de Achille Marozzo bolognese, mastro generale de larte de larmi*, Modena: Antonio Bergolae, 1536. John Hay Library, Brown University.

Agrippa rightly claims that his book will be different from those that preceded it, even if it was born from them, because, he says, those who desire honor from any science and art know that they will have it only when they combine theory with practice. Agrippa follows tradition to the extent that his first didactic illustration (fig. 49) was designed to prepare the reader for the discussion that follows on the movements of the duel, providing a visual armature on which to hang the rest of the book's discussion. Agrippa's model, however, is distinctly kinetic. Instead of a passive mannequin flattened for display, he shows four figures modeling the four basic positions, or guards, from which all further steps and moves will follow in combination with each other, or turned at different angles (fig. 54). Clearly labeled and with the natural origins of these fundamental movements thoroughly explained, Agrippa promises at the beginning to make the figures recognizable throughout the book by means of their identifying letters in alphabetical order, and to keep them always in proportion to the length of their swords. As rules for action, the emphasis on proportion makes them a radical departure in theory and in representation from the charts that produced *Fencing Man*.

It was also clear from the title of Agrippa's book, soberly lettered on the page facing his portrait, that it was composed for a different genre of reader from the one addressed in Marozzo's *New*

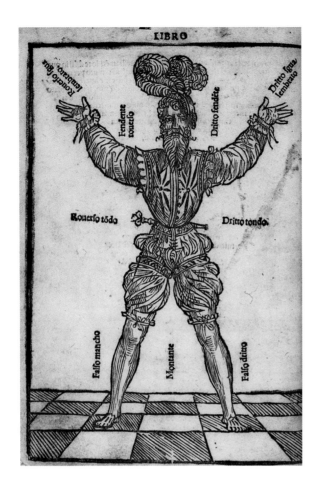

53   Achille Marozzo, *Fencing Man*, woodcut, *Opera nouva de Achille Marozzo bolognese, mastro generale de larte de larmi*, Modena: Antonio Bergolae, 1536. John Hay Library, Brown University.

Work, called, the Duel, or the Flower of Arms of singular combats both offensive and defensive. Marozzo's work included a table for calculating the appropriate degree of maiming in retaliation for the most expected insults, as well as the procedure for calling a duel. Agrippa's title prepares us for a discourse on natural philosophy. Instead of a teacher's manual recording the maneuvers of the duel peppered with anecdotes from years of teaching, Agrippa's book is presented as a hybrid work of persuasive theoretical discourse by a mathematically inclined philosopher whose familiarity with universal principles, rather than a pedigree as fencing master, gives him the right to theorize about this and many other things.

The right to speak, to participate in the publication of one's own practical knowledge, is clearly at stake in Agrippa's book. In the preface he tells us he is not formally educated, but before the first page of the treatise Agrippa posted an engraving (fig. 55) in which we observe him taking part in an animated public disputation with accredited scholars of philosophy. These disputations were designed to test students who had matriculated at a university in a variety of topics, and

54 *Four Principal Guards*,
engraving, Camillo Agrippa,
*Trattato di scientia d'arme, con
un dialogo di filosofia di
Camillo Agrippa Milanese*,
Rome: Antonio Blado,
1553. Typ 525 53.126,
Houghton Library,
Harvard University.

qualified them to hold the post of teacher or lecturer. In the fifteenth century the most common subject of disputations at Bologna was the nature of matter, one of the Aristotelian principles, and one of the subjects of the ensuing philosophical dialogue with Annibale Caro.[61] Here, the university scholars are identified by their togas and academic robes and backed up by a library full of books in which they are agitatedly searching for facts. Agrippa appears as a gentleman scientist whose research tools are, instead, the instruments that aid in knowledge derived from observation and experience. The compass, square, armillary sphere, and sword that occupy the place of books on Agrippa's side of the little seminar room are practical tools. Richly dressed in slashed leggings, fashionable codpiece, and decorative doublet, with his carefully trimmed beard and hair, Agrippa cuts a confident figure as a man of the world as he faces his unkempt academic opponent. His hand rests at the base of the armillary sphere that will reveal the working of the heavens, while his foot steadies the globe of the earth; his instruments seem like natural extensions of his own body. A man of action, Agrippa sits firmly balanced and still, tools at the ready, poised to jump

55   *Academic Disputation,*
engraving, Camillo Agrippa,
*Trattato di scientia d'arme, con
un dialogo di filosofia di
Camillo Agrippa Milanese,*
Rome: Antonio Blado, 1553.
Typ 525 53.126, Houghton
Library, Harvard University.

up if necessary. He seems particularly corporeal as he defends himself against the book-driven,
spirit-like fact-checkers who seem to inhabit every corner of the room, proof of the lively nature
of the men's debate, and showing that Agrippa is making them work hard. It is difficult not to
detect a challenge in the gauntlet thrown down on the floor like a signature at the right-hand
corner of the plate. If the purpose of such disputations was to qualify teachers, Agrippa is shown
here as surviving the onslaught: this is a pictorial certificate of his preparation as a teacher.

In showing that Agrippa's conception of natural phenomena was an idiosyncratic, Christianized
and humanist view of Aristotelian principles, Elio Nenci presents us with an early modern engineer whose conception of the world and literary style would have been very much in accord
with those Roman nobles and clerics for whom he would be working. Agrippa himself would
have had access to knowledge from books, and his native Milan was a particularly fertile place
for the cross-pollination of mathematics and pictures.[62] Part of Agrippa's Milanese heritage would

have been participation in a lively culture of armorers and military mathematicians, as well as image-publishing mathematically minded philosophers. Although he seems to have been about fifteen years old when he first arrived in Rome, his presence there is undocumented between 1535 and 1543, when he emerges as having fixed aqueducts, and he is again undocumented between the publication of the dueling treatise and 1575, when he emerges with the publication of his second book at Blado's shop, *Modo di comporre il moto nella sfera*, a book that detached the fencing treatise from the material of the philosophical dialogue, and concentrated on the latter.

The Milanese connection would have privileged such works as the Italian translation and commentary on Vitruvius by Cesare Cesariano, who worked at the court for a time. It was fully illustrated with detailed woodcuts of armillary spheres, Euclidean shapes, and Cesariano's rendition of *Vitruvian Man* (fig. 56). As Agrippa would do in the fencing treatise, Cesariano also used the

56   Cesare Cesariano, *Vitruvian Man*, woodcut, *Di Lucio Vitruuio Pollione De architectura libri dece, traducti de latino in vulgare affigurati, comentati, & con virando ordine insigniti*, Liber Tertius (n.p.), Como: Gotardus de Ponte, 1521. The Lessing J. Rosenwald Collection, Rare Book and Special Collections Division, Library of Congress, Washington, DC.

57   Luca Pacioli, fol. 6r, from Tractatus primus of *Divina proportione*, Venice, A. Paganius Paganinus, 1509. The Lessing J. Rosenwald Collection, Rare Book and Special Collections Division, Library of Congress, Washington, DC.

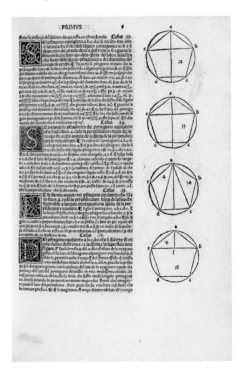

PRIMA

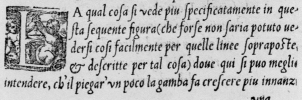

D' VN'ALTRA FIGVRA DI
Geometria. Cap. III.

A qual cosa si vede piu specificatamente in questa sequente figura(che forse non saria potuto ue-
dersi cosi facilmente per quelle linee soprapoSte,
et descritte per tal cosa) doue qui si puo meglio
intendere, ch'il piegar'vn poco la gamba fa crescere piu innanzi
vna

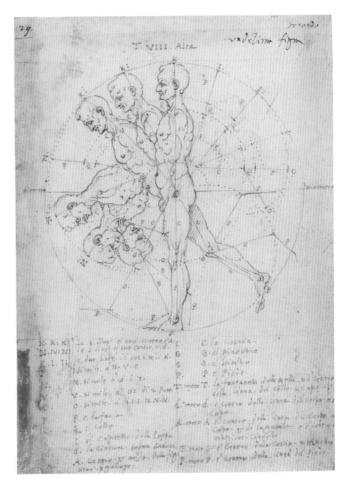

59   Carlo Urbino, Codex
Huygens (Urbino codex), sixteenth
century, fol. 22r. Pierpont
Morgan Library,
New York, 2006.14.
Purchased in 1938.

convention of adding a schematic diagram onto a picture in perspective to teach applied mathematics, showing an underlying principle at work.[63] Agrippa might also have been more aware
than most of Piero della Francesca's geometric and mathematical work as printed by the friar
Luca Pacioli, whose *Summa arithmetica* printed in Venice in 1494 would have provided Agrippa
with all the shapes he eventually encouraged his fencers to imagine in substitution for the flailing
enemies that might otherwise distract an undisciplined mind (figs. 57 and 58). Pacioli had also
teamed up with Leonardo in writing and illustrating the more famous *Divina proportione* of 1509.[64]
Images and ideas from Euclid and Vitruvius were available in Italian translation through the
medium of printed books, but in the sixteenth century the circulation of manuscripts and manuscript copies was still a vibrant business in spite of the grim predictions otherwise.[65]

The wide dissemination of the text and pictures in Leonardo's manuscripts proved this like
nothing else in the sixteenth century. They were never printed in the Renaissance, although they

60   Five figures in action, engraving, Camillo Agrippa, *Trattato di scientia d'arme, con un dialogo di filosofia di Camillo Agrippa Milanese*, Rome: Antonio Blado, 1553. Typ 525 53.126, Houghton Library, Harvard University.

were available to many as Leonardo's ideas and images turned up in copies and literary notices throughout the century. This is made particularly clear in the image of figures from the Codex Huygens, a manuscript collation of copies of pictures and transcriptions of text from Leonardo's writings on human proportion, theories of foreshortening, and how to show the movements of the human – and heavenly – bodies in motion in particularly cinematic ways (fig. 59). Some of them show the artist tracking the intermediate positions between lying down and standing up, both schematically and fleshed out as human figures.

Although Agrippa never claimed to have designed the engravings that were so important to his argument and that, more than any other single aspect of it, framed the book within a humanistic discourse, he never gave any other indication of the name of the artist or engraver. Unlike any of his other books, this one uses the individual human body as the central instrument through which Agrippa's grasp of geometry and even mechanics is displayed. While the more mechanical engravings from Agrippa's other books are confidently executed, these depictions of the human body reveal a painter's specialized training in anatomy, perspective, and proportion. Historically, the quality of the images has been judged so high that they have been attributed to Michelangelo and to members of Vasari's workshop, such as Stradanus. Lomazzo's attribution to the painter Carlo Urbino, admired for his excellence in rendering triumphs and battles, makes much more sense. Relatively unknown outside Milan and Cremona, Urbino was described in seventeenth-century sources as a student of Francesco Melzi, who had inherited Leonardo's manuscripts, brought them back to Milan, and made them available to people like Vasari and Lomazzo, who wrote of having seen them there. It has also been argued convincingly that the same Carlo Urbino was probably the artist who made the copies of Leonardo's drawings for the Codex Huygens.[66] In 1552, the year before the *Trattato* appeared, he was in Rome, where he made a drawing of the papal Swiss Guards.

As Claire Farago has pointed out, there are strong visual and textual similarities in the Codex Huygens and Agrippa's book, especially in the way figures are conceived as models to display theories and movement. She looks at Leonardo's drawings of men fighting, which were made throughout his life, and notes in them "the close connection between the variety of the pose and the sequence of movement."[67] Agrippa promises at the outset of his treatise that although correct and careful fencing looks difficult, anyone who wants to succeed at it will find it easier if he remembers that the profession is governed by points, line, tempos, and measure. The recourse to geometry and particularly to those four determining factors resonates closely with Leonardo's writing about motion, as do the treatises of other practitioners, such as Dürer, writing on proportion. Agrippa's immediate family included at least one other inventor and a bombardier, both professions that relied on a knowledge of mathematics and of drawing, and who would have taken an interest in Leonardo's work.[68] There is no shortage of evidence in his book that Agrippa might have had access, either directly or mediated through copies by Carlo Urbino and others, to Leonardo's manuscripts both in Milan and in Rome.

Agrippa's *Four Principal Guards* (fig. 54) are modeled by four basic figures in his first engraving, identified by letter, so that they can be treated separately and ultimately recombined in complex

61   *The Second Guard*, engraving, Camillo Agrippa, *Trattato di scientia d'arme, con un dialogo di filosofia di Camillo Agrippa Milanese*, Rome: Antonio Blado, 1553. Typ 525 53.126, Houghton Library, Harvard University.

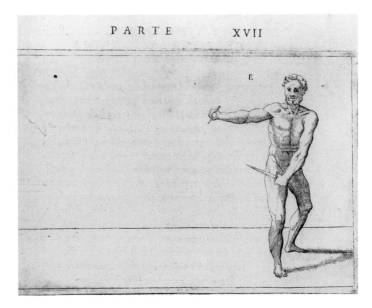

skirmishes later in the book (fig. 60). While the Leonardesque image in the Codex Huygens shows an uninterrupted sequence of motion, Agrippa's shows different men in a sequence of poses that originate from a move spontaneously made, Agrippa says, by sword-carrying men when provoked into anger. Here, however, the figures are deconstructed into the building blocks for more complicated moves that will come later. Agrippa explains their origins:

> Anyone who carries a sword at his side, stimulated by his own anger, or by some exterior provocation of words or deeds, after having taken the sword all the way out of its sheath, will be seen to be holding it up high in the form of a guard which, being the first that is made right away after removing the sword, is called the first guard. Subsequently, lowering the hand a little with the arm in the same plane as the shoulder, the second is formed. Then, lowering the hand with the sword in front of the knee he comes to make the third, and the last of these, carrying the hand with the sword behind the knee, will be the fourth. They are the *principles* because from them proceed and are formed the various other guards according to the most necessary considerations and occurrence of this profession.[69]

The sequence of movements that Leonardo seems to have investigated by drawing the ephemeral movements of a body in motion as it transfers itself from one stable position to another is deconstructed as a way of substituting memory and moderation for the uncontrolled angry impulse in Agrippa's *Four Principle Guards*. In his explanation Agrippa retains the notion of a natural, spontaneous movement stimulated by sudden anger and dissects it into four separate but connectable poses that the reader must commit to memory, to be resorted to in ideal form and in perfect control should exterior provocations arise.

The figures in Agrippa's book, carefully lit and with their shadows meticulously included, resemble toy soldiers put into the service of explaining themselves as they model their skills. Throughout his explanations Agrippa shows a remarkable attention to the fact that as figures turn in space, they appear to be very different through the new view of them that you will have. For "view" he uses the word "prospettiva." It sometimes seems that he is creating his explanations from the pictures as much as the other way around. In the case of the second Guard as seen from a position directly facing you (fig. 61), he precedes his explanation with an answer to the understandable anxiety about what you should most pay attention to when anticipating your enemy's next harmful action. He recommends keeping your eye on the hand holding the sword, since it is from this that the most terrible blows proceed. This inspires him to continue: "Finding yourself in the Third *larga di passo* far enough away from the enemy so that he cannot touch you, in case he tries to push away your sword by force, pull back your hand in the Second position as in this figure, which holds the sword in his hand in foreshortening, even though it is not visible for being in perspective."[70] The image relies heavily on a strong knowledge of how to effect foreshortening, or *scorcio*, in *disegno*, with an applied use of chiaroscuro to make it legible. The nude fencer faces the reader dead on in a slight crouch, with both knees slightly bent and one arm held out at the side grasping the hilt of a sword pointed directly at the reader's eyes. Therefore,

PARTE   XIX

chi benche fiano doi, non pero ponno uedere piu d' vn punto per volta, non potendo naturalmente andar' le linee loro, à Paralella, ma à Piramide, à finire in vn punto folo.

62   *An Act Denoted by 'G',* engraving, Camillo Agrippa, *Trattato di scientia d'arme, con un dialogo di filosofia di Camillo Agrippa Milanese*, Rome: Antonio Blado, 1553. Typ 525 53.126, Houghton Library, Harvard University.

all that is visible is a clenched fist seen in profile, holding the hilt of a sword whose foreshortened blade magically becomes invisible. This is an unusual way to demonstrate a fencing move, where the swords are always visible. One is reminded of Agrippa's promise always to keep the images of the fencers in proportion to their swords, something that seemed to lend assurance about the size of the sword. However, here we are not entirely assured that chimeras such as point of view might not be as dangerous as other deceptions. The engraver who incised Carlo Urbino's drawings has used masses of parallel hatching lines to emphasize the fencer's crouching lunge by shading the hollow of his concave belly and receding left leg, and the calf of the bent leg with which he approaches.

Optical theory is brought into play again in figure G from chapter XI, as it helps Agrippa to explain how to deal with someone who is trying to wound you with a left-handed blow (fig. 62). Agrippa's descriptions, and the words he uses to denote the movements of his fencers, are very much the same as those used by painters to describe points of view and the representation of objects in space. Therefore, he speaks of *superficie* to describe the outsides of geometric moves,

points, and lines, and even motions like *di sotto in sù*, which is the kind of perspective foreshortening used to paint a scene to appear as if you are looking up at it from below. After a series of
complicated passes Agrippa brings in Albertian optical theory to clarify that the good fencer must
remember to stay fixated on points in a description of a move in which the fencer fakes out his
opponent by unexpectedly stepping away from his line of vision: "as you see in the following
figure, with the many lines drawn to its back from the two points of the eyes, indicated in this
way to let you know that even if there are two eyes, they cannot look at more than one point
at a time, not naturally sighting along straight parallel lines, but in a pyramid, to finish in one
single point."[71]

It is this last startlingly impractical point, rather than the complex footwork, that is illustrated
in the accompanying engraving. It also bears a remarkable resemblance to Leonardo's statement:
"Perspective is a rational demonstration, by which experience confirms that all things send their
images [*similitudine*] to the eye by pyramidal lines." These lines are "those which depart from the
surface edges of bodies and coming from a distance draw together in a single point. A point is
said to be that which cannot be divided into any part and this point is that which, residing in
the eye, receives into itself all the points of the pyramid."[72] In his English translation of Agrippa's
book, Ken Mondschein smoothly integrates Agrippa's unique way of adapting the language of
Renaissance optical theory into a comprehensible discussion of feints and parries that describes
the movements of the duel. This is no doubt the way many fencers, interested in Agrippa's ideas,
would have read the text, straining at times to figure out his meaning, and reading it colored by
their familiarity with traditional fencing. Compared with Marozzo, or with other earlier fencing
books, one sees that terms like "tempo" or the idea of a wheel as symbol to contain movement
and describe position (much as we might say to move your foot to three o'clock) appear regularly
in more mundane sentences about fighting. Agrippa was taking full advantage of the meanings
these words might carry in another context and imported the language to a readership that would
understand his rhetorical move. In this, he was acting in the same way as Alberti when the latter
appropriated the language of rhetoric and geometry to ennoble the mechanical arts that used
disegno. Agrippa, probably taking his cue from the circles of Luca Pacioli and Leonardo da Vinci,
was using the new language of *disegno* as a conduit for demonstrating geometrical theories regulating the art of the duel.

In the second part of the book Agrippa joins his figures together in battle, showing how the
theories attached to the various steps and positions would work in the course of an actual skir-
mish.[73] He begins by summing up what he has already imparted to the reader, and says he will
move on to show how the principal guards may be used. He then reminds us that the letters A
through D posted at the feet of the figures signify the principal guards, and the succeeding letters
posted above signify the various attacks. To demonstrate this, he gives us the fencer whom we
last saw disappearing into an ever-enlarging pyramid (fig. 62) sinking his sword into his opponent
while simultaneously repelling his attack (fig. 60). The same figure turns up in various perilous
situations throughout the book, as do all the figures, demonstrating the contexts in which such

63   *Battle in the Forum*, engraving, Camillo Agrippa, *Trattato di scientia d'arme, con un dialogo di filosofia di Camillo Agrippa Milanese*, Rome: Antonio Blado, 1553. Typ 525 53.126, Houghton Library, Harvard University.

a position could be useful. The demonstrations in the second part of the book become increasingly complex, evolving into battles with up to four enemies at once, sometimes with the fencers wearing modern clothing, sometimes naked, and sometimes in Roman gladiatorial garb or on horseback. The treatise ends abruptly on an uncertain note with the admission that Agrippa has no experience of fighting on horseback; this is because, he says, he has been prevented from practicing it due to a handicap that he was born with.

As revolutionary as Agrippa's book promised to be, when it was reprinted in 1568 the publisher evidently felt there would still be a market for Marozzo's book among the less theoretically inclined swordsmen, and he reprinted that as well in the same year, updated with etched illustrations.

The Philosophical Dialogue

The treatise ended, we come to the philosophical dialogue promised in the title, headed *Dialogo di Camillo Agrippa / Annibale et Camillo*. The scene is the home of Annibale Caro, famous playwright, artistic advisor, deviser of emblems, and translator of Aristotle's *Rhetorics*, who had been among those who had encouraged Vasari to write his *Lives* at a conversation over dinner in the same palace, the Palazzo Farnese, in which he now lives in the service of Cardinal Alessandro.[74] Annibale has asked Camillo to come around so that he can give him what amounts to a bad reader's report, telling Camillo not to publish his treatise without first taking out "those geometric figures . . . so as not to confuse the minds of those who see them, who will then judge you to be something you are not."[75] Camillo thanks him for his warning and says he will pay attention to it – especially since the other evening he dreamed that he was mugged in the forum by certain philosophers who judged him to be presumptuous to write about these things, since he did not have a university education.

Inserted at the end of the treatise and signaling the beginning of the dialogue is an engraving of this particularly anti-academic scene, set in the Roman forum among the ruins of the engineering feats of antiquity (fig. 63). The swashbuckling Agrippa strikes one of the athletic lunges his nude models perform in previous illustrations. He, however, is dressed as a gentleman in slashed hose and a doublet trimmed with embroidery, while a strap slung around his hips allows a long sword to lodge against his prominent codpiece. Instead of taking on his adversaries in the measured choreography of a duel, a catfight takes place as the author is set upon by a gaggle of robed academics who pull his hair and derange his clothes, physically attacking him with the very instruments with which he defended himself in the disputation scene: one wields a compass, and a square lies trampled on the ground. The ragtag philosophers press in, brandishing instruments they do not know how to use, ridiculous with their wild beards, Phrygian caps, floppy doctoral hoods, and distinctly unathletic movements, their rear ends prominently displayed beneath their clingy togas in obvious insulting counterpoint to the powerful codpieces of Agrippa and the cadre of sword-bearing gentlemen who try to protect him from attack. Agrippa tells Caro that his carefully dressed and coiffed defenders are the friends and patrons who came to his rescue against the accusations of certain students of Euclid and Aristotle

> who were calling me presumptuous in wanting to discuss similar things, I not having studied. Then it seemed to me that with the help of many gentlemen who were my friends I defended myself . . . and I am telling you this so as to alienate any bad impression that anyone might have who should see these figures, and to show the world that even if I am not university educated, that naturally even I can speak logically about some things. And if you want to see the proof of this, take my book in hand, and find the figures that I made understandable using the letters of the alphabet, if, however, you are not bored by all this theory.[76]

At the root of the argument that we see as an academic disputation at the beginning of the treatise and as a mugging in the forum at the end is the question of how knowledge should be acquired and who had the right to present it.[77] Beyond their university affiliation, the philosophers are not identified in the text except as those students of Euclid and Aristotle who balked at crediting Agrippa's right to own natural knowledge through a grasp of theory as well as of mechanical craft, and therefore, in the context of the book, his right to be an author.

The argument about the trustworthiness of assumptions derived from practice (or art, or *techne*) which have been correlated with experience had a long history deriving, for these purposes, from Aristotle's *Poetics*, and one that was truly carried out as a kind of battle in itself in this period. Defined as knowledge gained from the habit of a particular activity, it was understood as being contingent on its particular circumstances and distrusted by the ancients as an uncertain and undependable distraction from *scientia*, or true knowledge.[78] Pamela Long shows that in classical Athens, "Xenophon and Aristotle explicitly separated material construction and the technical arts (*techne*) from the praxis of political and military leadership . . . the ancients thought that rulers and military leaders achieved success because of character traits such as courage and virtue, not because of technology or technique."[79] However, the Greeks claimed to value both the active and the contemplative life, theoretical knowledge tempered with practical knowledge.[80] Both the Sophists and the Cynics conceded importance to sense-derived experiential knowledge, deeming it a subject worthy of investigation. Julia Annas has shown that Plato's appeals to *techne*, or skill, in discussions of moral knowledge showed that practical expertise is Plato's model for knowledge, and that the kind of expertise gained through practice "demands a complete understanding of the relevant field."[81] As central as such thinking might have been in antiquity, it remained marginal in many authoritative circles, and little changed through the work of Renaissance humanists, many of whom were equally, if not more, attached to the authority of the book, despite lip service to the classical principle of checking in with the senses.[82]

However, there was a strain of humanist learning that took more interest in aspects of practice. As Long points out, there was a tradition in imperial Rome of gentlemen writing technical treatises or encyclopedias for members of their own class, and often for their own inner circles,

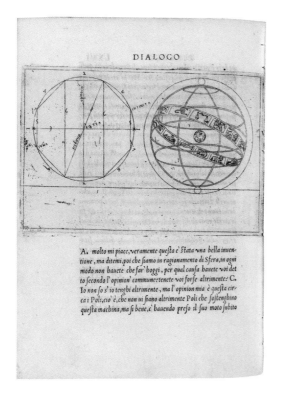

64 Armillary sphere, engraving, Camillo
Agrippa, *Trattato di scientia d'arme, con un dialogo
di filosofia di Camillo Agrippa Milanese,* Rome:
Antonio Blado, 1553, Folger Shakespeare Library.

on topics such as agriculture, military leadership, or natural history, which were considered suitable for holders of political office.[83] By the Renaissance, treatises on such subjects written by the people we would today call artists and perhaps especially architects, but including all kinds of craftsmen, became what Long has termed "trading zones" in which wealthy patrons and knowledgeable practitioners communicated and even collaborated across social boundaries: "A broad middle ground of communication, a trading zone of knowledge, developed between the wellborn who were seriously interested in the constructive arts and the skilled individuals who became their employees, clients, teachers and sometimes friends."[84]

When Agrippa constructed his polemic as a clash between effeminate university trained scholars in their academic robes and virile gentlemen in their fashionable finery and swords, he was not claiming that his practice alone gave him the right to write a book, nor did practice alone give him access to the truths about his subject that raise it to the level of, as his title claims, either natural knowledge or philosophy. Leonardo, who tirelessly observed and recorded the effects of nature, also recorded the process of losing faith in books, as when he realized in the course of an anatomical dissection that the heart was a muscle and not a kind of oven. Still, observation often misled him in murkier cases, when he misinterpreted what he saw in light of what he had read and knew about human anatomy from books.[85] Constantly checking as far as one could by

65 Tav. LXXXV, *Museum Mazzuchellianum: seu numismata virorum doctrina praestantium quae apud Jo. Mariam comitem Mazzuchellum Brixiae servantur / a Petro Antonio de comitibus Gaetanis . . . edita, atque illustrata; accedit versio italica studio equitis Cosimi Mei elaborata*, vol. 1, Venice: Antonii Zatta, 1761–3. Typ 725.61.414F, Houghton Library, Harvard University. Detail (below) Medal of Camillo Agrippa, r & v.

balancing experience with books and with nature, an inquiring person could approach true *scientia*. In his section on shadow and light in his notebook on painting Leonardo wrote: "I say here that the adversary who does not want so much science, for whom practice is enough in the imitation of nature, to him I respond that nothing misleads us more than to trust in our judgment without other justification, which is always proven by experience, the enemy of alchemists, necromancers, and other simple liars."[86]

Annibale Caro accepts the challenge to pick up Agrippa's book, and for the next three days and over the course of thirteen pages Camillo engages Annibale in technical demonstrations of how to make triangles, points, and lines, moving a compass this way and that to make hexagons and circles of every sort, including all the basics of geometrical measurement: how to form a square, divide a circle with its diameter, make pentagons, and so forth. At the end of the first day Camillo staggers home exhausted, promising that he will return to discuss ovals, octagons, and spheres, but first he wants to go home and write down what they have said in conversation – in other words, the passages we have just finished reading. Annibale tells him it is a good idea to write it all down while his memory is fresh, and anyway it is a holiday and the book could not be sent to the press on that account, even if everything else were in order. In contrast to Camillo's evident reluctance to make his work public without Caro's approval, Caro speaks of printing with a startling sense of immediacy, as if printing the written pages on two presses involved no more planning than making a photocopy.

The second day Camillo arrives with the "Latin about the material that I promised you," and immerses Annibale in a discussion of how to make an armillary sphere from an octagon, complete with the zodiacal band of astrological symbols, and a discussion of poles and equinoxes that he might have gathered from a popular primer such as Sacrobosco's *De sphaera* (fig. 64). Again, it becomes late and Camillo takes his leave promising to return, but by this time Annibale is becoming sensibly more eager to obtain knowledge from Camillo; he asks his friend to come earlier in the day than before so they can revisit the issue of whether or not to send the book for publication, which Camillo pretended to have utterly forgotten. The men spend the third day lost in conversation about the motion of the planets, the sun, the moon, and the stars, dealing with thorny questions about the center of the earth and the center of the world, about which there were then some doubt. They discuss the fixed stars (the zodiac) and those celestial spheres that move – among them, the earth.[87]

At the end, Annibale tells Camillo to include the dialogue with the treatise, to make any necessary clarifications to the pictures that are needed, and to get the book to the printer as quickly as he can. The printer in question, Antonio Blado, will turn up, perhaps already has turned up, as Barbagrigia, the benevolent pillar of the community in Caro's play *Gli Straccioni*, which takes place just outside the walls of the room where, in the fictive dialogue, he and Agrippa have been conversing.[88]

It may well have been Annibale Caro, the author of a treatise on the iconography of coins and medals, who devised the *impresa* on the reverse of a bronze medal of Agrippa made by Giovanni

Battista Bonini, a Milanese goldsmith working in Rome at mid-century in papal circles (fig. 65).[89] The medal bears a profile portrait of Agrippa on the front, and a curious emblem related to the philosophical dialogue on the verso. As an object, it confirms Agrippa's antiquary, courtly, and humanistic ambitions. Like the fencing treatise, the *impresa* shows two figures locked in a struggle for dominance. A caped and helmeted warrior in full antique armor, holding a lance (a weapon that, in the fencing manual, Agrippa admits has no rules of general use) and shield in one hand, uses the other to grab the forelock of a nude figure of Fortuna, who seems to be pulled off balance but steadies herself with one arm against his thigh, and with the other holds up a swelling sail for ballast that emphasizes the outline of her belly and breasts. In this way, her figure neatly fills the curve of the circular surface and underscores the central word of the appropriately ambivalent motto, "Velis nolisue": Whether you wish it or not. The naked and vulnerable woman, armed with a sail to harness the changing winds, and the armed warrior who uses his muscle rather than his military skill, seem in this instant to be oddly stable at the center of the bronze circle as if locked together in Fortune's spinning wheel.

Aby Warburg noted the ambiguity of the motto and interpreted the emblem as proof of man's violent and bestial nature, and the senselessness of trying to dominate Fortune (fate) with brute strength, in so doing, "destroying his means of transport for the price of a scalp."[90] But taken together with the fencing manual, the discussion of the direction of the winds, and the geometry of movement, and also bearing in mind Agrippa's later conception of himself at the center of a wheel of publications, the medal and the scope of his projects take on alternate meanings. The astrologer at the center of the microcosm, like Agrippa's warrior who tries to grapple with fate, was also attempting to master the forces of the universe. Fritz Saxl saw the Renaissance version of this as an assertion that Fortune is always weak and man the greatest force of all.[91] He ends ambivalently, finding in the stated relationship between man and the universe the real beginning of modernity: "The idea that the self is the prime object of our discernment and, on the other hand, the widening range of objects to be observed and measured, were the agents that led to a new explanation of the relations between man and the universe and through it to the complete break with the speculations of Late Antiquity."[92] But much of Agrippa's worldview was formed with the technologies that modernity would use to explain medieval versions of antique ideas.

If we compare the medal with Agrippa's engraving of the brawl in the the forum, we find out more about the importance of pictures to his notion of authorship. The engraving is a pendant to the frontispiece of the scholarly disputation; all three images, along with the engraved demonstrations of fencing in the treatise, are images of battle using wit, tools, and strength. The engraving of the brawl shows Agrippa pulled by his gentleman companions away from the squabbling and ridiculous academics who act in a landscape of unreconstructed ruins, a barren and silted up incomprehensible wasteland. Agrippa's world is robust with message-bearing obelisks and the clarity of architecture, weight-bearing columns attached to their entablatures, all of it knowable, weighable, measured, and pondered. On the medal, the gladiator stands in front of the door of a classicizing temple, firmly anchored to the ground. As she reaches towards her sail, Fortuna is caught off-balance with one leg crossed behind that of the weight-bearing leg of the soldier, her hand grasping his upper thigh, as if modeling a hold from the fencing treatise. "Velis nolisue" is visualized here as the domination of Fortuna through the force of the armed male, whose military weapons are pictured only as accessories, attributes of his nature and training. At the center of this small bronze wheel of the world, one that bears Agrippa's name and features on the other side, is the perpetual struggle of man for dominance over Fortune, a struggle in which man, by his nature, has the possibility to transform fate from a fickle force to a subservient mate. But can man really win this match? "Whether you wish it or not" cannot only be an admonition to Fortuna. Agrippa's vision of fate was a warrior locked in eternal struggle with the forces of nature, using his own nature as a weapon and a tool. In the dialogue he crafts with a fictional Caro, we see Agrippa trying to control the fortune of his authorship through the various conventions of publication, given visual expression in the engraving and in the medal.

The Defense: Gentlemen vs. Scholars

Can we say anything about these fictional gentlemen with the names of historical people who defended Agrippa's right to become an author, showing up for him in the forum of ideas as well as in the Roman forum? Caro, a Farnese courtier, was himself about to receive a lesson in the perils of publication and the vagaries of fortune. In the same year that Agrippa entrusted his fencing treatise to Antonio Blado with the literary Caro's enthusiastic encouragement, the real Caro published a serious and highly ornamented poetic appeal to the Muses to celebrate the transfiguration of the French kings into virtuous heroes and Olympic divinities. It was written in the form of a Petrarchan *canzone* and titled *Venite all'ombra de'gran gigli d'oro*; it found immediate praise in circles of Roman literati allied with the Farnese, who were firmly represented by the same lily and allied with the French. In 1554 a Modenese lawyer in Rome who did not think much of the verses sent a copy to the literary critic and grammatical historian Ludovico Castelvetro in Florence, who wrote and had circulated in Rome a sarcastic and biting response titled *Parere sopra la Canzone del Caro Venite all'ombra de'gran gigli d'oro*.[93] In this response he listed errors both of writing and *sentimenti* in Caro's work. Caro responded with another publication in reply (*L'apologia*), and the argument continued to the end of the decade, not only with verbal vitriol but actually resulting in the murder of a poet, in a volley of publication and violence aimed in part at defining what literature written in the vernacular should be.[94] At the end of the century the Inquisition hounded the members of the Florentine literary academy to act as knowledgeable censors in helping to expurgate heretical or unseemly passages from otherwise worthy products of the press, something the academicians had no wish to do. Unable to hold out any longer, they finally responded by turning their combined attention to one single book, expurgating Caro's poem with such thoroughness and single-minded intensity that they were relieved of the task of helping the Inquisition forever after.[95]

While it seems that Agrippa did have reason to worry, he occupied a social space between that of a freewheeling courtier without a court and bourgeois artisan in need of patrons, a space he calls that of the gentleman, somewhat below Caro or Vasari. The gentlemen who vouch for Agrippa do so in three ways in the *Trattato*. There was the desired upper class or noble patron to whom the book was publicly dedicated – the person under whose mantle, dedications often say, the book appears; there was the well-situated interlocutor in the dialogue, in this case Caro; and there were named gentlemen who popped up opportunely as if to provide evidentiary citations.

At the beginning of the dialogue with Caro, Agrippa mentions that he had often discussed the material in the book with Caro as well as with Alessandro Corvino and Francesco Siciliano. Corvino was a secretary of Duke Ottaviano Farnese and a friend of Caro's who had an antiquities collection that Caro acquired after his death. He was an enthusiastic letter writer, and his letters turn up in the papers of almost every important man in Roman artistic circles including Titian, Michelangelo, and Bartolomeo Ammanati. Pietro Aretino described him as a youth of incomparable intelligence, charm, and *virtù*.[96] Francesco Siciliano was a sculptor who made a statue of the Madonna and Child with putti for the church of Santa Maria sopra Minerva, as well as other sculptures for Roman churches.[97] Agrippa claims that he had spoken with those men about the use of lines and figures before choosing to go ahead and place them in his book. In their own time these men worked with a command of letters and perspective, of the active application of knowledge found in books, as well as the benefits of practice.

Other members of Agrippa's circle of gentlemen are named at the very end of the dialogue with Caro, who is worried that as good as Agrippa's reasoning is, there will be those who are not disposed to believe him. Agrippa answers:

> Certain people still did not want to believe that I could show, in a *Sfera materiale*, the course of the sun, that of the moon, its waning and waxing, the opposition between them, the interposition of the earth between them, the quantity of the zodiac and the other secrets of the heavens, and I showed you this with my *Sfera*, that I made, and you, Alessandro Ruffino, Iacomo del Negro, Hieronimo Garimberto, Francesco Salviati, and Alessandro Greco, with an infinite number of other *virtuosi* and *honorati homini*, have seen it.[98]

Alessandro Ruffino (d. 1579) was a Roman nobleman, the bishop of Melfi from 1548 to 1574, and before that a favorite secretary of Paul III Farnese. He owned a villa in Frascati and a palace in Rome near Piazza Navona that contained at least a few antiquities. Perhaps more importantly, he himself was a patron of a self-taught architect-engineer named Rafael Bombelli, who was active in hydraulic work and would soon publish a treatise on algebra.[99] Girolamo Garimberto (1506–75) was the bishop of Gallese, vicar of S. Giovanni in Laterano, and an active antiquarian with his own collection in the Palazzo Gaddi with a frescoed façade by Polidoro da Caravaggio, on the site of Montecitorio.[100] He also worked for Cesare Gonzaga at Guastalla, for whom he had procured a gallery of antiquities sent from Rome, and advised him and the Farnese on related matters.[101] He wrote a book on how to write, one on leading armies, one on the comportment

66 Whole page and detail with cell with address to Camillo Aggripa, Martino van Buyten, *Nuove forme di lettere*. C.2.11/47, Biblioteca Angelica.

of gentlemen as well as on cities, and in 1547 he published a book called *Della fortuna libri sei* that discussed historical and contemporary instances of the appearance and machinations of *fortuna* in the lives of famous people.[102] Francesco Salviati (1510–1563) was a Florentine painter in the circle of Vasari and the Medici in Florence, who spent increasingly more time in Rome. He was in Rome at the time of the publication of Agrippa's treatise, working on the decoration of the Sala dei Fasti Farnesiani in the Palazzo Farnese, the Farnese chapel in the Cancelleria, and the frontispiece for Antonio Labacco's treatise on architecture, among other commissions.[103] He also seems to have painted more than one portrait of his literary friend Annibal Caro, who had proven a benevolent intermediary in securing and facilitating commissions for the Farnese.[104]

Agrippa himself turns up in print, used by one of the literary gentlemen in much the same way in another example of Roman print culture. A Dutch printer/publisher working in Rome named Martino van Buyten engraved examples of the *cancelleresca* italic script for handwriting manuals, as well as the famously beautiful engraved lute music of Simone Verovio. He also published some broadsheet examples of fancy handwriting, particularly useful for impressive addresses at the beginning of formal letters (fig. 66).[105] This sheet was an engraved sampler in the form of a grid with elaborate decorative scrollwork and curlicues, as well as sample salutations for addressing letters. Like the page of Lodovico Curione's elegant handwriting manuals, which van Buyten also engraved, these examples offered some useful epigraphs by Verovio along with proper models for addressing flattering praise to high places: "The Most Illustrious and Most Excellent Lord, My Lord Cardinal Colonna," for example. The first cell of the grid contains van Buyten's own dedication of "various forms of lettering" to Giacomo Boncompagno. The last cell of the grid contains a different kind of dedication, disguised as another example of address, more informal and warmer than the others: "To my most dear, as a brother, Camillo Agrippa esq."

Agrippa was an architect and engineer, but also an active philosopher for his own busy circle of medallists, collectors, curial secretaries, cartographers, military engineers, inventors, engravers, and publishers. An avid polemicist, he was thoroughly steeped in the full flow of beliefs and superstitions common to most men of his time, as well as having studied Aristotle, probably through the works of Niccolò Tartaglia and other vernacular writers, perhaps through the manuscripts of Leonardo. He left purposeful notice of his intellectual relationships with friends, patrons, and mentors in his printed books, and the illustrations he commissioned for the *Trattato di scientia d'arme* make much of the richness and variety of that culture visible, inflecting his text as much as the self-conscious scattering of the names of his more famous acquaintances through its pages. His experience as an author, which he narrates in parts of his treatises and dialogues, points us to the formation of a community becoming adept at manipulating the possibilities that publication provided for the creation and control of reputation, a world that was accessible through print though conceived as local and knowable.

The claim that the road to knowledge and understanding was best attained by a grasp of theory brought to life in practice ("la Theorica vivificarla con la prattica," he wrote in his *proemio*), the gentlemanly and honorable subject matter of fencing, and the sum of Agrippa's interests – moving

67 Cornelis Cort, *Virgin and Child between Sts. Cosmas and Damian*, 1570, engraving after Michelangelo, Rafaello da Montelupo, and Giovanni Angelo da Montorsoli, British Museum.

obelisks, conducting water to a fountain, drawing bodies moving in certain ritualized ways – make him seem in some ways like the quintessential Renaissance man. But rather than see him as a poor shadow of Leonardo, we can understand him to have been skilled in an array of practices that were of interest to the members of the artisans' confraternity to which he belonged, the Compagnia di San Giuseppe, and to artists who first tried to stabilize their professions as liberal arts by founding an art academy in Rome, people for whom Leonardo's ideas were important. Exactly those skills that Agrippa was marketing were among the first requested curriculum of a proposed Academy of Saint Luke put forth by Girolamo Muziano in 1572.[106]

His personal battle to narrow the obstructive distance between theory and practice in the world of letters was visualized one last time in the engraved allegory illustrating the 1595 treatise on navigation, printed on the back of the following Letter to the Readers:

> Kind readers, the author of these navigations intends you to find some things that will please you, if however you take delight in them. Because delight makes man more diligent and avid in the arts and sciences, putting them into practice they will become more excellent in deeds than in words. Therefore, gentle spirits, look on my work with love, because the author gives

it to you with the hope and desire to satisfy you, hoping again that you will augment it: because there is no end to science, nor to art.[107]

The engraving (fig. 37) shows Agrippa again in disputation, but no longer confined to a musty room or subject to physical attack. Instead we see the author evidently on Parnassus in a court of natural law, laying out his arguments for the judgment of Nature herself. Parnassus, of course, was the home of the Muses, but also more immediately the artificial mountain and site of Agrippa's hydraulic triumph in the gardens of the Villa Medici.[108] Once again the author faces a crowd, but this time his audience is the Muses, and perhaps a few Liberal Arts, in the company of a rather marginalized Apollo. Agrippa, dressed as always in doublet and cape with his sword by his side, addresses Nature in the form of Diana of Ephesus, very much as she appears as a fountain at the Villa d'Este, here seated on a high bench in her fortified mural crown, rings of breasts, and a lion adorning each elbow.[109] She is flanked by two bearded men, the visual effect of the group being much like the figures of Cosmas and Damian with the Virgin and Child in the Medici chapel in Florence (fig. 67), but here with the suckling Virgin substituted by the figure of Nature, and the Medici doctor saints by figures of the *vita activa* (practice) and the *vita contemplativa* (theory).[110] The identities of these Michelangelesque figures are clarified by a decorative swag that runs like a curtain across the top and sides of the print, composed entirely of the sorts of weapons Agrippa has favored in the past: a string of books hangs at the side of the *vita contemplativa*, and a trophy made of pieces of Renaissance arms and armor frames the *vita activa*, who also holds a book. Nature is seated beneath a swag of instruments by which her laws are made known to men: compasses, squares, and an armillary sphere on one side, and musical instruments on the other. One of her sacred deer lies at her feet, staring benignly and with rapt attention into the eyes of Agrippa as he beseeches his master: "You understand, noble lady, the debate between the Muses and the regulated sciences, and the differences between us; and according to your wisdom may you judge our many disagreements so that we may live in peace."[111] In the almost half century since the appearance of the *Trattato di scientia d'arme* the terms of this debate seem to have become less violent, but otherwise not to have changed much. The author still sought a direct connection to readers hungering for information derived from experience, and seemed to feel that he had to continue to appeal to higher authorities about who had the right to educate the public. But by 1595 Agrippa felt he could portray himself as having unmediated access to Nature, who – pointing to him and inclining her crowned head towards him rather than towards the Muses and the Arts – seems to nod towards experience, thus rewarding in his person the unity of theory and practice. When Agrippa died in 1600 he was buried in the Lombard church of Santa Maria del Popolo, at the foot of the Medici gardens. In the notice of his death, though he never built a building or even wrote about such a thing, he is described in the elegant language of the university men, as "Magnificus Dominus excellens peritus ac sapiens Architectus" – Magnificent Master, excellent, skilled, and wise Architect.[112]

4

The Care of the Body in Pietro Paolo Magni's Manual for Barber-Surgeons

On a November day in 1587, Aurelia di Antellis, the recent widow of the engraver Adamo Scultori, met with the barber Pietro Paolo Magni in the presence of a notary. The purpose of the meeting was to close out the terms of a partnership (*societas*) "concerning the printing of certain books" that Adamo and the barber had entered into three years earlier.[1] The publication in question was the *Discorsi di Pietro Paolo Magni piacentino sopra il modo di sanguinare attaccar le sanguisughe, & le ventose far le fregagioni & vessicatorij a corpi humani* (Discourses of Pietro Paolo Magni of Piacenza on how to bleed, attach leeches and cups, perform massages and blistering to the human body), a quarto-sized book illustrated with eleven full-page engravings and an engraved architectural frontispiece that carefully and elegantly explained the role and techniques of a barber-surgeon in the art of letting blood and associated curative techniques.

The manual was published by Bartolomeo Bonfadino, and besides the engravings was further embellished with his signature ornamental capitals and friezes, and a single schematic woodcut diagram of how to score a patient's skin with a lancet when performing a particular procedure with cupping. Between the first edition of 1584, the date of the original agreement between the printmaker and the barber, and the second, dated 1586, a new frontispiece had been engraved that advertised corrections and new advice by the author, the language of the text had been upgraded into a more standard and authorial Tuscan, and certain passages were now emphasized by the word "nota" sprinkled judiciously throughout the margins of the text where it had not been before. Although the engraved illustrations were, for the most part, unchanged, a new emphasis on the hermetic relationship between the barber, physician, and client, and the insertion of text warning about the unreliability of female patients and practitioners, had the effect of changing the way the illustrations were read.[2] In consecutive editions the illustrations caught up with the text as they, too, were altered so that readers were guided to understand more definitively the authoritative role of the physician in light of a general mistrust of women, both

OPPOSITE PAGE   Adamo Scultori, Plate VI, Pietro Paolo Magni, *Discorsi di Pietro Paolo Magni* (detail of fig. 88).

as patients and healers, while the curative interventions of the barber – the stated matter of the book – remained unchanged.

This little illustrated book appeared in many editions, and as we follow its trajectory through the press over the course of the next century and more, we can see the dynamic relationship between text and illustration as it was manipulated in the hands of serial publishers. In spite of the costs entailed, they adjusted the book or its illustrations with each new edition to keep up with changes in the status of medical practitioners and in tolerance for the participation of non-professionals in the form of patients, relatives, and females in general. This began almost immediately after the first publication, with improvements to the text to make it more elegantly "Tuscan" and a corresponding reworking of the frontispiece. Each new edition worked both to secure and explain the barber's role in the treatment of ailing bodies, all the way into the eighteenth century. It is most notable that the techniques Magni describes, which depended on the treatment of the body according to theories of the humors, and which were described in terms of the relationship between patient and practitioner in sixteenth-century terms, remained relevant from the first edition of 1584 to the last, in 1703, in what became a medical best seller. The adjustments to the images in the book reflect the struggles over propriety and professionalism that marked the medical practices resorted to by most people, who still demanded bleeding as the remedy for an astounding variety of illnesses and conditions through the middle of the nineteenth century.[3] Although there is much to say about vernacular medical practice and the definition of the professions of barber and surgeon as the two roles began to detach from each other, what follows is about the publication history of what became an immensely popular illustrated book. It begins with the death of the artist who made the illustrations.

The Societas *and the Engraved Plates*

Adamo Scultori had died the previous May, and because Aurelia di Antellis was a new widow acting alone, a male relative – in this case, her brother-in-law, the architect Francesco da Volterra – also had to be present to conduct the negotiations for her. The meeting took place in the palazzo with the frescoed façade on the Via della Scrofa where the architect lived with his wife, the engraver Diana Mantovana, who was Adamo's sister. In settling the *societas* with the barber, Aurelia was acting as the mother and legal guardian of Adamo's heirs, Cesare and Felice, whom the notary described as the "infant sons" of the dead printer and publisher.[4] The engraved copper plates that reverted to the possession of the Scultori family, as well as the purchase this little treatise on bloodletting had on a reading public, represented part of the boys' patrimony. Those plates and copies of them made by others were put into service over and over again for the economic gain of the family and a variety of unrelated printers, publishers, and booksellers, as well as to edify would-be barber-surgeons.

The illustrations were referred to in each chapter of the book to clarify and demonstrate detailed explanations about how a barber was to approach, stabilize, and cut the veins of a patient. Besides clearly displaying the positions best assumed by patients and practitioners to assure successful procedures, in their style, their sobriety, and their obvious pictorial roots in mid-century narrative pictorial composition, the engraved illustrations implied a literary audience and set a decorous tone for the book. Later editions included the images in the same order and with the same numbering as the original publication, with the important exception of the replacement frontispiece engraved by Cherubino Alberti, on which Adamo Scultori's name no longer appeared.[5] The frontispiece of the first edition had been the single place in the book where Adamo's work was mentioned: engraved at the base of a column, under the figure of a barber-surgeon, were the words "Adam. Sculp. Exc." This inscription has been taken to refer to Scultori's work as the designer of either the frontispiece or all the illustrations. The document ending the *societas* does not entirely clear up the question of authorial responsibility for the engravings, nor of Adamo's role in their creation.

When Magni appeared before Adamo's widow in 1587 he had with him 620 of the books (noted in the agreement as Adamo's portion), valued at one giulio apiece, and thirteen engraved copper plates, which he was returning to the widow and which had been valued by the print and book publisher Pietro Paolo Palumbo, acting as an expert estimator of the value of the incised metal printing matrices, at 26 scudi.[6] The notary recorded the fact that the bookseller Bartolomeo Grassi had been applied to for expert valuation of the completed books, and it was his opinion that after calculating all the money advanced toward the project, Adamo's heirs were owed 66 scudi and 37 baiocchi, a debt satisfied by the addition of more books to their portion.[7] That the parties resorted to different outside experts in adjudicating the values of the book and the printing plates points up that the authors of texts, the printers of books, and the engravers of copper plates worked in contiguous circles that demanded different skills and, at their purest, were most conversant with different markets. Social and professional networks brought together the barber, the engraver and then his widow and children, the bookseller, the book publisher, and the architect in partnerships such as this one to create a product beneficial to all.

A New Frontispiece: The First Renovation

The new frontispiece (fig. 68), which also served as the title page, bore in the lower right of the image the monogram of the engraver Cherubino Alberti, who modeled it closely on the frontispiece for the first edition (fig. 69).[8] It announced the title of the book in an ornate cursive script rather than in the classicizing, lapidary lettering style Adamo had learned as a boy in Mantua. The information that the book was newly corrected by the author, and augmented with the addition of useful advice, appeared in a burst of calligraphic flourishes below a portal engraved at the center of the page, flanked by a mature, bearded physician on the left and a much younger

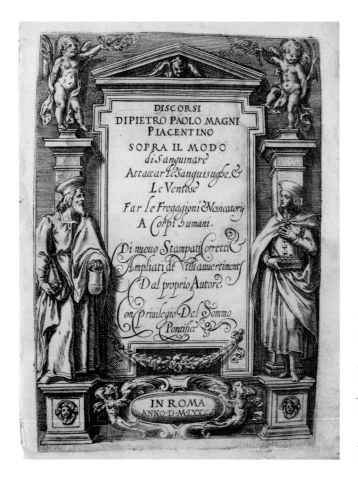

68   Cherubino Alberti, title page, engraving, Pietro Paolo Magni, *Discorsi di Pietro Paolo Magni piacentino sopra il modo di sanguinare attaccar le sanguisughe, & le ventose far le fregagioni & vessicatorij a corpi humani. Di nuovo stampati, corretti & ampliati di utili avvertimenti dal proprio autore*, Rome: Bartolomeo Bonfadino, 1586.

barber-surgeon on the right. Shoring up his professional credentials as a product of university training, the physician carries a large book under one arm and extends a urine flask toward the reader with the other, while pointedly turning his own attention to the text at his side in a pronounced contrapposto.[9] Across from him the young barber-surgeon, carrying a partitioned box that serves as the attribute of his trade, clasps one hand to his chest, marveling at the title text in turn. Both doctor and barber are elevated on plinths at the base of pillars topped by dancing putti, who cheerily wave palm fronds and laurel crowns. The classically clad doctor looks more like the statue of a prophet on a church door than a man to be encountered in the Roman street, an impression reinforced by the architectural details of fluttering ribbons, decorative harpies, and smiling masks that fill the bottom of the sheet.

The more prosaic physician in the original, and more sober, 1584 frontispiece, who is easily recognized as a version of the physician who appears prominently in every plate in the book, balances rigidly on his side of the portal wearing the formal, academic robe called a *toga*. Looking

69   Adamo Scultori, title page, engraving, Pietro Paolo Magni, *Discorsi di Pietro Paolo Magni Piacentino intorno al sanguinar i corpi humani, il modo di ataccare le sanguisuche e ventose è far frittioni è vesicatorii: con buoni et utili avertimenti*. Rome: Bartolomeo Bonfadino, and Tito Diani, 1584. RM182. M27, Boston Medical Library in the Francis A. Countway Library of Medicine.

out at the reader, he points sternly to the sensibly engraved title at center page, which does not yet advertise corrections and amplifications. The young barber is rather unchanged, but in the earlier edition he, rather than the doctor, is aided in his work by a book. The rectangular item dangling from his waist is a belt-book, containing bleeding charts and astrological figures that provided quick recourse to essential information – the proper and expected gear of a conscientious technician, which he must make do without by the time the second edition was printed.[10] In the original frontispiece both the physician and barber-surgeon are ennobled by the inclusion of discreet haloes, abstracting these modern figures as sixteenth-century versions of Sts. Cosmas and Damian, the twin doctor patrons of barbers' guilds from the medieval period onwards. Here, they are pressed into service for perhaps the last time in the task of uniting the two professions of barber and physician.[11] In fact, Magni's frontispieces were embellished, more dynamic versions of the iconic woodcut illustration used by the Vatican printer Antonio Blado for the frontispiece of the Roman statutes of the barbers and bath house workers (fig. 70), showing the haloed saints with urine flask and box, each holding the palm frond of martyrdom and standing on a grassy field. Above them, the basins that were their traditional corporate signs hang decoratively from nails attached to the sky.[12] These professions, as well as that of the barber and university trained physician or surgeon, were just then undergoing uncomfortable but irrevocable division. By the time a new frontispiece was wanted for the 1586 edition, the figures of physician and barber could no longer be comfortably thought of in terms of the indistinguishable equals connoted by the two often-conjoined saints. Rather than the halo that made a doctor godly, the physician of 1586 was ennobled with the philosopher's attributes of a weighty book, a flowing gown, and an oratorical contrapposto.

There were both professional and personal ties between Adamo's family and the engraver Cherubino Alberti, who came from a family of artists working in papal circles. One of Alberti's relatives, the painter Durante Alberti, stood as godfather to the son of Francesco da Volterra and Diana Mantovana in 1578.[13] Francesco, Diana, Adamo, and the Alberti Cherubino were all more or less active in the artistic organizations in the city to which printmakers, architects, and painters could belong, primarily the Compagnia di San Giuseppe, and aside from Adamo, who was no longer alive, the men were then active a few years later in the formation of the Roman artists' academy, the Accademia di San Luca, where Alberti lectured on the importance of decorum in painting.[14] But in 1586, when Alberti supplied the engraving for the second frontispiece, he had not yet become a famous master of baroque engraving as he is known today. At the time he was one of the more talented engravers of religious prints, title pages, and frontispieces for the book trade, and the publishers (whether Scultori, Magni, or Bonfadino) did well to apply to him for an introductory upgrade to the modern book on the practice of bleeding.

The Book

Both editions begin with the author's dedication to Ludovico Bianchetti, a canon of St. Peter's Basilica who had been a close familiar and *maestro di camera* of Gregory XIII until his death in 1585, and also of Magni's employer Cardinal Alessandro Farnese, and, like Magni, came from Emilia-Romagna.[15] They also both contain the same ten-year privilege from Gregory XIII to the author, which follows the dedication, along with a pair of sonnets praising the book and a conscientious table of contents.[16] A short *proemio* by the author follows in which he introduces himself as one who has spent his whole life in Rome letting blood from those who need bleeding, through cutting veins and applying leeches, as well as performing the cupping and massage that were also regularly ordered by physicians. Over the course of the next nineteen chapters he explains how a barber-surgeon can successfully perform the principal operations of bleeding by cutting veins with a scalpel-like instrument called a *lancetta*. These chapters are illustrated with the eleven numbered engravings (figs. 72–80), each printed on a full page separately from the text and tipped into the book. A short unillustrated chapter on the maintenance of leeches as well as

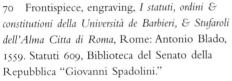

70 Frontispiece, engraving, *I statuti, ordini & constitutioni della Università de Barbieri, & Stufaroli dell'Alma Citta di Roma*, Rome: Antonio Blado, 1559. Statuti 609, Biblioteca del Senato della Repubblica "Giovanni Spadolini."

71 Printer's mark of Bartolomeo Bonfadino, woodcut, Pietro Paolo Magni, *Discorsi di Pietro Paolo Magni piacentino sopra il modo di sanguinare attaccar le sanguisughe, & le ventose far le fregagioni & vessicatorij a corpi humani. Di nuovo stampati, corretti & ampliati di utili avvertimenti dal proprio autore*, Rome: Bartolomeo Bonfadino, 1586.

how to attach them to veins (in the nose, neck, ears, arms, flanks, lower back, and anus), and how to judge how much blood each has removed, is followed by a discourse on the use of "ventose, o coppe" for cupping, a technique involving creating suction by heating glass cups placed on a patient's skin, through which – if the skin is scored – blood can also be drawn. This chapter is illustrated by the single non-narrative woodcut diagram. The last very short and unillustrated chapters are brief discourses on massage and *vessicatori*, or the technique of purging toxins by raising blisters on the skin through the application of plasters. When the 1586 edition was printed Bonfadino was no longer working in partnership with Tito Diani, and so the new colophon displays at its center Bartolomeo Bonfadino's magnificent woodcut printer's mark of a porcupine in an ornate frame, with the motto "Mordentes sauciabuntur": those who bite me will be wounded (fig. 71).

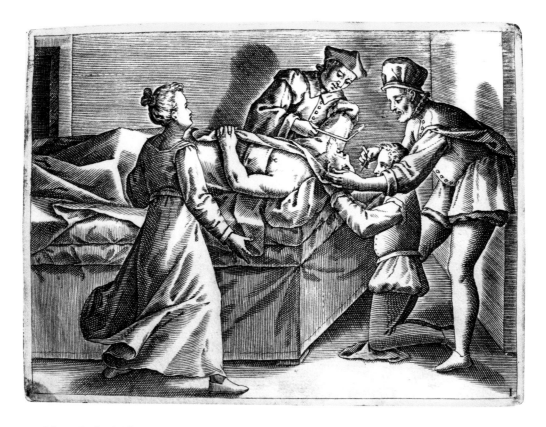

72   Adamo Scultori, Plate I, engraving, Pietro Paolo Magni, *Discorsi di Pietro Paolo Magni piacentino sopra il modo di sanguinare attaccar le sanguisughe, & le ventose far le fregagioni & vessicatorij a corpi humani. Di nuovo stampati, corretti & ampliati di utili avvertimenti dal proprio autore*, Rome: Bartolomeo Bonfadino, 1586.

The Images

The images in Magni's book draw from several different pictorial traditions to illustrate not only the techniques of, but also the tensions inherent in, the profession of the barber-surgeon. The engravings tell us more than the author does about the professional status and social position of men of this trade, once familiar and important among both the poor and the wealthy, in the hierarchy of Renaissance professionals. Like the manual of Magino Gabrielli, the subject matter and the illustrations present men and woman whose shared engagement with bodily fluids and close interaction between people and worms were a fact of everyday life. The settings are bed-chambers and small rooms in private homes that are probably much like the one in which Aurelia and the barber met to settle their association.

Whoever designed the pictures must have been working very closely with the barber, for there are no precise visual precedents for the ideas that the images have to convey. There were only a few readily understood visual notations for illness in the late sixteenth century. Most often, a mortally ill patient lay in bed, gaunt cheekbones and sometimes exposed ribcage testimony to the debilitating effects of loss of appetite, while at the bedside a veiled woman poised a hopeful spoon over a bowl of nourishment. In ex-voto paintings and woodcut images in manuals made to help people to a better death, saints and relatives gathered around the patient's bed or knelt to pray. Up to the late sixteenth century physicians were generally shown judging the strength of a pulse or the color of urine in a transparent flask, while surgeons and anatomists presided over opened cadavers. There was also a humorous tradition, much elaborated in the North in the seventeenth century, in which barbers and quack doctors were shown in disreputable circumstances and indecorous poses committing painful and dubious acts of healing with shabbily clad patients in obvious distress. But in medical treatises with any claim to seriousness, only midwives and barber-surgeons, among professional healers, were regularly portrayed as actively engaged with the messy, suffering human body, the midwives supporting and delivering the laboring mother, the barber-surgeons setting bones, sawing off limbs, and in general striking terror in any patient imagined in that position.[17] While the illustrations for Magni's book set an example for later Italian writers of treatises on the barber's art, as well as French and German barber-authors, in 1584 the barber and his illustrator had to look for solutions to representing bodies in meaningful and demonstrative interaction in situations for which there was no easily recognizable visual tradition. In cases like this, the authors and publishers who added images to their texts had to be as concerned with limiting the signification of the pictures as with extending the meaning of the text.

For Magni, it was necessary to characterize the intimate domestic spaces in which these very common medical procedures took place, with their intimations of personal drama (birth, sex, sickness, and death), as a proper workplace for a practitioner of the healing arts whose expertise was in need of new definition.[18] The engraved illustrations show the barber and his clients interacting with a stilted but engaged intensity that emerged from the unembellished engraving style, which was better suited to the didactic illustrations than to the self-promoting allegorical frontispiece. The engravings also carefully calibrate the elegance of furnishings and the procedurally important direction of light, turning each setting for bleeding into a diorama of well-informed practices in decent company, and successfully avoiding the many potential minefields of the barber's art that are vividly described in the book. The book is therefore remarkable not only for the fact that it is a professional treatise about a trade that was not held in much repute at the time, but also for the copperplate engravings that show the intervention of a sixteenth-century medical team into scenes of Roman domestic life at a moment of personal crisis.

The engraved illustrations are somber, formal, and seemingly narrative. In shadowy domestic interiors, work is demonstrated that seems at first glance to be familiar and ordinary, but almost immediately becomes incomprehensible, certainly uncomfortable; close looking fails to correct that sense entirely. The underside of a tongue, unbelievably, is about to be pierced, apparently with the full cooperation of its owner (fig. 73, Magni Plate II). A woman's foot is reverently inspected (fig. 78, Magni Plate VIII). The players in these compact tableaux part at the point where we seem to be standing, pinpointing the object of their attention in focused light. The staginess of these scenes and the dramatic lighting – which has been a primary focus of the engraver's attention – signal that there is a visual problem here, and that the engraver has been working at it.

Without reading the text it is difficult to explain, for example, the obvious mirroring of characters in the pictures: faces and gestures echo each other, like those of the two men in Plate VIII (fig. 78), respectfully kneeling at the feet, or the foot, of someone who, like her companion, wears the matronly headcloth of a married Roman woman. The two women are linked together in a single knot of emotional and physical attachment; one, standing, all active comfort and support, embraces the other, seated high on a pile of folded bedding, who leans into her breast. Their crossed arms emphasize their unity, obvious anyway in the oblique gestures of their tilted heads and the demure shadow falling equally over their faces and chests as they seem to hover in the background. But it seems to be the seated woman's right ankle, modestly exposed and tied with a tourniquet, suspended in mid-air over a basin set on a rectangular wooden bench, that is meant to attract our attention. This cannot fail to look like a religious ritual: the pillow under the knee of the kneeling man on the left, the altar-like bench, the decorative ewers at the center foreground, the kneeling, the solemnity, the priestly man in robes and beard standing dead center, the lit taper held to the privileged site. The sacrificial ankle shines brightly at the center of a pinwheel configuration of taper, supporting male hands, and the pointing finger of the central imposing figure of authority, who clutches to his chest a glove, which, uninhabited, continues to point down. This man is the only participant in the scene to stand alone and fully upright, intervening between the knot of female figures and the twinned, kneeling males at the base of the image. No one looks at him, yet events are clearly proceeding under his direction.[19] It is daytime, as we see beyond the arch of the balcony with its elegant balustrade in the next room, open to the landscape outside. But the bedroom in which this surgical procedure is taking place is removed both from natural light and perilous drafts.

Picturing Perils

From the text that these engravings illustrate we find that drafts and uncontrollable lighting are not the only dangers that threaten the close-knit stability of this chamber. Some threaten the patient – to us, most obviously, the possibility of procedural error – but for the living sixteenth-century woman, there also loomed the threat of charges of immodesty in having men in her bedroom, or of lying about the nature of her indisposition. There are also dangers to the barber-surgeon who cups her toes in his left hand as he prepares, under the watchful eye and explicit orders of the gloved physician, to cut a vein below the inner ankle called the saphenous vein in order to draw the amount of blood the physician ordered. The barber is the author of this book;

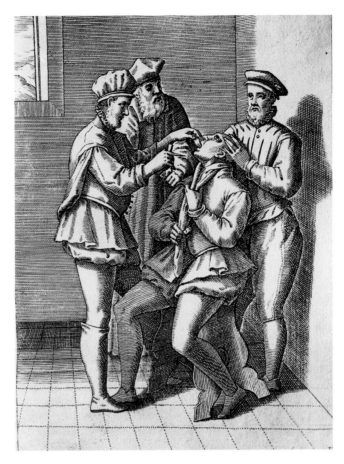

73   Adamo Scultori, Plate II, engraving, Pietro Paolo Magni, *Discorsi di Pietro Paolo Magni piacentino sopra il modo di sanguinare attaccar le sanguisughe, & le ventose far le fregagioni & vessicatorij a corpi humani. Di nuovo stampati, corretti & ampliati di utili avvertimenti dal proprio autore*, Rome: Bartolomeo Bonfadino, 1586.

he wrote it for the edification of men like the one who kneels across from him and helps to steady the patient's ankle, expertly lighting the bleeding site with a single taper in just the right place, stalwart, unmoved by whatever doubts the patient might be having, and learning by experience and through this book all he can from the barber whose double he is evidently becoming. The barber is a model of discreet affability, cleanliness, and sobriety: we find out in the text that he fears God, bad eyesight, poor lighting, dull tools, the trickery of women, and a bad reputation. When we, the readers, enter the scene, we do so in the position of the barber's apprentice, ready to learn all we can through pictures and words about a practice once so ubiquitous that its sheer ordinariness has allowed it almost to vanish from the medical landscape as we know it, leaving few traces.

The performance of phlebotomy was the part of the barber-surgeon's trade in which Magni, the son of a barber, took the most pride. He received his license to practice in 1558, which was granted in perpetuity in 1566.[20] He served the Farnese cardinals in Rome, but also tells us about bleeding and caring for the bodies of grooms, jewelers, children, and innkeepers. His treatise gives lively advice, which he tells us is the fruit of many years of experience, about how to deal with problems such as faint hearts, stout bodies, and subtle veins. But in these configurations of practitioners and patients we see something of the social status of barbers and physicians in sixteenth-century Rome, and the limits of social acceptability in gender and power relationships tested through picturing several aspects of this profession. We also see something more familiar acted out in these bedrooms: a serial distancing of the management of the body's cure, and an increasing necessity to relinquish active participation in that cure in order for the sick body to meet social norms of sickness, treatment, and the maintenance of health.[21] We can see the professionalization of medicine evolving before our eyes, expressed not, as perhaps more readily imagined, through changes in technology, but in the relationships of individuated actors – physicians, barbers and their assistants, patients and their families – to medical treatment as represented in pictures and in text.[22]

As a barber, Magni was active in the men's shave and haircut business. But the famous barber's symbol, the familiar red, white, and blue poles that still sometimes decorate establishments where we hope to see no blood at all, originated in the ancient sign for the trade that symbolized blood coursing through veins and arteries.[23] Charging far less for their services than doctors, and knowing something from experience about treating illness, barber-surgeons were often the only medical practitioners a poor person would ever see.[24] The intimate contact barbers had with the bodies of their clients was probably least suspect in the activities most closely related to medical practice: as Magni's title describes it, the practices of bleeding, attaching leeches and cups for cupping, and therapeutic massage were the barber's domain. As minor surgeons, barbers performed necessary and useful prophylactic cauterizations, removed rotten teeth, tended to certain wounds, and undertook curative bleeding when it was prescribed by a doctor. Magni wrote an entire book on cauterization intended as a sequel to the bleeding manual, published with Bonfadino in 1588, in which he describes further conditions that barbers were called to treat including toothache, leprosy, elephantiasis, and the bites of rabid animals.[25] As cosmeticians barbers were vulnerable to accusations of enabling the sin of personal vanity, making possible, for example, the scandalous deceptions of "certain old Ganymedes, shaving them under the chin and on the soft skin of their cheeks, so that the flies, attracted by the appeal of that honey, fly over them in the dark, and never stop their lascivious doings."[26]

Producing a Professional Space

Until the beginning of the seventeenth century the barber's trade in Rome was uncomfortably allied with that of the *stufaroli*, or public bath attendants, who were not only widely held to be ruffians who lived in rented rooms, but, in Rome, overwhelmingly German. The *stufaroli* presided over steamy saunas and therapeutic waters of varying temperature which were reputed to be the sites of "a thousand shameful and dishonest satisfactions of the flesh."[27] Needless to say, some of these establishments were entirely successful; at least one of them was known to be in continuous operation throughout three centuries. In these places the *stufaroli* were known to bleed clients who showed up with a variety of conditions including scabies, skin ulcers, and syphilis. Bernardino Ramazzini, after discussing the fame of Roman baths in antiquity, wrote that in 1700 there remained

> poor reminders of the baths, for the use of . . . those who have some cutaneous infection, e.g. scabies, psora, or the French Disease . . . where bathmen go through the form of washing them with tepid water and often they apply small cupping-glasses, scarify the whole body and draw off a considerable amount of blood; then when they have washed them, rubbed them hard and half killed them by inches, they let them go home. Too often patients and bathmen too do this without a doctor's advice.[28]

In their guild statutes of 1614 the barbers were finally able to proclaim success in shedding the hated alliance with the bath attendants, stating with relief that they were freeing themselves from all association.[29]

Magni wrote his treatise thirty years before that definitive change in the barber's guild. But the evolution of his manual through its several printings was guided by the necessity to distinguish his occupation from that of the bath attendants by defining it as closely tied to, and regulated by, university trained physicians and surgeons. At the outset of the second edition, Magni explains a few things about his audience and himself:

> Having spent by now almost my whole life in this beautiful city of Rome drawing blood from those in need of it, and in the use of leeches, and cupping, and massages ordered by doctors, and seeing that many people err in different ways in doing these things, with the greatest danger and damage to the patients, for this reason I have composed the present work, in which the barbers . . . can doubtless take no small pleasure. Because aside from warnings, and many secrets contained here . . . I have chosen the most clear and easy style that I could, remembering very well that I am writing for the barbers, to teach them how to draw blood, and not in order to speak well. And if my censors had considered this perhaps it would not have seemed strange to them that I have written some things less than *toscanamente*.[30]

By *toscanamente* Magni meant the polished Tuscan dialect that was becoming codified as the literary vernacular. He goes on to say that he has not been moved by the entreaties of some who begged him to honor the treatise by identifying himself as a surgeon, even though he has been admitted into the College of Surgeons in Rome, because the subject he is undertaking to explain – phlebotomy – cannot be spoken of without describing the profession of the barber, and in particular relating his own experience. And these things would have offended the surgeons who, without reason, greatly hate the trade.[31] It was a commonplace in medical practice that surgeons were not accorded the same respect as physicians; Vesalius disapprovingly noted that the physicians

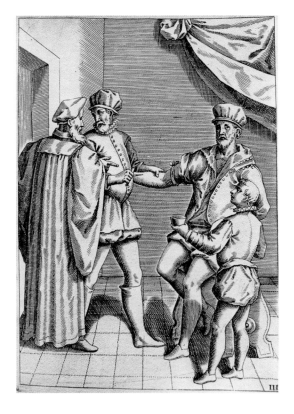

LEFT 74 Adamo Scultori, Plate III, engraving, Pietro Paolo Magni, *Discorsi di Pietro Paolo Magni piacentino sopra il modo di sanguinare attaccar le sanguisughe, & le ventose far le fregagioni & vessicatorij a corpi humani. Di nuovo stampati, corretti & ampliati di utili avvertimenti dal proprio autore*, Rome: Bartolomeo Bonfadino, 1586.

OPPOSITE PAGE 75 Adamo Scultori, Plate IIII, engraving, Pietro Paolo Magni, *Discorsi di Pietro Paolo Magni piacentino sopra il modo di sanguinare attaccar le sanguisughe, & le ventose far le fregagioni & vessicatorij a corpi humani. Di nuovo stampati, corretti & ampliati di utili avvertimenti dal proprio autore*, Rome: Bartolomeo Bonfadino, 1586.

"regard them as less than their own servants."[32] At Magni's level the prejudice was magnified through the extension of that disdain from surgeon, a rank he could brag about having attained, to the even more lowly barber, the trade he practiced.[33] He was adamant, however, in insisting that in this book he was writing as a barber for other barbers, and for the betterment of the profession.

Although bleeding was an expected aspect of medical practice, most doctors did not do it themselves. Instead they had the important supervisory role of prescribing bleeding as a treatment when a patient was ill; a barber was called in to perform the task.[34] Gianna Pomata notes, with reference in particular to Magni, that the barber's job was to care for every necessity of the body and the regulation of its humors; as such he worked within a culture of the body where no procedure was too repugnant to carry out, as opposed to those whose work had a higher status and were removed from actual contact with the ill body.[35] Similarly, the botanist Leonhart Fuchs, writing in 1542, noted with disapproval that many doctors were loathe to learn about the medicinal properties of plants, considering it beneath their dignity. This caused patients to put themselves in peril by being forced to rely on the uncertain knowledge of illiterate empirics for medicines.[36] The partitioned box that the barber-surgeon carries in the frontispiece to Magni's book is prob-

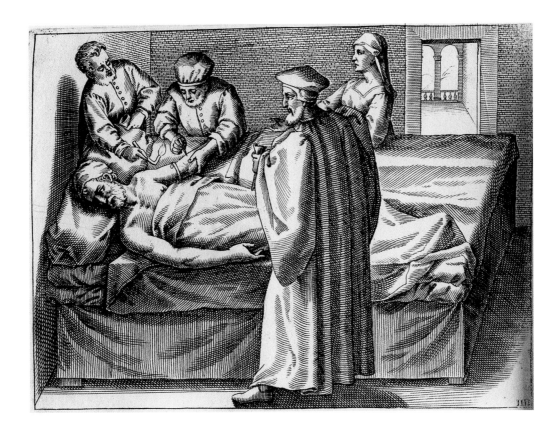

ably a box of medicines or simples, relating to the activities of the saints Cosmas and Damian, who were sent to their martyrdom when they were caught in the woods while collecting herbs for medicines (simpling).[37] However, although the frontispiece distinguishes the barber from the physician through the attributes of administrative medicine box and diagnostic urine flask, Magni and his colleagues would not have been allowed to prescribe medicines themselves, which fell to the doctors. Formal measures to separate the occupation of barber from the rank of physician varied; at the University of Paris, for example, graduating physicians in Vesalius's time had to take an oath that they would not practice barbers' work.

In Magni's illustration of the bleeding of the sciatic vein (fig. 77, Magni Plate VII) we see the physician presiding in watchful inaction during the procedure, which required first soaking the leg up to the knee in a barrel of warm water. In order to enter their guild, barbers had to demonstrate knowledge of anatomy including, specifically, the location of all the veins in the human body, and how to draw blood from those veins either by incising them or by applying leeches or cups.[38] But the barber apprentice would have learned most of the practical aspects of his trade by assisting at procedures with a more experienced barber, as we see the apprentice doing in Magni's illustrations. This is in large part the practical point of Magni's illustrated book.

76   Adamo Scultori, Plate V, engraving, Pietro Paolo Magni, *Discorsi di Pietro Paolo Magni piacentino sopra il modo di sanguinare attaccar le sanguisughe, & le ventose far le fregagioni & vessicatorij a corpi humani. Di nuovo stampati, corretti & ampliati di utili avvertimenti dal proprio autore*, Rome: Bartolomeo Bonfadino, 1586.

Magni was evidently aware, as most people in the medical profession would have been, of the *Fasciculo di medicina*, a fifteenth-century medical book from the circle of the University of Padua reissued in several editions. It is famous for its woodcut illustrations: one of them shows an anatomy lesson about to begin with the public dissection of a corpse (fig. 81), providing an interior frontispiece for a section of the book devoted to the anatomical text of Mondino dei Luzzi, a fourteenth-century anatomist.[39] In this illustration a hierarchy of student/teacher/worker seems firmly in place, but as Jerome Bylebyl has shown, the didactic function of the dissection is actually dispersed throughout the figures in this representation. As a lecturer in surgery recites the material in the text from a seat on a lectern above, medical students and other physicians gather around the cadaver below, which is the shared object of their, and our, attention. A non-academic surgical assistant, the dissector, is most closely associated with both the text – which instructs the reader in exactly this procedure – and the corpse, which he bends over to begin the dissection. An academic demonstrator (*ostensor*) on the right is ready to point to the parts discussed in the text as they are revealed, and he will also explain whatever there is to know about them.[40] We, the readers, join the students located on the other side of the anatomy table, hearing with the ear of our minds the voice enunciating Mondino's *Anatomy* in the form of the text we are studying. Vesalius was quite clear about the inutility of this "detestable ritual" which, in its division of labor, reflected the fragmentation of the practice of medicine that resulted in the physicians never practicing the dissections, and the dissections being performed by barbers who had no access to the Latin anatomical texts.[41] When Vesalius took over the chair of anatomy at the medical school in Padua in 1537, he claimed to do what had previously been the work of the dissector himself, and mentioned that as a student in Paris he had impatiently waved the barbers aside in order to demonstrate the proper way to dissect the muscles of the arm.[42] The professional space of the barber-surgeon was clearly outside the academy, but in picturing it in the palace interiors of well-to-do Romans, it was also clearly not in the bath houses.

Teaching, Learning, and Medical Illustration: Magni

Before Vesalius's famous anatomy book appeared there were already in circulation certain sheets with paper doll-like figures printed on them from which students were supposed to make a kind of early version of the Visible Man and Visible Woman.[43] Vesalius, too, provided a set of ideal bodies as models for the study of anatomy, figures which appeared in a publication he called the *Epitome*, also published in 1543.[44] Models like these, designed to be cut out and pasted to stiff parchment with the overlays of veins and muscles attached above the forehead, belonged to a genre popular with medical students at universities as well as among the literate, but not Latin-trained, barbers. The sheets containing these models were available in unbound sets ready to be made up, or could be purchased cut out, hand-colored, and glued together, ready to use. Magni shows his familiarity with such pedagogical tools in the first illustration from his book, a didactic plate he calls the "Universal Figure" (fig. 72).

By 1584 Vesalius's figures had been appropriated by other artists plumbing the market for printed anatomical illustrations, most notably, for Magni, the Spanish doctor Juan Valverde. The illustrations for Valverde's anatomy book, which went through several editions, were made by the French engraver Nicholas Beatrizet, who, like Adamo Scultori, worked for the Roman print market.[45] Both men had an intimate knowledge of Roman antiquities, which, along with religious scenes, were the most important subjects of the sixteenth-century print trade in Rome. The transposition of modern medical models onto antique bodies was one of the most noticeable differences between the medical books of the Renaissance, including the illustrated editions of the *Fasciculo*, and the schematic charts and diagrams of the Middle Ages.

Vesalius took special pride in the pictorial nature of his illustrations. He followed Galen in recommending that the body to be used for dissection be derived from perfect proportions and ideal, unindividuated features, that it "be as normal as possible according to its sex and of medium age, so you may compare other bodies to it, as if to the statue of Polykleitos."[46] Magni's "Universal Figure" provides such a model body in order to demonstrate "the sites in which the veins of the

LEFT 77   Adamo Scultori, Plate VII, engraving, Pietro Paolo Magni, *Discorsi di Pietro Paolo Magni piacentino sopra il modo di sanguinare attaccar le sanguisughe, & le ventose far le fregagioni & vessicatorij a corpi humani. Di nuovo stampati, corretti & ampliati di utili avvertimenti dal proprio autore,* Rome: Bartolomeo Bonfadino, 1586.

OPPOSITE PAGE 78   Adamo Scultori, Plate VIII, engraving, Pietro Paolo Magni, *Discorsi di Pietro Paolo Magni piacentino sopra il modo di sanguinare attaccar le sanguisughe, & le ventose far le fregagioni & vessicatorij a corpi humani. Di nuovo stampati, corretti & ampliati di utili avvertimenti dal proprio autore,* Rome: Bartolomeo Bonfadino, 1586.

human body can be cut with a lancet."[47] Magni, too, is clear about the limits of specificity in his mode of teaching: "My teaching is about ordinary bodies and not those in which by some defect of nature the veins are hidden, as in too much fat, or coldness, or other causes."[48] The normative body, as represented in this manual, is in fact the Greek model of perfect male proportionality, presented and appropriately coiffed in the mode with which a barber like Magni would be so familiar.

The "Universal Figure" was printed with unlabeled dart-like lines pointing to the pertinent bleeding sites. One owner of the 1584 edition labeled these points with letters and penned in an extra line to indicate the jugular vein (marked "V") which Magni specifically did not point out, not wanting anyone to be tempted to bleed it (fig. 72).[49] In drawing attention to that illustration, Magni wrote: "But before I come to talk in detail about the sites of bleeding, I wish to propose the brief figure, which I call Universal, because in it are seen and depicted all the ordinary sites together," simplified to show only those veins large enough to be cut with a lancet, mapped onto the intact body of the living patient. One of the illustrations of the fifteenth-century *Fasciculo* was a blood-letting chart (*Flebothomia*) that used a floating human body with its tongue exposed to demonstrate the usual bleeding points by means of snaking lines to sur-

VIII

79   Adamo Scultori, Plate VIIII, engraving, Pietro Paolo Magni, *Discorsi di Pietro Paolo Magni piacentino sopra il modo di sanguinare attaccar le sanguisughe, & le ventose far le fregagioni & vessicatorij a corpi humani. Di nuovo stampati, corretti & ampliati di utili avvertimenti dal proprio autore,* Rome: Bartolomeo Bonfadino, 1586.

rounding captions in small, chart-like boxes (fig. 84). The comprehensive and intricate Vesalian diagrams of all the blood vessels in the human body (fig. 85) were of little practical use to the barber, whose job, Magni goes on to reveal in the next part of the book, has more to do with stabilizing the intact but trembling body of an ill, embarrassed, and frightened human being, in low-light conditions and in the presence of concerned relatives, than it does with the demonstrative study of an obedient cadaver. Yet we find unexpected echoes of the strange dispersal of information and attention in Ketham's introductory anatomy scene in Magni's text and illustrations.

The staged and static quality clearly portrayed in each of Magni's scenes makes visible the tension produced by the intersection of several skills, desires, and conditions: (1), the visible fears and textually revealed skepticism of the patients and their family members, which are suppressed in the presence of (2), the elevated status of the expensive, Latin-literate, professionally aristocratic physician, whose orders make him a presence in the room whether he attends the procedure or not, and are necessary to legitimate the actions of the barber (3), who desires to carry out that

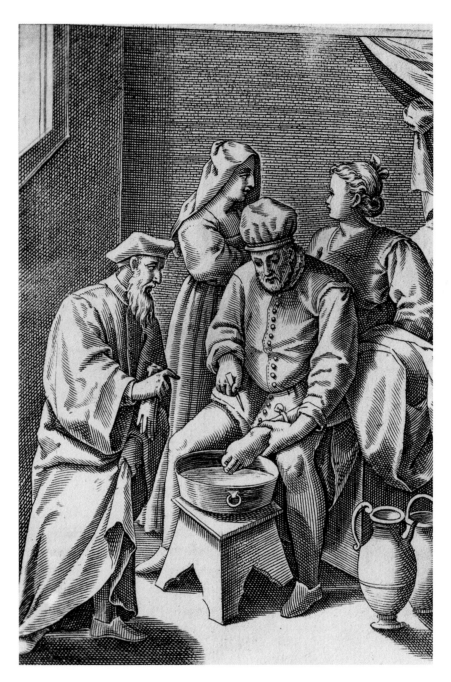

80   Adamo Scultori, Plate X, engraving, Pietro Paolo Magni, *Discorsi di Pietro Paolo Magni piacentino sopra il modo di sanguinare attaccar le sanguisughe, & le ventose far le fregagioni & vessicatorij a corpi humani. Di nuovo stampati, corretti & ampliati di utili avvertimenti dal proprio autore*, Rome: Bartolomeo Bonfadino, 1586.

81   *Anatomy Lesson*, woodcut,
Johannes de Ketham, *Fasciculus
medicine*, Venice: Joannem & Gregori
de Gregorijs fratres, 1500. f★AC85.
H7375.Zz522k, Houghton Library,
Harvard University.

specialized part of his job in which he is doing the most good for mankind and which, if performed well, will bring him honor. Here, this complex relationship is represented in the form of
a treatise, the sort of text usually employed to ennoble a liminal trade or pastime, which brings
us to a final point of tension: that barber-surgery was actually one of the lower-ranked trades,
socially and professionally, not only because of its overwhelmingly practical rather than theoretical
status (although Magni does take pains to show the connection to medicine by way of surgery,
itself the lowest branch of medicine), but also because of its service-oriented nature. In spite of
this, Magni felt it reasonable to produce a sumptuously illustrated technical manual in which he
emphasized both his own close connection to medical knowledge and his creativity, born of
experience, in problem solving. For this treatise, writing in the vernacular, he enlisted the artistic
skills of an engraver trained to depict religious monuments of the High Renaissance and Counter-
Reformation in the visual vernacular of Roman commercial engraving.

Teaching, Learning, and Illustration: Scultori

Before turning to the plates individually and together as a pictorialization of problems in the art of phlebotomy, it would be best to say a few words about the training of the person whose name appears on the first frontispiece of which he was, most likely, the engraver. Adamo Scultori, whose name appears as the publisher of the engravings in the 1584 edition and nowhere else, had learned the art of reproductive engraving at a young age from his father, Giovanni Battista Scultori, an engraver and sculptor at the Mantuan court who was proud of his son and did his best to find him patrons in the circle of the papal court. Still as a boy he had engraved Michelangelo's figures from the Sistine Chapel, works that his father had sent to Cardinal Antoine de Perrenot Granvelle, an active and knowledgeable patron of arts. Both Adamo and his sister Diana went to Rome in the papacy of Gregory XIII, and besides being a busy printer and engraver Adamo also seems to have been active in publishing. Like Cherubino Alberti, he made engravings for the print trade after paintings and sculptures by the most well-known artists, and also provided, in at least one other case, the frontispiece for a book.[50] He is not known as a designer of images, mostly seeming to engrave prints after works by others. "Adam Sculp. Exc." only tells us, insofar as we understand the cryptic inscriptions on early modern prints, that he was the owner of the plates. Adamo's last name, Scultori, which came from his father's occupation as a sculptor and engraver at the Gonzaga court in Mantua, was written in Latin as "de Sculptoribus," so the inscription could mean "Adam engraved and published this," or it could mean "Adamo Scultori published this," and it has been interpreted both ways.[51] Book publishers could, and did, contract out the printing of illustrations to other printers, and since all the engravings in Magni's publication are blank on the verso, they could have been printed apart from the text very easily. The publishers of the first edition, Bonfadino and his partner in those years, Tito Diani, worked this way on other publications, for example, when they farmed out the cutting and printing of the woodblocks for the illustrated *Herbal of Castor Durante* to Leonardo Parasole in 1585.[52]

Oddly, given the subject matter, those very early images from the Sistine Chapel may have provided Scultori with a good model for the somber staging of the scenes involving silent, studious clusters of people in darkened rooms. The small engravings he did as a boy of Michelangelo's *antenati* and the Prophets and Sibyls from the Sistine ceiling are handled with similar intensity of focused light in rectangular, room-like coffers, against backgrounds shaded with a screen of evenly spaced parallel lines (fig. 86 and 87).[53] The monumentality of Michelangelo's frescoed figures was noticeably diminished in the boy's engravings, and his father apologized to Granvelle for the fact that the prints had not turned out better. Of all the images by Michelangelo, the *antenati* and the Prophets and Sibyls are the least heroic in gesture and stance. The ancestors sit and wait, engaged in mundane tasks like combing their hair, reading, or spinning yarn, while the Prophets, once the startling color of the frescoes gives way to the black and white of the prints, seem primarily to be craning around in their compartments to catch enough light to make out their revelatory texts. Only the "Universal Figure" shows a godlike man, who, if not for his

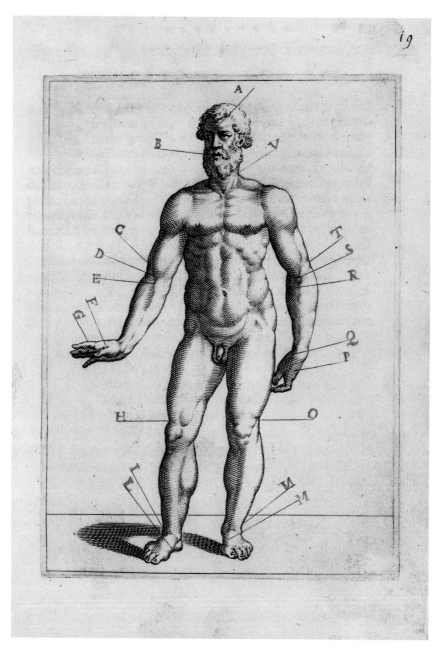

82   Adamo Scultori, *Universal Figure*, engraving, Pietro Paolo Magni, *Discorsi di Pietro Paolo Magni piacentino intorno al sanguinar i corpi humani, il modo di ataccare le sanguisuche e ventose è far frittioni è vesicatorii: con buoni et utili avertimenti.* Rome: Bartolomeo Bonfadino, and Tito Diani, 1584. RM182. M27, Boston Medical Library in the Francis A. Countway Library of Medicine.

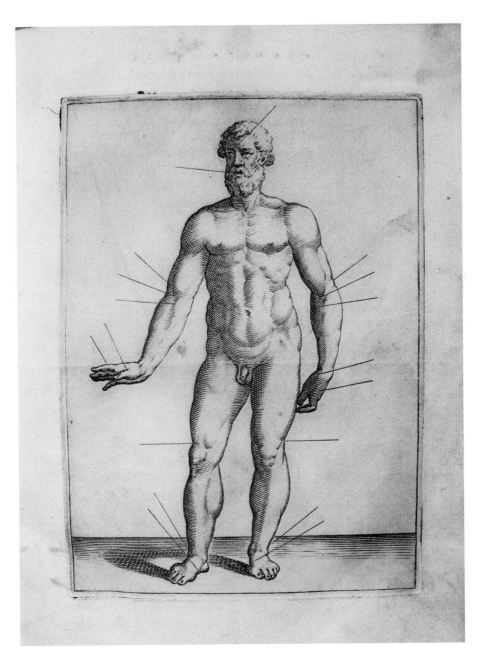

83   Adamo Scultori, *Universal Figure*, engraving, Pietro Paolo Magni, *Discorsi di Pietro Paolo Magni piacentino sopra il modo di sanguinare attaccar le sanguisughe, & le ventose far le fregagioni & vessicatorij a corpi humani. Di nuovo stampati, corretti & ampliati di utili avvertimenti dal proprio autore,* Rome: Bartolomio Bonafidino, 1586.

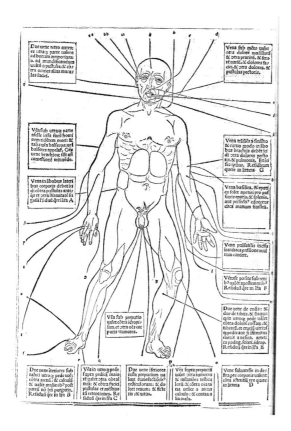

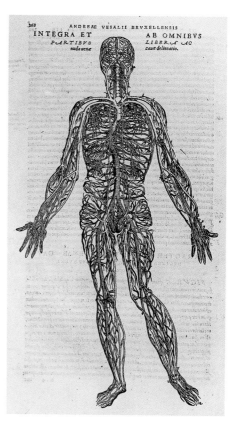

84 *Vein Man*, woodcut, Johannes de Ketham. *Fasciculus medicine*, Venice: Joannem & Gregori de Gregorijs fratres, 1500. f*AC85.H7375.Zz522k, Houghton Library, Harvard University.

85 *Vein Man* (and all the veins and arteries), woodcut, Andreas Vesalius, *De humani corporis fabrica*, Basel: J. Oporinus, 1543. Rare 2-S QM21.V37 1543, John Hay Library, Brown University Library.

exceptionally modern hair and beard, could be Hercules with the addition of a club and a handful of apples. In these engravings the same solidity of form and the sense of a motionless tableau that, in the frescoes, communicated helpless waiting or heavenly lighting here give Magni's scenes a feeling both of modernity and gravity that suited the purpose of the book. On the one hand, it was a good way to communicate the importance of focused light in finding the right vein, the stillness of the bodies to be bled if the procedure was to go without fatal mistakes, and the seriousness of the work at hand. On the other hand, it lent an elegance and even sacrality to the trade that Magni was looking to enhance in the eyes of future patrons and of the city, removing it from the bath houses and making it welcome in the most reputable palaces of Rome.

Women and Children

Although patients believed that letting a good amount of blood was necessary in order to achieve relief from the symptoms of illness, the barber acknowledges that patients also at times feel he poses certain risks to their well-being. Magni includes a chapter in which he discusses ways to overcome the paralyzing fear and timidity of some of the patients he treats, describing tricks barbers can use to distract the patients from their own self-consciousness. He tells of the charming subterfuge by which he reassures small children who are afraid of being bled, letting them marvel at the leeches swimming around in a vase of water that allows him to display the quickness of hand for which good barbers were famous, bleeding the patients before they know what is happening.[54] A patient's fear has other, practical implications for the barber. Besides making him twitchy, rendering it more difficult to cut the vein properly, it is bad for business: one time in Venice, Magni relates, a barber killed a man accidentally by cutting an artery along with a vein, throwing the whole city into such a terror that for a long time nobody dared to have themselves bled, no matter how great their indisposition.[55]

86 Adamo Scultori, Plate 40, studies after Michelangelo, *Figures from the Sistine Chapel (The Antenati)*, *c*.1547, engraving, 142 × 113 mm, British Museum.

87 Adamo Scultori, Plate 59, studies after Michelangelo, *Figures from the Sistine Chapel (The Antenati)*, *c*.1547, engraving, 137 × 100 mm, British Museum.

A short chapter at the beginning of the book takes up the issue of possible timidity on the part of the barber himself, who has to act with confidence and purpose not only under the eye of the lord physician from whom he takes his orders, but on occasion under the eye of his social superiors when he is called to bleed princes. In those cases Magni recommends remembering that the prince is also only a man like any other, and that the barber should do his best always to pretend he is bleeding a poor man, proceeding slowly and calmly. To aid in this, no one should speak to the barber or otherwise distract him while he is at work, an injunction that is reinforced by the still, empty rooms, sober lighting, and emphasis on immobility that translate pictorially into silence. The pleasant expression on the face of the barber, exuding a sense of serenity without levity, helps to set the scene in which a confident expert will discreetly manage and succeed in bringing into balance the humors of a frightened patient of any gender or class. The author concedes that in every profession it is possible to make mistakes, but in the case of the barber these mistakes are more serious, and so should not proceed from ignorance. This is one of the reasons for writing the book, and for showing clearly in the figures how to practice the art.[56]

Such considerations also appear within particular descriptions of procedures, as in the unillustrated chapter explaining how the male barber should apply leeches to a woman's private parts. In the case of a modest woman subjected to humiliating exploration in curing anal fistulas through the application of leeches, he recommends draping her completely with cloths so that her identity is unknown and no one can be embarrassed if they later meet:

> It seems necessary for me to recount what happened to me a few times, being called to perform this operation on a modest woman, when the lord physicians had ordered leeches applied to said part. For modesty's sake she did not want me to apply them, so I advised her relatives to cover her face and everything else that should be covered, so that neither would she recognize me, nor would I recognize her, and so in this way she was satisfied to receive this great aid. Her relatives knew, and she herself recognized, that any hope for her life depended on that particular remedy. I am saying this because few perform this operation as it should be done. And while some women can be found who make a profession of this and other similar things, one should not have them do it, not even if everybody trusts them. Because many barbers find that in this procedure they commit many errors, how much more would be committed by women [practitioners], who are far removed from the barber's art?[57]

This entire anecdote with its warning about trusting female practitioners appears in the second edition, appended at the end of chapter XII, "Del modo d'attaccare le sanguisughe al fondamento," constituting one of the additional *avvertimenti* mentioned in the new title, emphasized with the word *Nota* printed in the margin.[58] The new warning against female practitioners and information about modest and fearful female patients throw the maleness of the orderly group working around the patient into relief; it signals a low point in the licensing of women, who nevertheless continued to be trusted in the homes of the ill, particularly where treating women was concerned.[59] While the original medieval decrees that all practitioners, male or female, should be licensed did not distinguish between men and women, the licensing procedure quickly began to exclude women from the practices of physic, or internal medicine. They were routinely excluded, by the mid-sixteenth century, from barbers' guilds.[60] Evidence is mixed as to whether women were preferred to men in the treatment of female patients on grounds of propriety or modesty, but by Magni's time women's shame and modesty in discussing their health problems with men, or being treated by them, was a topos of medical writing.[61] Equally a topos by this time was the charge that women practitioners were ignorant of the best medical practices, a charge used by university trained physicians against all practitioners below them in rank, although it was particularly sharp when aimed at women.[62]

The chapter meant to allay the fears of barbers in dealing with their superiors, "Avvertimenti per dar animo a' principianti," also ends with a warning, added in the second edition, to be alert when dealing with women who are somewhat knowledgeable about medical procedures:

> But it seems to me that it should be the duty of the physicians to judge all of these things and that they should leave them in writing – not only that the sick person must be bled, but from which vein, and whether the cut should be wide or narrow, and how much blood should be collected. I say this not to give orders to the lord doctors, who advise them to do these things, but for those who do them, who are in truth very many, about whom I am rather unhappy because these mistakes cause many inconveniences that I see every day. And the physicians have as much more to do as there are inexperienced and worthless modern barbers (even if I always respect the good, wise, and careful ones, because I am not speaking against them here), who bleed people according to their fancy and without a doctor's orders and often for the worse, because they do damage to the patient. It will also be good to do this in order to remove the opportunity for those women who procure an abortion, which is unconscionable, because there are many who know that bleeding the *safena* can easily provoke one.[63]

Rather than describing the barber as subservient to an imperious doctor, Magni conceives of him as a conscientious practitioner whose skills and experience are required by princes, and who gratefully operates under the guidance of a knowledgeable physician. The viewer's attention in the pictures is therefore evenly apportioned between the patient's fear and pain, the barber's confidence and skill, the watchful presence of the university trained physician, the helplessness of the family member who summoned the doctor in the first place, and the apprentice whose earnest desire to learn well is the pretext for the book (but also magnifies the position of the barber). The participants of this group form a nexus of activity, bound together in these somber pictures in silent and studious concentration. While there is no visual mode in which the author can be singled out as a modern medical hero, the male barber is pictured as confident, trustworthy, and discreet, as he is described in the text.[64] In the cluster of solemn actors, only the figure of the concerned relative, always female, seems to have no specific task (fig. 88). With the addition of the new passages in the text warning about ignorant female practitioners and sly female patients, these previously anodyne figures begin to take on a new and somewhat threatening demeanor.

88   Adamo Scultori, Plate VI, engraving, Pietro Paolo Magni, *Discorsi di Pietro Paolo Magni piacentino sopra il modo di sanguinare attaccar le sanguisughe, & le ventose far le fregagioni & vessicatorij a corpi humani. Di nuovo stampati, corretti & ampliati di utili avvertimenti dal proprio autore*, Rome: Bartolomeo Bonfadino, 1586.

Of the ten images of bleeding, four of them show the male patient sick enough to be lying unclothed, in bed, attended by a female nurse or family member. Only two of the images show a female patient, and in both she is fully dressed and sitting up. She, too, is attended by a female advocate who seems also to act as a chaperone in the intimate chamber full of men. So in six of the ten illustrations (Plates I, IIII, VI, VIII, VIIII, and X) a woman is included who never participates in the procedures except to relate to the emotional state of the patient. In Plate I (fig. 72), which is the first of the procedural illustrations, this woman's agitation is clearly communicated by her helpless gesture and flying gown. She is probably not as distressed, as we might be, by the sight of a loved one about to have a vein opened in his forehead, but more by the illness from which he suffers. The text tells us that such a procedure is indicated when "the patient seems out of his mind, as often happens, either from craziness and delirium or melancholic humor . . . it will be necessary to hold the patient's head firmly using both hands placed at the temples, by some person who is both strong and discreet, and he can place himself at the head of the bed between the legs of the barber, as shown in the figure, and sometimes it will also be good to tie up the [patient's] feet and the hands, or to hold down the whole body."[65] Accordingly, the young barber apprentice kneels to stabilize the patient's head between his hands while the barber leans over him, bracing his back against the wall, poising his lancet by the light of a taper held steady by the physician, who helps with the lighting in three of the images. Neither the patient lying flat with his head hanging off one end of the bed, nor the little apprentice on his knees at its head, rigid with apprehension, nor the barber, wedged against the wall with his sleeves rolled up, the lancet held awkwardly and visibly in mid-air, nor the formally dressed physician in hat and robe, could be mistaken for figures from any known mythological or religious scene. Only the agitated wife or daughter who has nothing to do and looks on helplessly might possibly be found, with a change of costume, at a scene of martyrdom or other disaster. Otherwise, the artist had very specific work to do and was forced to make up his characters without relying on the familiar images of saints and pagan gods from the Renaissance print seller's stock.

The engravings therefore portray the different and highly specific postures recommended by Magni for all the people involved in an operation (barber, physician, assistant, patient, and even the hands-off position of the attendant relative), assigning each a unique task in the tense and dimly lit room, and requiring that all the thoroughly modern and recognizable figures display their part in the operation with great clarity. Magni refers the reader to the images as a clarification of the written instructions: "and so that you can better understand this operation, look at Figure I that will show you how you have to do it"; "And wanting to better understand the way you have to do it, look at Figure VI and you will not be able to err"; or more specifically, "I made this Figure III that is placed in front [of this chapter] just so you can see the distance from the thumb of the left hand to the place where you have to cut the vein, and the length of the scalpel that has to be held uncovered in the right hand, so that you can do this operation well." This usually comes at the very end of the chapter, neatly referring the reader back to the beginning of the chapter, where the illustration precedes the text and announces the procedure to be explained. Plate X (fig. 80), alone, refers the reader to the illustration throughout the explanation, because the subject of the passage is an embellishment of the operation described earlier, and consists largely of recommending a new way to grasp and approach the patient that allows a right-handed barber to work on the left side of a patient in the most stable way. From the point of view of the engraver, this was a highly original assignment.

However, if the images in this book are placed side by side they can be grouped to show us how the engraver went about composing the pictures most efficiently, and in a way that presented a variation on the same familiar characters throughout, creating a "storyboard" effect. The following short technical examination of the images is of interest to see how a businesslike engraver could modify essentially similar compositions and figures to come up with plausible variations that both encompass a range of types of apprentice and patient while all appearing to belong to a single authorial practice. We see the reuse of the paired foreground figures of barber and physician or barber and assistant; the barber, kneeling with reverence and elegance, is distinguished by padding or a pillow for his knee, and a medley of vessels provided for his use (fig. 89). The horizontal Plates I, IIII, VI, and VIIII (fig. 90) show the possible positions for bleeding a patient lying down in bed, so in each of them the nude, male patient lies on the bed parallel to the long edges of the plate, attended by a female relative. In Plates VI and VIIII the attending relative is shown in the background, on the far side of the bed, leaving the foreground free for the physician and barber to confer. In Plate IIII (fig. 75) the barber and assistant work from the far side of the bed as the relative watches from a discreet distance by the doorway. Because Plate I shows a patient whose head has to hang over the pillow, the bed is necessarily pushed away from the wall, and the engraver has varied the monotony of the images by reversing the direction of the bed and by switching places with the female relative and the physician. If you reverse the image it becomes clear that they are variations of each other as the pictures of seated men were also variations of the same group of figures.

Fig. 91 shows vertical Plates II, III, V, and X. Except for Plate X, all the vertical plates show a clothed male patient seated in a chair in a small room with a squared pavement and a single opening. The sense of variation depends primarily on simple backdrop elements: a high window and dark shadow cast on the wall in Plate II, a dramatic drapery swag as backdrop in Plate III (fig. 74), and a tremulous question mark of a feather on the patient's hat in Plate V (fig. 76). There is a great economy of facial types: now the patient shares the physiognomy of the barber, who seems to be bleeding his twin in Plate III, now the doctor, as in Plate V. Otherwise the engraver rotated the circle of attendants – physician, assistant, and barber – around the seated man like a tiny carousel. Plate X retains the general composition of the other vertical plates showing a seated figure, but disconcertingly the seated figure at the center of the ring of participants is the barber. The swag of fabric behind the patient in Plate III (fig. 74) has been converted into a bed curtain and the doorway at the left into a window, the assistant has been replaced by an attendant relative, and the physician – heretofore either looking on with approval or offering minimal assistance – leans into the picture and points vigorously at the proceedings. We will come back to this.

Returning to the image with which we started, it is now clearly an odd feature of the picture that a family member is holding the patient in her arms, which is not seen again in this book; also unusual is the sex of the generic patient (fig. 78).[66] The whole picture, as we saw, seems reverent and respectful; the genuflecting barber and the physician focus their gazes earnestly on the discreetly uncovered ankle. The privacy of the patient is further protected with a veil of cross-hatched shadow. She is removed from the bad effects of direct light and air, and in order to enhance the flow of blood she is elevated, as if seated on a throne, like an image of the Virgin in a slightly decentered *sacra conversazione*. But in this world and this setting, the woman is not a virgin – we saw right away that she, like the companion who cradles her in her arms, wears the headcloth of an honest married woman, an item of clothing legally forbidden to Roman prostitutes, and not yet adopted by young unmarried virgins.

89   Kneeling figural group of barber-surgeon and physician, Plates VII, VIII, and VIIII of Magni, *Discorsi di Pietro Paolo Magni piacentino sopra il modo di sanguinare attaccar le sanguisughe, & le ventose far le fregagioni & vessicatorij a corpi humani. Di nuovo stampati, corretti & ampliati di utili avvertimenti dal proprio autore*, Rome: Bartolomeo Bonfadino, 1586.

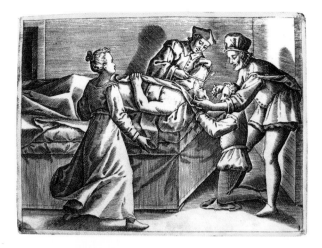
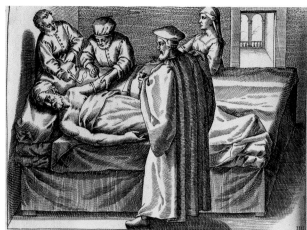

90   Horizontal images of bleeding bedridden patients, Plates I, IIII, VI, and VIIII of Paolo Magni, *Discorsi di Pietro Paolo Magni piacentino sopra il modo di sanguinare attaccar le sanguisughe, & le ventose far le fregagioni & vessicatorij a corpi humani. Di nuovo stampati, corretti & ampliati di utili avvertimenti dal proprio autore,* Rome: Bartolomeo Bonfadino, 1586.

The *safena* vein, Magni tells us, is bled to provoke menstruation and to cure post-partum ailments, and for that reason the operation is almost always performed on women, although occasionally men suffering from pain in their testicles or melancholia would be bled from that site.[67] It is a delicate operation, and it is in the description of this procedure that Magni discourses at length about the importance of paying attention to your light sources. It also requires holding the foot steady in mid-air, and in the last plate in the book, Plate X (fig. 80), Magni shows what is undoubtedly a more stable, though far less delicate, version of how to do this. Plate X demonstrates the same operation, performed for all the same reasons, but shows the patient being bled from the left, *sinistro*, foot. This is a very difficult procedure, Magni says, because you basically have to be born left-handed in order to pull it off successfully.[68] However, he is proud of a method he has developed for stabilizing the foot and allowing the barber to proceed using his right hand, and wants us to see how it is done. In so doing, the artist, who would have known exactly what he was about, gives us not only a picture of a plausible pose for a difficult task, but one that is the mirror opposite both procedurally and in terms of decorum of the medically identical regimen, bleeding from the right foot. In Plate X the patient sits bolt upright, alert and fully dressed, but unlike the woman in Plate VIII she is not very far elevated above the medical men. She wears no modest headcloth; instead the artist gives her an elaborate hairstyle and fancy bow. Instead of leaning trustingly into the arms of her companion, this patient turns her face into full light to address the more honestly attired woman at her side. Not only is the companion not offering support and maternal comfort to an obviously indisposed friend, but the two women appear to be chatting with each other. The barber, rather than genuflecting reverently at his patient's feet, wedges himself between her legs, draws her leg over his, and grasping her bare foot in his lap, prepares to prick her ankle with his lancet.

Few positions could be more immodest for a woman. Few visualizations of female indisposition could be less decorous or less discreet, and we have seen that the sentiments of modest women in the presence of the doctor was something Magni was very aware of elsewhere in the text. The engraver, familiar with mythological subjects from the print trade and from growing up around the works of Giulio Romano in Mantua, would have known that the gesture of a woman's leg slung over a man's was a visual euphemism for the act of love, while the image of the woman in Plate VIII, in modest shadow, supported under the arms by a concerned companion, would have reminded anyone of nothing so much as a birthing scene.[69]

Early in his treatise Magni mentions one of the reasons barbers should act under the supervision and on the order of physicians: to remove the possibility of women who know that bleeding the *safena* vein can provoke menstruation, that they might take advantage of the procedure to procure an abortion.[70] In Bologna it was, in fact, prohibited to bleed a pregnant woman, and even to draw blood from a woman's foot at all. The more commonly recognized reason to bleed a woman from this vein was to alleviate illnesses that occur after the delivery of a baby, and so the honest Roman matron in Plate VIII is shown in the time-honored position of a woman giving birth, firmly associating her with the right and proper use of Magni's procedure. The fact that she is being bled from the honorable right side of her body, and her counterpart from the sinister left, is another fact that would put a Renaissance reader on guard for abnormalities of nature. In Plate X the barber proceeds unassisted, in part because of the stability of the position, in part, perhaps, because we are to understand the nature of this operation as clandestine. The barber, still under the eye of a physician (who points with emphasis from the left much like the pointing finger a student might use to annotate the margin of the book), works in direct light from the open window, precisely what Magni warned us against in his earlier description of this very same operation on the other foot. Besides the eyesight of the barber, the health of the

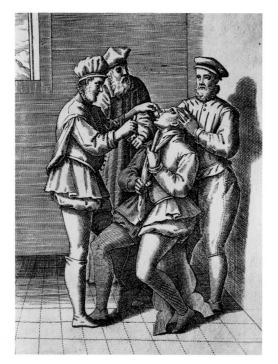 

ABOVE AND OPPOSITE PAGE 91 Vertical images of bleeding seated patients, plates II, III, V, and X, engraving, Pietro Paolo Magni, *Discorsi di Pietro Paolo Magni piacentino sopra il modo di sanguinare attaccar le sanguisughe, & le ventose far le fregagioni & vessicatorij a corpi humani. Di nuovo stampati, corretti & ampliati di utili avvertimenti dal proprio autore*, Rome: Bartolomeo Bonfadino, 1586.

patient is also compromised by this exposure to the elements. But we are not interested in the fate of this woman's soul nor, aside from the impact it would have on the barber should the operation go wrong, in her health: it is the barber's reputation that is at stake. Aside from the schematic woodcut diagram on the position of bleeding points for cupping, it is the last picture in the book.

The visual evidence suggests that the woman in Plate X is neither modest nor honest, and if we read the text we know that the barber is not working under optimum and wisely controlled conditions, although he is still under a physician's direction. There is no apprentice pictured as absorbing the lesson to be learned, as if the picture was both teaching and not teaching, perhaps registering one of Magni's promised warnings as much as demonstrating a procedure to carry out. But the very perplexing question remains of why Magni, who took such care to produce an upstanding image of the barber in his writing and in previous pictures, would include an image that could be understood as lascivious or illustrative of clandestine practices in a treatise he hoped would show the barber's trade in the most respectable light.

 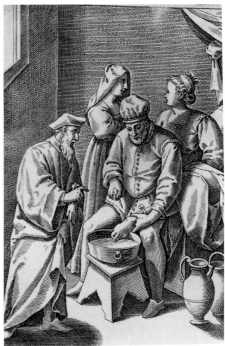

Erotic imagery had already been used in the service of anatomical illustration, notably in the woodcuts illustrating Charles Estienne's *De dissectione partium corporis humani* of 1545, which had adapted female figures from a series of plainly erotic engravings of the *Loves of the Gods* by Jacopo Caraglio.[71] The woman in Plate X, with her oddly balletic turned out ankle, seems to be some version of Caraglio's Deianira seated on the lap of Hercules, most likely by way of one of Estienne's woodcut illustrations showing the veins in the outer layer of the womb (figs. 92 and 93). Magni's illustration makes use of connotations of the erotic as well as of the pregnant woman in both these images, which seem to remain, palimpsest-like, carried in the visualization of a woman receiving an operation related to sexual activity or the reproductive system. It was also Magni's only chance to be pictured as the protagonist of a procedure that he had himself invented, claiming centrality through mastery and invention, the lack of a visible assistant both underlining the ease of the procedure and focusing attention on the figure of the barber.[72] The artist, encountering the uniquely visual problem of portraying a female body in such a pose, used the only kind of body that could be pictured in such a position. The illustration evidently caused no one else any problem, and was copied in later editions without significant changes.

Magni's illustrated treatise in the vernacular attempted to stabilize more than just the members and humors of the afflicted. There is evidence in the many existing copies of his books, studiously annotated and underlined in fading brown ink, and in the frequency with which it was reprinted

92 *Figure of female anatomy*, woodcut,
Charles Estienne, *De dissectione partium corporis
humani libri tres*, Paris: Simon de Colines,
1545, p. 270. Hay Lownes 2-size QM21.E82x,
John Hay Library, Brown University Library.

and updated, that his advice was a serious resource for aspiring barbers. "Nota bene" one reader
penned vigorously in the margin next to the warning to remove hot towels exactly seven minutes
after applying them to the area to be bled, so as not to cook the blood in the veins. Since it was
not the place of the barber-surgeon to exhibit the degree of technical knowledge usually depicted
in medical texts, we see the author (via the engraver) struggling to display a familiarity with anatomical illustrations, while unable to take his cue from the pictures that emphasized the overly
complex system of all the veins and arteries (fig. 85), or smiling skeletons in landscapes, and flayed
cadavers of prisoners still bound and garroted (fig. 94).[73] He wrote for and about barbers, but used
the illustrations to deflect the authoritative center of his work in order to emphasize the authority
of the physician, the connection with whom definitively separated his activities from those of the
hated bath attendants. The object of attention was ostensibly the ill and suffering patient whom
the barber bled by order of the physician. Technically, the barber's area of expertise in bleeding was
confined to the points he indicated on the "Universal Figure," picked out by the skilled pictorial
engraver in focused light on less idealized bodies in the engravings that illustrate his book. Aside
from that, his success and his expertise was located in putting the patient at ease not only to prevent

93   Gian Jacopo Caraglio after Perino del Vaga, *Hercules and Deianira*, engraving, British Museum.

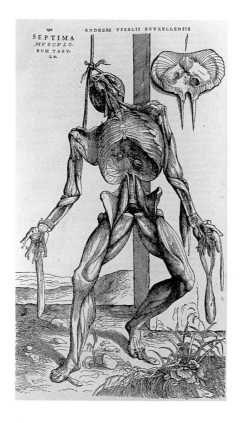

LEFT 94 *Hanged Man* and anatomical figures, Andreas Vesalius, woodcuts, *De humani corporis fabrica*, Basel: J. Oporinus, 1543. Rare 2-S QM21.V37 1543, John Hay Library, Brown University Library.

OPPOSITE PAGE 95 Engraved author portrait and frontispiece, woodcut, *Discorsi di Pietro Paolo Magni piacentino sopra il modo di sanguinare, attaccar le sanguisughe & le ventose, far le fregagioni & vessicatorij a corpi humani. Di nuovo stampati, corretti & ampliati di utili avvertimenti dal proprio autore*, Brescia: Bartholomeo Fontana, 1618. RM 182.M19 1618, Pilcher Collection, Taubman Health Sciences Library, University of Michigan.

accidents with the lancet, but so that he would be successful, would perform his operations with grace and honor, and presumably would be called back. Magni says he wrote his book in Italian so it would be useful to barbers, but he fitted it out with accomplished and expensive engravings by someone familiar with the modes of Counter-Reformation painting as well, so that his book – if not a Latin medical text – at least had little in common with the broadsides or popular pamphlets that could so easily have signified vernacular Roman literature.[74]

Magni's illustrations, then, had to disperse attention across an entire field of linked actions and intentions. Whereas anatomical prints were often detailed views of a single aspect of the violently exposed, dead dissected body, Magni's showed the dexterous management and care of the whole, active living body amid a group of other individuals. At the beginning of his treatise, when describing all the responsibilities of barbers, Magni says no barber can style a man's beard without judgment and making it conform to his face, having respect for and taking into consideration his mode of dress and the quality of person that he is.[75] Representations of caring for the living body, among the many attentions both medical and cosmetic that Gianna Pomata has discussed as the "culture of the body," had to take into account the entire web of social attachments and relationships into which the barber entered in the course of his work.

The Future of the Book

At the beginning of this chapter we saw Aurelia di Antellis settling up with Magni in the matter of the return of the plates, and determining the family's proper share of the finished books. Because the book would not be comprehensible without the illustrations, whoever owned the plates and a single, copiable version of the printed text had the means to print the book again with the least expense; in this way the enterprise could theoretically continue to be profitable for her family. In practice, it was not necessary to own the plates, only a convenience. Rome was full of talented copyists who could engrave new plates from the old models, and this is what happened once the ten-year privilege protecting Magni's financial and authorial interest in the book had expired. At some point it seems that the family sold the plates, which eventually went to Brescia, Bonfadino's hometown, where they turn up in a reprint of Magni's book by a publisher named Bartolomeo Fontana in 1618 (fig. 95).[76] Fontana also produced an edition of Magni's treatise on cauterization (1588) the same year. In his edition of the bleeding manual, the words "Con privilegio del sommo/pontifice" and the date at the bottom of Cherubino Alberti's frontispiece were scraped off and re-engraved to display Fontana's name and printer's mark. The woodblocks that illustrated the cauterization book with pictures of instruments and bandages, which could

LEFT AND BELOW 96 Title page and Plate VIII, woodcut, Pietro Paolo Magni, *Discorsi di Pietro-Paolo Magni piacentino sopra il modo di sanguinare, attaccar ventose, e sanguisugue* [sic], *far le fregagioni, e vessicatorij a' corpi humani, novamente ristampato con sue figure*, in Bologna: nella stamperia del Longhi, 1703, Biblioteca d'arte e di storia di San Giorgio in Poggiale, Bologna.

never have belonged to the Scultori family, seem to have made their way to Brescia as well, and appear unchanged in the new edition with the exception of the frontispiece that records the new publisher's information. Fontana put to use the woodcut author portrait of Magni, that originally appeared only in the 1588 cauterization book, twice, including it also to introduce his edition of the bloodletting book.

Aurelia eventually married a French printer working in Rome named Cristofano Blanco, who specialized in printing and selling single-sheet prints, and in 1613 the sons, by now grown, gave 300 scudi worth of their inheritance in Adamo's plates to Blanco as their mother's dowry. The plates for Magni's book are not listed among them, signifying that either they were no longer in the sons' possession or Blanco was not interested in the production and sale of this illustrated book.[77] In that same year – also the year the barbers' company succeeded in liberating itself from the *stufaroli* – the images for Magni's book were re-engraved for a new Roman edition at the behest of the Roman bookseller Pietro Fetti, printed by Jacopo Mascardi. Fetti dedicated the book to the barbers' guild, writing that he wanted to reissue "the book for the common good, as much for the practitioners of this art as for the sick. And considering that I could not do better than to direct this to your *Collegio*, I willingly dedicate it to you as much because it is the subject of your profession, as to defend the work, thinking it would be useful to join the fame of the author with your enjoyment, and especially because this is a book that has often been requested of me."[78] The plates seemed to stay at Mascardi's shop, where the book was printed again in 1626 with a dedication by Giacomo Marcucci, who, although the title was still "Discorsi di Pietro Paolo Magni . . . ," referred to it as "my book about surgery" in his dedication to the papal surgeon, Prospero Cecchini.[79] In all, the book was to go through eight Italian editions and a French translation. The original plates were used for the first two Roman editions of 1584 and 1586, and the 1618 edition printed in Brescia. They served as models for copyists from the first Mascardi edition of 1613 through the last editions printed in Bologna in 1674 and 1703, which appeared with crude woodcut figures (fig. 96).[80]

In the editions printed by Mascardi in 1613 and 1626 the text is barely altered, but the peculiar balance and tension between authoritative figures has been changed by a new emphasis provided in the updated illustrations. Reading the text and looking at the pictures in the first editions, it is possible to take one of several subject positions. The images can be viewed sympathetically from the point of view of the barber, the patient, or the apprentice, even if the physician is slightly too abstract to evoke this sort of response. From the first, Magni has shown a certain irritation with interfering family members in several parts of the book, and goes so far as to say that if they are pressed into service as assistants – for example, steadying an arm or a leg – the barber will not be able to do a good job, as their empathy with the patient is so great that at the first cry of pain they will drop the limb and ruin the entire operation.[81] In spite of this Magni has allowed a female family member to be represented among those present during the operation, as a visible comfort to and chaperone for the patient, and because in some instances, he mentions, they are pressed into service to provide cloths or approve the process. In all likelihood, and

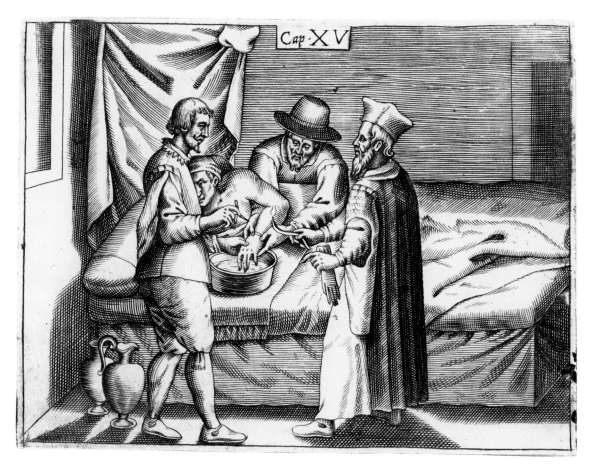

97   Plate VI, engraving, Pietro Paolo Magni, *Discorsi di Pietro Paolo Magni piacentino sopra il modo di sanguinare, attaccar le sanguisughe, & le ventose, far le fregagioni & vesicatorii a corpi humani. Di nuovo stampati, corretti & ampliati di utile avertimenti dal proprio autore . . .* , Rome: Iacomo Mascardi, 1626. Rare Books RM182 .M27 1626, Boston Medical Library in the Francis A. Countway Library of Medicine.

according to the anecdotes Magni likes to include, they were an important if problematic feature of the early modern sickroom. In all cases this person noticeably refrains from assisting in the procedure in any way. Instead, it is the disinterested barber's apprentice or else the doctor himself who assists at the operation, and the relative is shown standing slightly apart, comforting or distracting the patient. By 1613 even this was no longer acceptable. In the later version of Magni's Plate VI (fig. 97), the lady simply vanishes, while the authority of the doctor is clarified and reinforced with the addition of a pair of highly visible gloves, front and center. In the copy of the 1626 edition I used, the reader took special note of Magni's warning not to have family members present at operations, underlining the relevant phrase to make it, perhaps, as emphatic

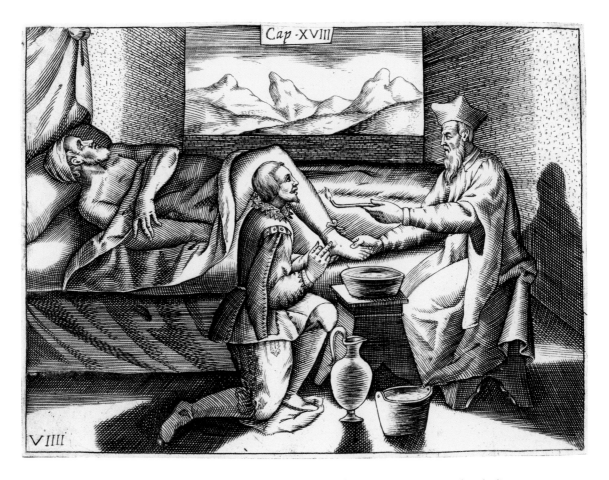

98   Plate VIIII, engraving, Pietro Paolo Magni, *Discorsi di Pietro Paolo Magni piacentino sopra il modo di sanguinare, attaccar le sanguisughe, & le ventose, far le fregagioni & vesicatorii a corpi humani. Di nuovo stampati, corretti & ampliati di utile avertimenti dal proprio autore . . .* , Rome: Iacomo Mascardi, 1626. Rare Books RM182 .M27 1626, Boston Medical Library in the Francis A. Countway Library of Medicine.

as the new illustrations seemed to warrant. Magni's Plate VIIII is more dramatically changed, as the lady turns into a bizarre landscape seen through a window cut into the wall (fig. 98). This breaks at least two of Magni's cardinal rules. Magni is very clear about lighting:

> I have also seen and heard it said that many err in [operating] in the dark; I say that if they are in natural light, as near a window or a door, they have placed the patient completely contrary to the way one must occupy one's hands, so one would make a shadow oneself, and because of this they would not see what they are doing. The same thing also happens by candlelight, which seems to me to be a great mistake, and I would not in the end be surprised

if they do not do a good job, for this is an operation of great importance, and not one to try in the dark. I have also heard when listening to the doctors that there are some who try to perform this operation with the light of two or three lit candles, which is the greatest error, and the poor things have not worked out that one light contrasts with another, and they will not see what they are doing . . . and in spite of a candle being lit, they also leave a window open, and this too is a mistake, because the bigger light fills up the smaller one, and breaks down vision.[82]

Aside from that, Magni says it is placing an ill and bedridden patient in the gravest peril to subject him to fresh air from an open door or window. Magni's advice here did not seem to override pictorial considerations, for the engraver of the sixteenth-century editions did include windows in two scenes, resulting in operations performed by natural light (Plates II and X). But the engraver of the 1626 edition placed the patient in terrible danger and risked sacrificing the barber's reputation simply in order to picture healing as a male-only profession, something which became increasingly the case from this time forward.

Like the Marcucci dedication, the dedications of the Bolognese editions refer to the book as the effort of the producer, calling it "my little work." By 1674 Magni was, of course, no longer alive, and the engraved plates for the finely tuned illustrations in respectable quarto had been reduced into much more schematic woodcuts in octavo-sized books that preserved changes in background and hat fashions that first appeared with the Mascardi editions. Shading is rendered in dark blocks, and facial expressions are inferred through the positions of the body. The landscape in Plate VIIII was further simplified to a blank window, but the female relative in the sixteenth-century editions did not return (figs. 96, 98, and 79). However, the book was still a reasonable moneymaker in the eyes of the Bolognese publishers, who went to the trouble of having the woodblocks made for the latest editions. Dedicated by one Giovanni Battista Vaglierina to a well-known local surgeon named Antonio Bosio, it was printed at a shop specializing in burlesques, Latin grammars made easy, and printed madrigals and love songs. The connections between artist, designer, and engraver that had originally prevailed had been turned into something else – the book had taken on a life of its own, and had become "a thing in itself." These were the words that Adamo Scultori's father had used writing to Cardinal Granvelle in the 1560s anticipating a negative reaction to his son's engraved versions of the paintings on the Sistine ceiling, blaming the degeneration of the image on the inexpert copies that were available to the boy in Mantua from which he was forced to work.

The original engravings for Magni's book walked a fine line between claiming a social space and a professional space for the best sort of barber-surgeon, one whose honor depended on the legitimacy of a physician who had to be present in the illustrations as watchful and abstract, conferring the legitimacy of an embodied prescription while maintaining the desired sense of distance. The artist worked religious imagery in the most personal vein, bringing the emotional sense of familiar groupings from scenes like the birth or death of the Virgin to a state-of-the-art medical practice that could not be mistaken for that of the charlatans, the bath house men and wise women who might also deal with living and suffering bodies.

Appendix: Editions of Magni

1 *Discorsi di Pietro Paolo Magni piacentino intorno al sanguinar i corpi humani, il modo di ataccare le sanguisuche e ventose è far frittioni è vescicatorii: con buoni et utili avertimenti* (Rome: appresso Bartolomeo Bonfadini, and Tito Diani, 1584).

2 *Discorsi di Pietro Paolo Magni piacentino sopra il modo di sanguinare attaccar le sanguisughe, & le ventose far le fregagioni & vessicatorij a corpi humani. Di nuovo stampati, corretti & ampliati di utili avvertimenti dal proprio autore* (Rome: per Bartholomeo Bonfadino nel Pellegrino, 1586).

3 *Discours de Pierre Paul Magni, . . . touchant le saigner des corps humains, le moyen d'attacher les sangsues et ventouses et de faire frictions et vessicatoires . . . traduits nouvellement d'italien en françois* (Lyon: per J. Leretout, 1586). Noted in Brunet, 1299, Bibliothèque interuniversitaire Sainte-Geneviève (Paris).

4 Venice, 1606?. Reported in Giuseppe Borghini, "Salasso e l'opera di Pietro Paolo Magni Piacentino", *Piacenza sanitaria* 4, no. 9 (1956): 3–30 (26).

5 *Discorsi di Pietro Paolo Magni piacentino sopra il modo di sanguinare attaccar le sanguisughe, et le ventose, far le fregagioni, et vessicatorij a corpi humani. Di nuovo ristampato ad istanza di Pietro Fetti libraro in Parione* (Rome: per Iacomo Mascardi, 1613).

6 *Discorsi di Pietro Paolo Magni piacentino sopra il modo di sanguinare, attaccar le sanguisughe & le ventose, far le fregagioni & vessicatorij a corpi humani. Di nuovo stampati, corretti & ampliati di utili avvertimenti dal proprio autore* (in Brescia: per Bartholomeo Fontana, 1618).

7 *Discorsi di Pietro Paolo Magni piacentino sopra il modo di sanguinare, attaccar le sanguisughe, & le ventose, far le fregagioni, & vesicatorii a corpi humani. Di nuovo stampati, corretti, & ampliati di utili avertimenti dal proprio autore* (Rome: ad istanza di Iacomo Marcucci in Piazza Nauona: per Iacomo Mascardi, 1626).

8 *Discorsi di Pietro Paolo Magni piacentino sopra il modo di sanguinare attaccar le sanguisughe, e le ventose, far le fregagioni, & vessicatorij a corpi humani* (in Bologna: per Gio. Recaldini, 1674).

9 *Discorsi di Pietro-Paolo Magni piacentino sopra il modo di sanguinare, attaccar ventose, e sanguisugue [sic], far le fregagioni, e vessicatorij a' corpi humani, novamente ristampato con sue figure* (in Bologna: nella stamperia del Longhi, 1703).

MAGGINO HEBREO VENET. INVEN. AETAT. SVÆ XXVII

5

Courts and Other Theaters: Magino Gabrielli's Dialogues on Silk

Whereas Camillo Agrippa's publications were meant to be linked together in a series pointing to the same inventive author, a now rare illustrated dialogue on a new invention for how to produce silk twice in the same year focused attention on a single inventor who, as it turns out, was really a corporate entity of inventors, merchants, authors, engravers, and multiple inventions and enterprises. As Agrippa's dueling manual was printed in Rome but dedicated to the Medici in Florence, the silk invention that was the subject of this book was also printed in Rome and offered simultaneously to the pope and to the Medici duke, Francesco I.[1] Whereas Agrippa dedicated his many books to different possible patrons, the book on silk attempted to address a cosmology of patrons in a single publication, meant to flatter all of them at once. *The Dialogues of M. Magino Gabrielli, Venetian Jew, on the Use of His Inventions Concerning Silk* (fig. 99) was dedicated to the reigning pope, Sixtus V. It explains a new invention that would guarantee an abundant crop of cocoons twice a year, doubling the expected harvest of sixteenth-century silk growers.[2] Lavishly illustrated and expensively printed, most circumstances of the book's printing are unknown. Except for one, the artists who cut the woodblocks are still anonymous, and how Magino came to use the heirs of the Roman printer Giovanni Gigliotti as his publisher is yet to be discovered. It is not even entirely clear who wrote the book, or whether most of the blocks were cut in Magino's native Venice or after he came to Rome. But much else about the story of the named author and his ambitions for this publication is abundantly clear.

The topic of the book, the technique of raising silkworms, was subject to some of the same sort of pastoral imagination in the Renaissance as the life of music-making shepherds, and the literary and pictorial forms in which this process was expressed ran from Vergilian ode to agricultural treatise to painted wall decoration. The illustrations of Magino's *Dialogues* represent the process as entirely refined and delightful, at the extreme upper end of artistic compositions related to particulars of the silk industry.[3] The emphasis on the universal benefits of open trade and

99   Title page, Magino Gabrielli, *Dialoghi di M. Magino Gabrielli, Hebreo Venetiano: sopra l'utili sue inventioni circa la seta*, Rome: Heredi di Giovanni Gigliotti, 1588. Capponi II 74, Biblioteca Apostolica Vaticana.

modern technology in the *Dialogues* also makes it an early representation of certain arguments for considering the industry of Jews as beneficial for inter-religious harmony and, above all, as profitable for European court cities. The book presents the unconventional idea of a healthy economic order in which unity focused on common commercial profit furthered the public good, a modern principle that Magino claimed was desirable for people of all religious beliefs.[4] The world of international commerce described here was a neutral sphere, able to accommodate, even override, the prejudices that damaged the possibilities for peaceful coexistence in political and social realms. In the pages of this book, Magino made good use of the relatively newly established conventions of print culture to represent himself both as a literary author and as an inventor with access to the secrets of nature.[5] Through words and images the book promoted his business interests in an atmosphere that more usually obstructed rather than enabled Jewish economic ventures. Arguing that his invention would be useful and delightful to all of humanity, this otherwise marginal early modern subject took advantage of printed text and pictures to construct and deploy an operable identity for a Jewish entrepreneur within an overwhelmingly Christian society.

The incubation of silkworms at the most delicate stages of their life cycles was an instance of cross-species symbiosis much like that of shepherds and their flocks, and an example of the intimate access to the resources of one body to profit from the resources of another that characterized so much of premodern labor. Unlike the sheep whose coats provided necessary wool, and whose outdoor existence in the company of rough shepherds was most idyllic in the imagination of those who had never experienced it, the more privileged silkworms were born and raised in comparatively luxurious surroundings. Coddled into existence between the breasts of young women whose natural warmth maintained the developing worms at the recommended temperature, these valuable creatures were happiest when nursed with wine, raised to the sound of joyful music, and eventually hand-fed with fresh-picked mulberry leaves from which the damp morning dew had been carefully removed. As they began to venture from their first nests, they were carefully shaded from the scorching temperatures of the midday sun or an open hearth, and protected from exposure to nocturnal damp in the marble palaces and arcaded villas that were their preferred homes. Although women of any class or occupation could and did supplement their incomes by loaning their bodices to this task, and most of them were either nuns, widows, or poorly paid wives and relatives of vendors or artisans, the story of sericulture is a decidedly patrician tale as it unfolds in the images of this book.[6]

After making a demonstration of his invention to the pope's satisfaction in the full, leaf-wilting heat of summer, in August 1587 Magino attained his initial goal: permission to put his new process, for obtaining two crops of silk cocoons in a single year, into practice at the Villa Montalto, the papal villa near the Baths of Diocletian.[7] There, Sixtus founded a silk workers' colony over which Magino may have presided for just under two years — the documentation for what actually took place there is sparse. According to Luca Molà, who has uncovered the origins of Magino's interest in this invention in the context of the development of the silk industry in Venice, the process Magino was recommending worked in China and subtropical India, but never actually worked in Italy until the Japanese bred a more resilient mulberry tree in the late nineteenth century.[8] But for this short time, the commercial utopia described in the book seemed to be realized in the rarified terrain of the papal villa. Although the well-being of the silkworm and its miraculous regeneration is the subject of the dialogues, the creature also becomes a surrogate subject for the industrious, luxury-providing Jewish inventor, whose ingenuity makes the world a better place. In the text and its illustrations, the character of the Jewish inventor Magino discourses at length on the value and utility of the humble and defenseless worm, thereby also aiming to convince readers of the value and desirability of a vulnerable people.

Apart from its sumptuous subject matter, the folio-size book is a luxurious object in its own right. Its many elaborate full-page woodcut illustrations and text are printed on fine writing paper, and its entertaining dialogues are staged in rich palaces and healthful, thoughtfully planned gardens that are identifiably situated at the different Italian courts where Magino and his business partner wished to install their process. The pictures show an ideal aristocratic world that includes Christians, Jews, Moors, animals, plants, men, and women, working together in well-ordered palatial settings towards the common goal of increasing the amount of silk in the world. These Edenic printed gardens are presided over first by the face of Pope Sixtus V, printed on the back of the title page in the book's only copperplate engraving, and then, throughout, by the figure of Magino, the Jewish inventor. The woodcut illustrations provide a vision of virtuous industry, which tempers the otherwise overwhelming impression of luxury by showing honestly occupied hands and earnestly inquiring minds, focused under the instruction of a capable master identified in almost every case as "the Inventor." Only once in the captions for the illustrations (fig. 112) is the presiding courtier identified by the religious difference that provided the tension that structures both the work and the enterprise. In this figure, set in Naples, the hatless, bearded courtier in lace ruff and embroidered suit who steps forward to offer a handful of fresh-cut mulberry leaves to a young girl seated at his feet is expressly identified as Jewish: "The Jew who teaches how to feed [the worms] after each mutation."[9]

It should be said at the outset that the historical Magino Gabrielli was indeed a talented and ambitious young entrepreneur who, over the course of the ten years following the appearance of the book, went on to a career as an inventor, businessman, political leader of the Jewish community in Livorno, and diplomat in Italy, France, and Germany, but he was not the inventor of the silk process that the book declares him to be.[10] While still in Venice Magino entered into a partnership with a nobleman from Lucca, Giovan Battista Guidoboni, who had been registering patents with the Venetian Senate for instruments pertaining to sericulture since 1569.[11] He registered the very processes Magino declaims in his *Dialoghi* in 1586 and 1587, and drew up a partnership with the Jewish merchant in which Magino would travel to make the process known through family connections in other cities, and at foreign and Italian courts. He would obtain further privileges and patents, sell equipment to be used in the silk process, and report back often to Guidoboni.[12] The *Dialoghi* seem to be an instance of Magino presenting the invention in his own name. Guidoboni sued to dissolve the partnership in June 1588, less than a year from the time Magino received his Roman privilege and published the silk book.

That same year Magino received papal recognition for another invention, which he mentions in the book, for making a transparent and clear crystal that he proposed to use for windows, mirrors, and bottles, clarifying the glass using potash.[13] Later the same year he received a papal monopoly for making crystal bottles (*misure*) for the sale of wine and oil in legally mandated measures that would have to be adopted by all the Roman *osti*, or people who kept *osterie*, popular establishments for local fare and wine, in order to prevent the sorts of fraudulent measuring or substitution of inferior wines that had evidently been encouraged by the use of opaque terracotta measures.[14] In fact, much of the time and money Magino spent in Rome between 1587 and 1589 was dedicated to promoting his own invention having to do with the manufacture of clear glass, not only for windows and bottles but also for a dozen gilded chandeliers decorated with papier-mâché "cherubini" and hung with crystal drops.[15] The clear glass windows were eventually consigned to the workshop of Domenico Fontana, who was responsible for installing them in the new papal palace at the Vatican, at the church of St. John Lateran, and in the pope's showcase project, a hospital (or prison) for poor beggars at the foot of the Ponte Sisto, for which Magino also provided large mirrors.[16]

Magino supplied the windows to the Holy See gratis, along with a large sum of money, in return for the pope's granting of the monopoly on the glass measures, the ability to live outside the ghetto, and the notice that both privileges would devolve also to Magino's heirs. The papal edict forcing the *osti* to buy all new glass wine bottles – a portion of which money would go to Magino – was, unsurprisingly, highly unpopular. It was met by those involved with much hostility, including a last-ditch and unsuccessful reminder to the pope that they should not be forced to do business with Jews. An account of their displeasure made its way into the Roman *avvisi*, the manuscript newspaper that circulated to the Netherlands and elsewhere with the news and gossip of the Roman court, which records the terms of the edict in spite of extreme efforts on the part of the *osti* who "curse every hour the inventor of this novelty, Magino *hebreo*."[17] As Sixtus lay dying the following year, lightning struck the papal coat of arms posted over the door of Magino's house, which was widely taken as a sign that Sixtus's misplaced liberality and Magino's impertinence were to be rewarded with the removal of his privileges regarding the monopoly on the carafes, a sentiment enacted by the succeeding pope, Innocent IX.[18]

In September 1589 Magino entered into bankruptcy proceedings, although later that year he rented a house in the rear of the new ghetto along the Tiber.[19] He finally left Rome in the summer of 1591 amidst the ire of the Roman *osti*, at which time he became the consul appointed by Duke Ferdinando de' Medici to represent the merchants of the Jewish Levantine nation in Pisa and Livorno.[20] He began again in Pisa as an artisan and entrepreneur, setting up a glass workshop as he had in Rome. In 1595 he transferred the workshop to Livorno, where he had already established businesses selling silk and wool cloth and making gold leaf, but again fell into debt.[21] In 1597 he was sent to France as an ambassador for the Jewish community at Lorraine, and then he traveled with Abramo Colorni, the Jewish court engineer and alchemist, to Stuttgart and Württemberg in 1598. He appears in payment records of the papal *fabbrica* supplying glass to the papal court in 1598, and after that there seems to be no further documentary record of his activities in Rome.[22]

Magino possessed an intricate understanding of Italian commercial law and court etiquette; he was particularly good at calibrating the operating space between aristocratic greed and a ruler's sense of public responsibility. As for his literary skills, there is no proof of his exact role in writing the extravagant book that bears his name as author. He did see the manuscript through the press in Rome, as he portrays himself doing in the text, just as he actively pursued the business of getting the silk invention underway in the Papal States. Whatever his role in the writing of the book and the invention of the process it declaims, issues of authorship, intellectual property, and invention are at the forefront of the concerns the book promotes. In it, the person of Magino Gabrielli, *hebreo venetiano*, is given voice and shape, and the book acted as a passport and a cover for Magino during two years of intense financial and artisanal negotiations in several cities simultaneously. Whether he was the composer of the rather professionally written *Dialogues* may never be known; he was not, in the end, the inventor of the process they describe. He is, however, presented in the book as its author, chief interlocutor, and the solitary inventor of a very profitable silk process. It is this representation of an author and inventor in word and image, as well as its formulation in this very rare illustrated book, that will concern us here.

The *Dialoghi* are crafted with as much attention to the order and composition of the entire illustrated book as to the exposition of the silk process. Theories of writing and reception are discussed alongside theories of sericulture and trade, and in the process the book deals with delicate questions about religion that address important aspects of doing business with non-Christians. It seeks to allay fears about entering into economic relationships with people of different religious beliefs, anxieties that were borne out by the actions of the Roman *osteria* owners when they asked the pope to exempt them from doing business with a Jew in order to avoid an unpopular tax. The very deliberate order of the book begins with a ten-page *proemio* and continues with three dialogues, the last two of which are interspersed at regular intervals with full-page woodblock illustrations. This order was closely related to the way the book was supposed to be used, both legally and in the practice of silk making, and several times during the *Dialogues* Magino warns his interlocutors not to interrupt the order he has planned.[23] However punctilious Magino was about obtaining privileges and permissions, maintaining order in the printing of the book itself seemed to elude him in the chaos that resulted from the rush to get the book into print before the beginning of the next coordinated cycle of silkworm-hatching and the leafing out of the black mulberry tree that provided their food. This history, too, is found inserted among the precise instructions for his silk process, where asides about the printing process add a soupçon of comedy and provide opportunities to bring out the characters of the inventor, his interlocutors, and even the silkworms themselves. It also becomes a way for the author to reveal both his knowledge of the publishing process and, in the promises to correct errors in the text and pictures in future editions, his plans for the future of the silk project. Printing, trade and its protections, religious tolerance, and particulars of the silk industry are all subjects of this book.

The title page opens the *proemio* and bears a notice, printed above a large woodcut of the papal coat of arms, that everything pertinent to the material in the treatise is demonstrated in "charming narrative pictures." The typeset page with its woodblock centerpiece is, very unusually, printed on the other side of the full-page portrait engraving of Sixtus, which had an independent life as a single-sheet print (fig. 100). This meant that the sheet was printed twice on two different kinds of presses, as was Camillo Agrippa's fencing book that combines type and engraving on the same page. The portrait is the only engraving in the book, and rather than acquiring the engraved plate, it would have probably been more economical for the publisher to buy already printed engravings and print the title pages on their reverses. It was originally planned to produce the book in an edition of 1,500 copies.[24] For a book with a small print run this would have been a better investment than purchasing or commissioning an engraved plate and printing it on a press different from the kind used for woodcut and type. Publishers of single-sheet portrait engravings could count on markets such as this for useful images of famous people.

The engraved portrait of Sixtus was copied from an official model used to produce standard papal portraits in all art forms, including sixteenth-century media such as wax and tapestry.[25] As was customary for this kind of portrait, the pope appears in the *camauro* and *mozzetta* of his religious office, rather than in the imposing tiara that would also signify his temporal position.[26] The elements of the papal *impresa* have been dispersed into the four corners of the engraving, casting the papal body as a cosmological figure, the center of a *mappa mundi*, surrounded by signs that defined his individual self in terms of his family lineage.[27]

In a popular print of a type made to show the most remarkable events and accomplishments of the papacy, parts of the Sistine *impresa* (the lion with the three mountains from Sixtus's place of origin, Mont'Alto, and the pears signifying abundance, playing on the pope's family name, Peretti) were allegorized in a decorative way that enlivened the didactic image (fig. 101).[28] These elements, the lion, the pear branch, the three mountains, and a star, coalesced into the Peretti coat of arms, a personal constellation seen shining over the pope's shoulder. Although they were also bound into books, the *Res et gesta* print was a type made for late sixteenth-century popes and usually sold as single-sheet engravings to pilgrims and tourists.[29] In this genre of print the pope is pictured surrounded by the major accomplishments of his reign, his deeds forming the basis for his status as a good ruler. Sixtus's memorable works pictured here not only included building churches, canonizing saints, and moving the city's ancient obelisks (engineered by court architect Domenico Fontana), but also effecting an increase in abundance and bounty for the people. These accomplishments are allegorized in the lower right section of the print through the image of a lion shaking pears from a tree to feed a flock of sheep, a scene explicitly labeled *abundantia*.

The benign image of the all-providing Sixtus therefore regards the reader from the left-hand side of the open book, while seeming to receive the words addressed to it on the right. There, in an open letter to the pope, Magino states his gratitude for the important privileges conceded to him: "so that my intentions can be better understood, with great expense I have represented [*figurati*] in the present book all the instruments and practices that are used."[30] Following the conventions of Renaissance books of secrets, and adapting the protestations of unworthiness that were standard in courteous dedications to his particular situation, the author expresses relief that he will no longer have to hide from the world the secrets that will so greatly enrich it, and hopes that his work will be accepted:[31]

> I therefore humbly petition Your Highness to accept my work, looking not at the baseness of the person who offers it, even though he be of the law of Moses, but (as is your way) towards the universal profit in the introduction of such a rich practice that, by the grace of God, seems to have been reserved for your most happy [*felicissimo*] reign, under which we can wait so much more comfortably, much more so than in times past (thanks to your Holiness), we all enjoy tranquility in peace, uncorrupted justice, and great abundance.
>
> Most humble and unworthy servant, Magino Hebreo[32]

100   *Sixtus V*, engraving, Magino
Gabrielli, *Dialoghi di Magino
Gabrielli, Hebreo Venetiano: sopra l'utili
sue inventioni circa la seta*, Rome:
Heredi di Giovanni Gigliotti, 1588.
Capponi II 74, fol. i verso,
Biblioteca Apostolica Vaticana.

After playing on "Felice," the Pope's given name, Magino also flatteringly invokes the image
of the lion and the pears to describe his reign. The reference to living more commodiously might
have had to do with Sixtus's plans to enlarge the crowded Roman ghetto, which gave the Jews
a welcome feeling of permanence in the city while they were waiting for the Messiah to appear.[33]
Yet in an immediate sense, it referred to the increase of wealth that would attend the adoption
of Magino's invention.

Indeed, among Sixtus's first stated concerns after assuming the papacy was to bring industry
into Rome to relieve, or at least make less visible, the pressing problem of poverty in such a busy
city. Until that time, there was no major industry apart from tourism, the sale of its antiquities,
and, eventually, the manufacture of carriages. In 1587 Sixtus founded the Ospizio dei Mendicanti,

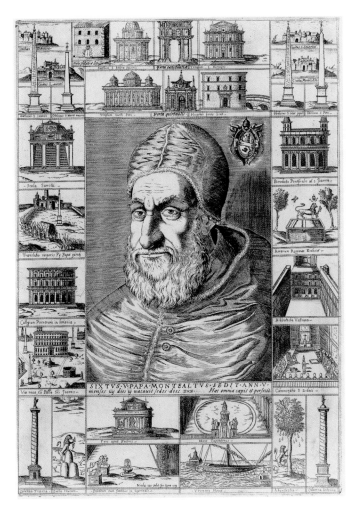

101   Nicholas van Aelst, *Deeds of Sixtus V*, etching with engraving, Rome, 1589, © The Metropolitan Museum of Art, The Elisha Whittelsey Collection, The Elisha Whittelsey Fund, 1949.

the pauper's hospital built by Domenico Fontana at the Ponte Sisto in order to remove beggars from the city streets, but he was also interested in ways to expand employment.[34] Like his predecessors, he looked to the textile industry as a natural possibility for Rome, with its good water supply and many wool-producing sheep grazing in the nearby countryside. Shortly after he took office, it was announced that the pope had resolved to find a means for the poor to be able to live by their work: he would introduce and support the wool and silk trades in Rome.[35] He commissioned a public tub with running water for washing raw wool placed near the Trevi fountain, and – demonstrating a particularly practical approach to the immovable monuments of pagan antiquity – planned to turn the Colosseum into an immense wool-and-cloth-making facility.[36]

This was the aspect of Sistine patronage to which the author made a point of appealing in his dialogues and preface. The dialogues affirm that an increase in the production of silk was good not only for the court, whose members would have more of it to wear, but also for the army who used it for flags, for the banner makers who made the flags, for the farmers who grew the mulberry trees, and for the women who tended the silkworms; in short, one did not have to be among the few nobles who were qualified to wear the fruit of the worm in order for one's condition to be improved by producing it, and the concept of universal profit appears many times in Magino's text.[37]

The letter to Sixtus is immediately followed by a poem in Hebrew in two columns, surmounted by the papal coat of arms, and signed with Magino's Hebrew name: "Meir Sarfati, poet."[38] Following this imposing page is one titled: "Translation of the preceding Hebrew canticle by Messer Magino di Gabriel hebreo venetiano, in praise of Our Lord Sixtus V, made into *ottava rima* by Signor Sebastiano Tellarini" (fig. 102).[39] The poem elaborates each of the elements of the Sistine *impresa*, interpreting them in the light of the pacification of hatred and the expression of just rule. Sixtus is called a new Moses, who provides his thirsty people with water from dry rock and leads them to the sumptuous vineyard by the *terme* (or baths), which Magino includes among Sixtus's greatest urban improvements, bringing to mind the monumental fountain built by Domenico Fontana featuring a Sixtus-sized Moses (fig. 103) that displayed the pope's munificence in revivifying the newly named Acqua Felice. The author also invokes the cycle of nature and mechanics of the cosmos when he inserts himself into a particularly Sistine cosmological order. Sixtus's coronation took place on 1 May 1585; Magino uses this detail to explain that his own name results from his birth on the first of May (*maggio*), and this, along with the fact that May begins the silk cycle, is pointed out here and later in the dialogues as evidence that he is the pope's true and natural servant.[40] The poem gives him an opportunity to mention his other invention, too, and to highlight his own industry and work, not for his own enrichment, but for his desire to improve the state of all human beings: "And only thinking of human delight / Among Alpine forests, and wild branches / I explored so much in mountain and plain, / That I finally discovered a deep virtue, In the grasses that I collected with my own hand. / I could make from them an artful mixture: from which / comes a clear oil, and crystal, that will make / So many palaces gleam like diamonds."[41]

The poem is followed by a second letter from Magino addressed to the reader, in which he apologizes for being late in publishing his secrets to the world. On the back of this letter is a decorative woodcut frontispiece that bears Magino's portrait and a sonnet about silk (fig. 104).[42] The author's portrait at first looks unindividualized and unimposing; however, the inclusion of an author portrait announces a learned treatise, and with it Magino fashioned the cultivation of silkworms as a scientific process. Magino used the image twice, once as frontispiece to the book, and once, with modified text, as a kind of divider between the descriptions of the first and second harvests. It shows Magino as both author and inventor, in the manner of Vesalius, and before that role was assumed by Domenico Fontana, in the frontispiece portrait of his treatise of 1590 that

102   Hebrew Canticle (*proemio*, fol.ii verso–iii recto), Magino Gabrielli, *Dialoghi di Magino Gabrielli, Hebreo Venetiano: sopra l'utili sue inventioni circa la seta*, Rome: Heredi di Giovanni Gigliotti, 1588. Biblioteca Nazionale di Napoli.

publicized his particular ingenious accomplishments in moving the Vatican obelisk.[43] Both pro-emial portraits in Magino's book make use of Latin texts, the language of monuments and also of the legal profession, to locate the identity of the person portrayed within a meaningful group. Sixtus, with his official costume, easily identifiable features, and lapidary inscription in the margin is installed among both the Peretti family lineage and that of all the popes. Magino is identified in the writing around his portrait – "Magino, Venetian Jew, Inventor, his age 27" – and ennobled by the appearance of a crown at the top of the page, as well as two coats of arms that winged putti kick into play at the bottom corners.[44] The crown at the top has a fleur-de-lis in its pattern, and along with the same element on the *scudo* at the right suggests nobility as well as symboli-cally stating Magino's family name, "Sarfati," being Hebrew for the word "French."

Just as the pope is identified as the high priest of Christendom by his uniform, Magino was expected to display his identity as a Jew by exhibiting the red or yellow hat that Venetian Jews

103 Domenico Fontana, Leonardo Sormani, and Prospero da Brescia, Fontana dell'Acqua Felice, 1585–8.

were required to wear; he would likewise have been prohibited from wearing the black hat that would have proclaimed him a Christian.[45] In the sixteenth century there were circumstances under which Venetian Jews might have been excused from wearing identifying clothing. A charter of 1548 provided that "while traveling from one place to another, they shall not be obliged to wear the yellow *baretta* to prevent many troubles that could ensue"; they were also allowed to wear the identity-concealing black hat until the second day after their arrival at their destinations.[46] Although Magino wears no hat in the tiny frontispiece portrait, he is more fully revealed in the theatrical illustrations that accompany the dialogues (figs. 106–114). Here we see that with his stylish, starched linen ruff, his slashed and embroidered outfit, and his rakish cape with its small, peaked collar, he is dressed and even coiffed in a manner that makes him indistinguishable from any Elizabethan courtier, lacking only the gentleman's sword and a visible coat of arms. In many of the illustrations he holds a hat in his hand, or else it rests on a table nearby. Is the hat yellow, as it should be? The prints are discreetly black and white, so all we know, in an inconclusive way, is that the hat is definitely, and correctly, not black. In the frontispiece, however, rather than a securely signifying hat, it is the circlet of Latin text that snugly surrounds the youthful face in the portrait, taking the place of the cultural iconography Magino's contemporaries would recognize in a color image.

In Rome, at least, Magino was not allowed to appear without an identifying sign, although there is evidence that he asked for such a privilege. The concession in which he was given the monopoly on making glass carafes also allowed him and his family to live outside the ghetto (for the cost of 25 gold scudi a year). Ever since 1555, when Pope Pius V proclaimed in his bull, *Cum nimis absurdum*, the creation of a ghetto for the Jews in Rome for the first time, all Jews in the city were forced to live in a crowded gated area close to the Tiber, the fish market, and what had been the ancient theater district.[47] Magino's concession also exempted him from paying the punitive dues to the "Università delli Hebrei" and all the other rules inhibiting the actions of Jews in the city, but it explicitly states that he and his family will, however, have to wear "the sign usually worn by the other Jews."[48] The merchant's identity as a Jew is neither shown stitched onto his clothes in the many images of him that appear throughout the book, as were the circles of yellow braid that Roman Jews had to wear, nor reversible or removable as were the colored hats worn by the Venetian Jews. Magino, *hebreo*, declared himself a Jew in the designation after his name by which he appears as the author of the book, in the many parts of the discourse in the *Dialogues* in which the issue of religion is addressed, and with the name and special ritual touching of the pen "in the style of the Jews" that the notaries report took place whenever he initiated the many legal documents that accumulated in the records of notaries all around the city in the years he was in residence.[49]

On the frontispiece of his book Magino's portrait nests inside a mulberry wreath that provides a framework for putti and adult female angels to act out the infinitely repeatable cycle of the hatching of worms and the sprouting of leaves that culminates in the production of silk thread, which we see rising like steam from a vat below. Magino's image seems to coalesce above the vat, as if brought forth by the ministrations of the angel stirring the cauldron of cocoons with one hand while turning a crank that unwinds them onto a large bobbin with the other. Magino's fancy embroidered jacket unravels into and takes shape from just these threads, as though the always significant clothing of the Jew not only begins to stand as the excluding sign of his religious identity, but also identifies him with the nature and culture of silk production, which he models in his portrait cameo. An angel on the right exhibits a fringed cloth decorated with the Sistine *imprese* of star, mountain, and pear, gifting the project's ultimate patron with a monogrammed example of its finished product (fig. 105). The metaphor of regeneration and the cyclical nature of the silk process is emphasized at the top with the image of a phoenix flying towards a glowing sun. This mythological symbol of rebirth forms the central jewel of a crown below, held aloft by a winged cherub and angels. The crown is poised to confer royalty on the natural wonder

of the conjoined life cycles of the black mulberry tree and the silkworms, a triumph of nature as worthy of admiration as the renewal of the golden phoenix from its ashes.

The trick in all the courtly sericulture literature of the sixteenth century was to transfer a sense of nobility to the fiercely determined, vulnerable, but infinitely valuable and downright miraculous actions of the caterpillar that spins glossy silk to make its own tomb. From this sepulchral cocoon it is resurrected as a winged creature whose single concern is to leave eggs for its own regeneration and then again to die, only to be reborn as the cycle begins over again in marvelous, clockwork synchrony with the sprouting of its favorite food, the leaves of the black mulberry tree, the *gelso moro.* A portrait of the noble, self-sacrificing worm at the moth stage, made for the frontispiece of an engraved series on the art of sericulture by Johannes Stradanus that appeared at about the same time as Magino's book, shows something of this in its characterization of the triumphant *animaletto* as determinedly fierce and dragon-like (fig. 115).[50] In Magino's frontispiece, however, the putti are offering their circlets of leaves and cocoons as if they were wreaths of victory, and the crown, although supported by angels with signs of Sixtus, hovers above the head of the author. Magino is depicted here as a member of a nation whose people have risen yet again from the ashes of their destruction, and someone who has unique access to the secrets of this other natural

106   Prima Figura, Magino Gabrielli, *Dialoghi di Magino Gabrielli, Hebreo Venetiano: sopra l'utili sue inventioni circa la seta*, Rome: Heredi di Giovanni Gigliotti, 1588. Capponi II 74. Biblioteca Apostolica Vaticana.

107 Figura Seconda, *Della staccare dell'ove . . . (Roma)* [not repeated in third dialogue], Magino Gabrielli, *Dialoghi di Magino Gabrielli, Hebreo Venetiano: sopra l'utili sue inventioni circa la seta*, Rome: Heredi di Giovanni Gigliotti, 1588. Biblioteca Nazionale di Napoli.

R O-

A Inuentore che ordina quel che si doue fare.
B Distribuisce le pezze grandi piene di seme.
C Si mette dette pezze d'oua in petto per asciugarle.
D Piglia le pezze grandi per compartire il seme in più pezzette piccole, perche nasca tutto à vn tempo.

M A.

E Il Sole.
F Piazza, e Chiesa di S. Pietro.
G Obelisco d'Augusto, dedicato dal gran SISTO V. alla Croce Santa.
H Palazzo di sua Beatitudine.

D 2

OPPOSITE PAGE 108 Leonardo
Parasole, Figura Terza, *Del governo del seme (Roma)* [not repeated in third dialogue], Magino Gabrielli, *Dialoghi di Magino Gabrielli, Hebreo Venetiano: sopra l'utili sue inventioni circa la seta*, Rome: Heredi di Giovanni Gigliotti, 1588. Biblioteca Nazionale di Napoli.

RIGHT 109 Figura Quarta, *Del modo di conservar la Notte . . . (Firenze)*, Magino Gabrielli, *Dialoghi di Magino Gabrielli, Hebreo Venetiano: sopra l'utili sue inventioni circa la seta*, Rome: Heredi di Giovanni Gigliotti, 1588. Lessing J. Rosenwald Collection, Rare Book and Special Collections Division, Library of Congress, Washington, DC.

cycle. With only his own *ingegno* (as he makes a point of saying in the *Dialoghi*, and Sixtus confirms in his privilege) he has penetrated the secrets of nature to become someone whose vivid scientific acumen is the subject of what follows.[51] The poem under the crown explains the organization of the frontispiece: "From the right side of the sublime circle [the author puns on sublime circle, *cerchio eccelso*, and mulberry wreath, *cerchio di gelso*] / Learn how the worm's food grows, how it makes silk, and below, how it is taken whole from its cocoon / See how the happy angels of beautiful cloth wind, dye, and weave it in the other part; / Above, by the crown and its fitting motto, you know that this stands above every other art."[52]

110 Figura Quinta, *Quando i Vermi cominciano . . . (Venezia)*, Magino Gabrielli, *Dialoghi di Magino Gabrielli, Hebreo Venetiano: sopra l'utili sue inventioni circa la seta*, Rome: Heredi di Giovanni Gigliotti, 1588. Biblioteca Nazionale di Napoli.

Like author portraits in other scientific books, the frontispiece was also one of the newer conventions of print culture. Magino uses it to claim a degree of nobility for the worm (comparing it to the mythical phoenix) equal to that of the luxury thread it produces. It is a nobility in which his own image takes part in the position of inventor and man of science, as well as with the addition of the two small coats of arms that flank the sublime mulberry wreath at the bottom of the page. The image is also ambassadorial: Magino's face on the frontispiece, promising to reveal the secrets of silk production, is his side of a pact with the pope aimed at generating a profit for both of them. In fact, the portraits of pope and merchant are two sides of the same coin; together they bear witness to the value of a joint enterprise for which the book is the plan, the proof, and the tool.

111   Figura Sesta, *Come si allevino i Vermi (Milano)*, Magino Gabrielli, *Dialoghi di Magino Gabrielli, Hebreo Venetiano: sopra l'utili sue inventioni circa la seta*, Rome: Heredi di Giovanni Gigliotti, 1588. Biblioteca Nazionale di Napoli.

The frontispiece faces a Latin version of the papal privilege, which appears in Italian on the following page and was registered with the papal secretaries on 4 July 1587.[53] On the face of it, this is completely ordinary; many sixteenth-century books included their printing privileges up front; the privilege was a guarantee of doctrinal purity to booksellers and buyers, and warned off would-be pirates. But the document in Magino's book was not actually issued to allow and protect the sale and printing of the book, as privileges included in books generally are. It was instead a privilege granted to Magino to practice the art that the book describes. Therefore, the legal license is juxtaposed with that bestowed by nature, and allegorized on the frontispiece.

The privileges in two languages constitute two full pages of impressive legal text. Unlike privileges conceded to Christian subjects, which often began, "To you, so-and-so, beloved of God

LEFT II2   Figura Settima, *Sopra lo allargare i vermicelli (Napoli)* [not repeated in third dialogue], Magino Gabrielli, *Dialoghi di Magino Gabrielli, Hebreo Venetiano: sopra l'utili sue inventioni circa la seta*, Rome: Heredi di Giovanni Gigliotti, 1588. Biblioteca Nazionale di Napoli.

OPPOSITE PAGE II3   Figura Ottava, *Del governo di Vermicelli fino all'ultima muta . . .* (Turino), Magino Gabrielli, *Dialoghi di Magino Gabrielli, Hebreo Venetiano: sopra l'utili sue inventioni circa la seta*, Rome: Heredi di Giovanni Gigliotti, 1588. Biblioteca Nazionale di Napoli.

[*diletto di Dio*]," Magino's privilege began, "To you, Magino di Gabrielli, Jew, living in Venice, and yet to know the true way and how to abide by it."[54] Instead of the usual greeting and benediction with which indulgences began, Sixtus confirmed what otherwise would have been simply a pretty conceit in Magino's dedication when he asked the pope to excuse the "baseness" of the donor, "even though he be of the law of Moses." It then goes on to recognize, from the demonstration Magino had recently made, that he had found a way to raise and keep silkworms so the silk was greatly improved in both quality and quantity, and could be procured with less arduous work at lower expense than before, and that the church could collect twice the amount of taxes on it. Therefore, the pope granted the inventor and his heirs the right to collect a percentage of any money earned by anyone in the Papal States by means of his invention for sixty years. Magino would also receive a third of any fines levied on those who transgressed the terms of the privilege,

TV-

A Donzella che sbruffa i vermicelli con malua-
 gia, ò con aceto & acqua rosa per rinfrescar-
 gli, e guarirgli.
B Inuentore che dimostra come si debbano go-
 uernare i vermicelli.
C Gentildonna che mette vna incannucciata sopra i vermicelli per diradarli.
D Graticci di canne accomodati l'vn sopra l'altro.

RINO.

E Profumo che assicura, e conforta i vermicelli.
F Moretto che co'l sonare il tamburo opera che
 i vermi non siano offesi dal trono, ne da altro
 strepito.
G Sterco & auanzi della foglia de' vermicelli
 che si mette à piedi de' Celsi per ingrassargli.
H Residenza del Serenissimo Signor Duca.

114   Figura Nona, *Come si devono mettere con nuova inventione i Vermi a làvorar la seta (Genoa)*, Magino Gabrielli, *Dialoghi di Magino Gabrielli, Hebreo Venetiano: sopra l'utili sue inventioni circa la seta*, Rome: Heredi di Giovanni Gigliotti, 1588. Biblioteca Nazionale di Napoli.

as well as a portion of any silk earned from a second annual harvest, the technical secret that was the subject of the book.

Furthermore, the pope specified the conditions under which Magino would make his invention known, which constituted another form of publication: "Therefore" the papal secretary wrote, "so that our subjects can more easily learn how to practice these new inventions, you have promised us that on all the Fair days, in the *piazze* and in the public places, the principles of how to practice this new invention can be read in print, so that the above mentioned subjects can easily practice the same art with little expense and to great advantage."[55] It was also mandated that the patents and privileges for the practice of the silk trade that had been granted to Magino were to be read out loud everywhere that he wanted to practice the invention. Therefore, between 1587

115   Philip Galle after Johannes Stradanus, *Silkworm at the Moth Stage* (TOP), detail of frontispiece, *Vermis Sericus* (BOTTOM), Antwerp: s.n., 1580? Hay Lownes SK25 S8x 1650, John Hay Library, Brown University Library.

and 1588 he spent a good deal of time in notarial offices arranging with lawyers and agents for the privilege to be sent to him in whatever city he or his agents found themselves, so that after negotiating with the authorities for the necessary permissions, the printed privilege might be read and made visible.[56]

The legal protections and administrative requirements of the invention were also reinforced visually on the title page of the book, which takes the same format as a legal pronouncement. By the last quarter of the sixteenth century most laws and even court judgments had both a printed and a vocal form of publication, posted in the form of the printed banns, or *bandi*, in particular parts of the city, primarily in the Campidoglio, the seat of the city's civic government. We saw that a *trombetto* was in charge of posting the *bandi*, and in notarial documents this person's signature is found at the bottom of the notarized pronouncement, or on the back of an archived example of one of the printed *bandi*, confirming that it had been posted in the usual places.[57] When Magino exhibited his privilege or had it read out loud, or when he read his printed book out loud in the *piazze*, he was actually publishing in this particularly early modern format.

The title page of the book – the first page that the reader would encounter – confronts the reader in the form of proclamations like the *bandi*, with its ornate papal seal signifying at the same time allegiance to and the approval of, and possibly an order from, a very high authority. The same format – a title in all capital letters and its rather smaller explanation over a papal coat of arms centered on the page, with the name of the publisher and date of publication at the bottom – was used for the late sixteenth-century printed *motu proprius*, or printed papal edicts (*editti*) (fig. 116). Magino's title page is both rather formal and also completely formulaic for products of the press that printed the book, the heirs of Giovanni Gigliotti, who did not print many large or illustrated books.[58] Magino's differs from the others only in that it was clearly copied from the papal edicts that would have appeared on walls in the "usual places" all over the city, substituting, in the space where the description after Sixtus's name bearing the words "Papae V" would be, the designation "Hebreo Venetiano."

After the Latin and Italian versions of the important papal privilege, we arrive at the *Dialogues* promised in the title. The dialogues take place between Magino, as the author and inventor, and three Italian Christians who have been waiting to hear about the new process first hand. The second and third dialogues contain the "delightful narrative figures" mentioned in the book's subtitle that demonstrate the practice and instruments of the art of silk. The first dialogue, which is entirely unillustrated, introduces us to Giulio Cesare, a Neapolitan merchant, and Horatio, a Roman businessman and man of letters, as they pass the time before Magino's arrival discussing many things about the usefulness of silk, how it takes part in a history of practical inventions that have helped mankind, and the virtues of ethical business practices in general. They say that Magino has promised to come by with a copy of his new book, and so the expectant reader enters the dialogue's vestibule to join the merchant and the businessman in waiting for Magino. It is here that the author lays out a philosophy about open trade and the commonalities of all religions.[59]

The literary nature of the book was introduced in the *proemio*, which included poetry, the official language of legal briefs, the conventional flowery format of the courtier's dedication, and the epistle to the reader. This is heightened in the *Dialogues*, which are brought forth in a far more theatrical way than was usual for treatises in dialogue form (like Agrippa's, for example). They are written with a strong sense of literary symmetry and a purposeful use of puns and humor. The treatise departs from the usual back and forth, student–master prompting of an academic dialogue and instead is launched through contrast, argument, and proof, and ornamented with many courtly pleasantries, digressions, and amusements. A sense of being present at an actual play rather than reading a literary dialogue in studious solitude is emphasized by the style and placement of the full-page woodblock prints, starting with the unfolding of the silk process in Dialogue Two. Each of the images places the action against different architectural backdrops not

SANCTISSIMI
D. N. D. SIXTI
P A P AE. V.
BVLLA

Elargitionis Indulgentiarum,& applicationum,ijs, qui Chriſtiano-
rum, qui inferuntur, Imperatorum numiſmata deuoto
animi affectu penes ſe tenuerint, aut alijs fidelibus
accommodauerint,vel tandem Eccleſias,in
quibus ea fuerint aſseruata,
viſitauerint.

ROMÆ Apud Hæredes Antonij Bladij Impreſſores Camerales.
M. D. LXXXVII.

116   *Editto* of Sixtus V, 1587 (compare
with title page, fig. 99), ASR Not.
RCA1078.295r.Bull.2. Courtesy of the
Ministero per i Beni e le Attività
Culturali di Italia.

only identified by their iconic settings but also labeled in a key below, which signal Magino and Guidodoboni's plan to disseminate the silk process in all of the Papal States and other provinces of Italy. Therefore, the settings against which the process is staged include Rome, Venice, Florence, Turin, Milan, Genoa, and Naples. The dialogues themselves develop character, contain jokes, anecdotes, and asides, and include descriptions of the interlocutors, who are, in two cases out of four, illustrated in the images.

The interlocutors also ostentatiously participate in the authorship of the treatise, although Magino is at every turn acknowledged as the inventor, the teacher, the master of silk, and the author of the treatise. When he is asked by Horatio at the beginning of Dialogue Three if he minds if the two gentlemen enter in the middle of his explanation of the process to Horatio's wife, Isabella, Magino tells them not only does he not mind, but their inquisitive presence is sure to draw him out further about the process in ways that he would not have bothered with otherwise, and therefore their interruptions will actually help him to bring the book to perfection.[60] The text is further noteworthy for constant references to the circumstances of its own production, to the work involved both in getting the book through the press and in the composition of a literary dialogue, as well as the conditions under which the book is to be sold and read. After Dialogue One the action proceeds from the pictures, which move the story along, and are the focal point for the discourse in which Magino teaches the art of sericulture that he promised the pope he would do.

The first dialogue opens with the appearance of Giulio Cesare at Horatio's house. While Magino's preceding letter to the reader begins with an apology for being late in getting the book to press, the dialogues begin with Cesare apologizing for arriving early at his friend's house; he blames it on his great eagerness to hear about Magino's marvelous invention. He says that he will wait happily in conversation with the learned Horatio until Magino appears because, as Horatio tells him: "The invention has already been printed and illuminated with the most beautiful pictures that represent what he clearly tells us to do in terms of this process, and his very useful secrets are universally published in Italian and in Latin so they can also be learned beyond the Alps, and I certainly believe that he will bring a copy of his book today to my wife, because he told her he will not return without bringing her one here."[61] Happily waiting is not, here, a neutral subject: Jews were in the irritatingly quasi-permanent condition of waiting for the enlightenment that would result in their conversion. The hope that they would convert, thereby adding great numbers of Christians to those on earth and hastening the end of days, was one excuse stated by popes and Christian rulers for keeping Jews around at all. It was one reason the Roman ghetto had been built and Jews were allowed to continue to live in the city (finance was another, but it occupied a completely different theoretical realm).[62] The book includes many references to Magino's religion both in the pictures and the text, beginning on the title page and continued throughout the volume. The waiting that was a characteristic of Jews contrasted sharply with the intense activity Magino talks about (and which is borne out by the many documents he signed in these years, cluttering the files of notaries in several different offices) in getting his book to press and licensing his invention; the waiting and the hurriedness are counterpoints that the author plays against each other with humor.

Soon after Cesare enters the scene impatient to hear about silk, he tells Horatio: "the inventor seems to me to be one of the greatest and best regarded Jews that were by chance born from the wise King Solomon on down, with the exception of the saints and those beloved by God [*diletti di Dio*] who came from those people."[63] Horatio agrees, referring Cesare (and the reader) to the papal privilege a few pages before as evidence of the fact, although that very same standard expression, *diletto di Dio*, is pointedly not used in the initial boilerplate language of Magino's privilege. Mention of the privilege spurs Cesare to narrate its principal provisions about the usefulness of the invention to all people, and about how Magino has promised to make his invention known in all the provinces that could make use of his secrets to increase their silk production with little expense and with great ease. "But the thing that seems most incredible and of much more importance," Cesare says, "is that he has discovered the secret of – and is obliged to and most liberally wishes to teach to everyone, and to publish to all of Christendom – the way to reap a second harvest of silk in a year with not much more trouble than the first . . . and this is what is contained in the privilege."[64]

Theatrically speaking this seems a little bit stiff, perhaps, and like stacking the deck since the reader has already encountered the privilege twice in its legal form. But in fact, the privilege stated that in order for Magino to gain from his inventions in all the Papal States, he had to promise that not only the silk technique but also the "main points of the exercise" would be read aloud in public places. Including the main points about the terms of the exercise that were stipulated in the privilege was certainly part of the deal, and having them narrated by Cesare was a good deal more attractive to an audience than a straightforward reading of the legal text, no matter with how much pomp or trumpeting. This also removed the performance a bit from the realm of that of the *trombetto* and set it rather neatly into that of the charlatan – *trombetto* and charlatan being two of the local figures who could at any time be seen declaiming written texts in public.[65]

Charlatans, according to one (probably not atypical) physician in 1632, were "those people who appear in the square and sell a few things with entertainments and buffoonery."[66] On the other hand, charlatans were literate vernacular practitioners, very often of medical practices, who sold their wares and advertised their services through the use of public spectacle, generally involving humor and theatrical effects. Their practice took shape in Italy in the fifteenth century, and by the time Magino was looking for all possible networks to publicize his businesses there were both medical charlatans and merchant charlatans. Of the two kinds, David Gentilcore says, the merchant charlatans were more likely to print up chapbooks, some of them beautifully designed and illustrated, to promote their businesses. Those chapbooks often copied the form of papal edicts and *bandi* and were sold along with their remedies, as the title page of Magino's book imitated the look of papal bulls (fig. 116). The merchants were also the most likely to obtain privileges for their activities.[67] Sometimes charlatans accompanied their merchandise and the chapbooks advertising their methods with a public display and declamation of those licenses and protections under which they were operating.[68]

In this way, too, Magino's practice, either by papal decree or through his own choice and particular concatenation of skills and experience, mirrored many aspects of that of the charlatans who mounted platforms in civic squares to declaim their miraculous cures and sell their wares. Magino, too, sold products along with his book, as he promised his Venetian partner he would do. These were items he claimed would last a very long time and were authenticated by identifying stamps. The author's promised procedure would be to attract crowds by holding up the "lovely narrative illustrations that demonstrate all the practices and instruments one needs in the silk trade," accompanied by the declarative reading out loud of the papal privilege. The charlatans and mountebanks, whose most public realm of activity was the street or piazza but who were also called to perform in modified ways at princely courts, drew audience attention through the performance of music and little plays that were the sixteenth-century predecessors of the *commedia dell'arte* performances.[69]

It seems to me that the inclusion of the many expensive and accomplished woodblock prints and the engraving, and the length and the luxurious folio format of the book, more than anything else set Magino's work apart from even the more elaborate charlatan chapbooks that promised equally desirable and improbable results. Magino was a part of the economic and cultural world that included charlatan markets, performances, and sales, as well as their use of the written and spoken word. His activity either expands even the more generous notions of charlatan life and work or – probably more accurately – can be understood to include the business practices of charlatans as one possibility and opportunity for this inventor, merchant, political leader, and (would-be) courtier. It is also true that in Venice it would have been difficult for Magino to produce a printed book, especially one advertising itself as emerging under Jewish authorship and in the printing of which he had to be present: a Venetian edict of 1571 explicitly stated that no Jews could work at a press or print books, and that if they issued books under the name of a Christian they would also be punished.[70] Oddly, Magino's enterprise was a case of the opposite: a Christian invention audaciously presented as Jewish.

The printed privilege displayed at the beginning of the book was therefore an important component of Magino's practice. Printed in two languages up front, it was available to those who could understand written and spoken Italian, and also appeared in impressive legal Latin comprehensible to those outside Italy. Its components are further summarized within the dialogue by Cesare, so that at the first moment of the reader's entry into the dialogue, Cesare sends the reader back to the *proemio* as if to ensure the most important parts of what that sheaf of documents contained was clearly understood. The religion of the author, the legal status of the invention, and the legal protection that the aged ruler of all of Christendom afforded the young Jewish inventor are brought to the forefront of the reader's consciousness.

The dialogue continues with Cesare and Horatio discussing the prehistory of remarkable inventions that will lead naturally to the one to be revealed in Magino's book. The first invention Horatio speaks about is religion, "the greatest gift that man receives from the Supreme Creator, because by its means we reflect on divine things, it pacifies us and unites us with God, from which we follow and to whom belong our lives and our comfort, justice, knowledge, manual labor, and finally eternal happiness."[71] Horatio goes on to list the inventors of many things, most of them invented by the gods on Olympus who seem to be neutral deities, neither Christian nor Jewish. After a brief discourse on luxury he moves on to the invention of laws, which gives him a chance to place Moses and Aaron at the beginning of the conversation, but from there proceeding to the Christian world to mention Constantine, Ferdinand and Isabella, Charles V, Phillip II, and then Christopher Columbus (whose invention of the new world was truly remarkable because while most inventions happen in this world, his happened in a new one).

The discussion reprises other Renaissance collections of the origins of great inventions and the nobility of certain trades, with many flattering references to the courts Magino hoped would accept his business, and also specifies what constitutes a real invention.[72] The classification of trades was part of a popular preoccupation of ordering and reordering that popped up in games, songs, and treatises, the most recent and most popular being Tomaso Garzoni's *La piazza universale di tutte le professioni del mondo*, first published in Venice in 1585 and containing a dedication in the 1587 edition to Abramo Colorni, a Jewish engineer in the service of Alfonso II d'Este in Ferrara, who would eventually become Magino's business partner.[73] One of the odder inventions that Cesare particularly wants to discuss calls the reader's attention back from the annals of history to an awareness of the craft of the book being read, as will happen several times before the *Dialogues* are finished. After talking about music, the interlocutors pass to the invention of writing dialogues, noting that the proper number of interlocutors should ideally not be more than three.

The number of interlocutors in the dialogue we are reading has so far only been two, but a third appears shortly when the reader is presented with the confusing first appearance of the already published book, presumably containing the *Dialogo Primo* that is unfolding before us. Contrary to most theories of dialogue form, the author at this point appears in his own dramatic work to take the subject in hand.[74] Magino arrives with a freshly printed copy of the book we thought we were already reading. This disorienting moment is framed as an epiphany, and begins with an apology for being late:

M: Good day gentlemen, please excuse my lateness, but I was delayed at the print shop because up to now I could not have sent the illustrations.

H: You are a thousand times welcome, and you could not have arrived at a better time because Sr. Cesare and I have been discussing your ideas for almost two hours, and when you came we were saying that if the little worms had spirit or understanding they would never be able to tire of honoring and celebrating you, their biggest illustrator. All the rest of us that, thank God, have intellect, have to hold you in admiration and esteem as the public benefactor of all of Christendom, and in particular of the many faithful whom you employ with notable utility in your business; is this not so Sr. Cesare?

C: It is so true, M. Magino, also Sr. Horatio has been discoursing on many beautiful inventions, and at length on many inventors, and he has concluded that your inventions are wonderful, like those that bring usefulness to everyone: because the princes get to make more deposits into their accounts; the people are therefore wealthier, not having to pay to get silk from the East and from the infidels; the merchants and those who follow the arts augment their traffic. Many private persons also do better by way of your works, as poor women earn their livings better with this than with other practices, and those who have mulberry trees can sell the leaves again in the summer for the second silk harvest, for all of which I say and affirm that you are a more than mediocre inventor among the greats, and a man of sublime ingenuity; and we are both waiting for you with the same desire that you wait for the Messiah.[75]

Magino's arrival into his own dialogue, directly from the site of the production of the book, is accompanied by references to printing problems as well as to his problematic ethnicity. The tension of coming face to face with the Jewish author is discharged with a good-natured joke that makes the famous "waiting" status of the stubborn Jewish people no more significant than that of any Roman inconvenienced by a temporary layover in any of the world's real or literary vestibules. Magino made a point of mentioning it in the dedication when he thanked Sixtus for enlarging the ghetto, and making a place where his people could wait more commodiously. This kind of sentiment, repeated more than once in the dialogue along with statements about how great a blessing Magino's invention is for Christianity, highlights the important fact that Jews did not consider themselves, nor did Sixtus consider them to be, either infidels or heretics in light of official papal policy.[76] As Sixtus's privilege said, they were simply ignorant, "yet to know the true way," but came from the same religious stock as Christians, which was one reason why they were tolerated in Rome. It was important for the point to be rehearsed again at the moment that the speaking author appeared to take control of his dialogue.

This is also an early instance of something much like what Wayne Booth termed the self-conscious narrator, who interrupts his narration to discourse about the difficulties involved in its creation, a literary style (he also calls it "the facetious style") that seemed to appear first in fiction in the seventeenth century with Cervante's *Don Quixote* and which flourished especially in the middle of the eighteenth century as epitomized in *Tristram Shandy* or in *Tom Jones*.[77] Like Magino, Booth's self-conscious narrators call attention to their own act of creation, with the result that the reader is also brought into awareness of the act of reading.[78] The artifice of fiction seems to be broken apart, and the reader is aware of the written work in the moment of its composition, and of the interaction between reader and work that, in this case, has the effect of making the author appear to be an active, purposeful individual in consummate control of the thoughts and emotions of the reader. However, like most practical dialogues of the late sixteenth century, Magino's is not quite fiction and not quite fact, as we have seen and will see. It is instead the ornamentation of fact (a process for raising silkworms) with many literary devices that include, but are not limited to, book illustration. The collaborative moment of printing itself, as in the visit to the printer's shop that resulted in the author being late for his own book, is presented several times here and in other sixteenth-century dialogues as a risky, somewhat dangerous limbo where mistakes can and will be made. The time that text and images were in a printer's shop would be relatively short compared to the time spent in the creation of its contents in most instances, but not only was it a period when others would become involved in translating the author's intentions into set type and engraved lines, it was also the pregnant moment just before the author becomes an Author. It was a liminal period between private and public pronouncements, and a moment when events are outside the author's control.

The second and third dialogues explain the process announced in the first one, and constitute the illustrated portion of the book. The second dialogue contains a full-page woodblock print before every two or three pages of text that explain the information presented in the image. Magino is examining the book's images with Horatio's wife, Isabella, the only other interlocutor in this dialogue besides the author. The main elements of the woodblock images of sumptuously dressed, industrious women, children, and animals working under Magino's

117 Figura Ottava from the first harvest (Dialogo Secondo), and Quinta Figura from the second harvest (Dialogo Terzo), Magino Gabrielli, *Dialoghi di Magino Gabrielli, Hebreo Venetiano: sopra l'utili sue inventioni circa la seta*, Rome: Heredi di Giovanni Gigliotti, 1588. The Lessing J. Rosenwald Collection, Rare Book and Special Collections Division, Library of Congress, Washington, DC.

supervision in elaborate palaces and gardens are signed by letters of the alphabet. They take up about three-quarters of the page, and are anchored at the bottom by a two-part key that matches the letters to terse identifications of people, objects, and setting. Dialogue Three, which explains the second harvest, uses the same illustrations as Dialogue Two but entails a slightly different version of the process (fig. 117)[79] Therefore, while the images remain the same, the information in the keys below is adjusted accordingly, and the fact that there are fewer images in the third dialogue reflects the shorter process. For that dialogue Magino and Isabella are rejoined by Horatio and Cesare.

Concordance of Figures in the Second and Third Dialogues

Dialogo Secondo	Dialogo Terzo
Prima Figura (fol. 19) (Rome, Castel Sant' Angelo)	Prima Figura del secondo raccolto (fol. 61)
Figura Seconda (fol. 23) (Rome, Villa Montalto)	Does not appear in the third dialogue
Figura Terza (fol. 27) (Rome, St. Peters)	Does not appear in the third dialogue
Figura Quarta (fol. 31) (Florence)	Figura Seconda, del modo del far nascere l'Ova de'Vermicelli la seconda volta (fol. 65)
Figura Quinta (fol. 35) (Venice)	Figura Terza (fol. 69)
Figura Sesta (fol. 39) (Milan)	Figura Quarta (fol. 73)
Figura Settima (Naples)	Does not appear in the third dialogue (and missing in the Vatican Library copy of the book)
Figura Ottava (fol. 47) (Turin)	Figura Quinta (fol. 79)
Figura Nona (fol. 53) (Genoa)	Figura Sesta (fol. 85)

The *Dialogo Secondo* is subtitled: "In which M. Magino, inventor, teaches Signora Isabella the true way to exercise the art of silk, and its secrets, and furthermore to show her, in different delight-fully drawn images, everything that appertains to said practice."[80] Magino and Isabella begin by agreeing that Isabella should ask questions which Magino will reframe if necessary, but will cer-tainly answer to her satisfaction. The questions Isabella asks proceed from her close examination of each image, and sometimes the chapters are set up or introduced by her impatience to know what is happening in the image she has paged ahead to view. Magino begins with the first figure (fig. 106): "In this you will see portrayed from life [*ritratto dal naturale*] all that I have told you up to now, and we will observe the same order in all the lessons that follow, and underneath each figure will be the declaration of the letters of the alphabet, and of the contents: summarily, however."[81] Although this works well, Isabella's role in the dialogues equally seems to be to point out those places where Magino's ideas or technology fail him, framing the relationship between the characters and setting the tone of the dialogue with a sly but never mean-spirited banter. She assumes the character of an intelligent soubrette, whose acute observations are the pin that pricks her teacher into fuller investigations and analysis of his topic. For example, having studied the Prima Figura (fig. 106), she observes that certain important information involving color is lacking in the black and white woodcuts: "I: I remain very satisfied having seen this figure, and it seems to me a beautiful invention to be so diligent about putting everything into one picture . . . but I want to know which cocoons are the best . . . because there are yellow, white, orange, and pale green ones as you know."[82] The learned and generous Magino compliments her on her question before answering it fully, and they talk for two pages that lead seamlessly to examination of the next picture. The pictures are presented to Isabella and the reader as memory aids and as a way to understand the process as fully as possible; the dialogue takes shape through the conversation between Isabella and the author as they discuss the woodblock figures. To move on to the information in Figure Two, after having named all the different words for cocoon he can think of, Magino declares the subject exhausted and gives his student a pause to reinforce her understanding of the process by reviewing it in the picture. Isabella rejects the inference that she might have forgotten anything he previously told her:

> M: Now you should look closely at the second figure, so you do not forget how to prepare the seed, and after that we will continue to discuss how to accommodate it to make it grow.

> I: Well done, I will happily look at the figure, but in order to learn more, rather than because I might have forgotten your order.

> M: Here it is.[83]

The first step of the process (fig. 106) is demonstrated in a loggia with Corinthian columns ornamented with classicizing grotesques and Sixtus's coat of arms. The loggia affords an iconic view of the fortified Castel Sant'Angelo in Rome and its bridge, with its famous statues of Sts. Peter and Paul. The view encompasses another example of industry in the image of a little mill on the right whose wheel is churned by the water of the Tiber river, and we see that there are gardens just beyond the loggia that are leafy with trees. A joyful fountain close to the balustrade takes the form of a nude woman standing on a ball, who pours vases of water into a basin from a raised arm that rather softens the identical gesture of St. Paul, who appears behind her with his raised sword. She seems to rise triumphant from the outstretched hand of the young bearded man who enters the scene stage right, labeled "A," and identified in the key below as "the inventor who teaches how one must set the cocoons to make seed." The second picture (fig. 107), the one that Magino presented to Isabella above, takes place in a loggia overlooking the papal villa, the Villa Montalto. While Isabella learns the process, we do too. But we also have occasion, in studying the pictures, to rehearse points raised while we were waiting in the previous dialogue: the worthiness of the undertaking, its delightful and healthful nature, as well as its universality are demonstrated through the different *mises-en-scène*, and its benefit for Christendom through the constant visual connections to the pope.

The Villa Montalto (designated "K Vigna della Santità di SISTO V.") is where Magino asked to situate his silk factory and the most likely location for the experimental demonstration to the pope and related illustrious company, in August 1587, that was decisive in granting him the privilege for the invention. Like all the backgrounds in the images, this one was taken from another print and modified for use as proof that Magino's process was affiliated with the courts of the most powerful rulers and potentates, as he says it will be. The image shows young women laying the fertilized seed, or silkworm eggs, onto white cloths that they gracefully spray with mouthfuls of wine. These cloths are then tied up into damp little packets and heated in the bodices of young virgins until the eggs hatch, illustrated in the third picture (fig. 108). Magino has here arrived at one of the most important and delicate moments of silk production, and one that caused it not to be tried in the puritanical American colonies until the eighteenth century, when Benjamin Franklin discovered a stove in France that would replicate the gentle heating action of a virgin's bosom.[84] Marcus Hieronymus Vida, who was also the archbishop of Alba, described it poetically:

> Next how to hatch them, and by what device;
> The art is curious and the method nice . . .
> Be thou advised: thy Bosom be their Nest,
> Wrap'd close beneath thy Garment let them rest;
> Nor need'st thou blush beneath thy Breasts to hide,
> If on their glossy web thou buildst thy Pride;
> . . . When, by thy Bosom warm'd, the fertile seed,
> Bursts into life, and shows the teeming Breed,
> New forms they take, and from their Cells remove,
> And round their wrapping Lawn begin to rove.[85]

Magino is more practical about the farming out of these small bundles to several surrogate mothers:

M: A young woman will put it in her bosom to dry for a day or two, because, if by chance she be one of those women who does not clean much, or who sweats a lot, being accustomed to that from the beginning the seeds will not take offense when put to hatch from that *tufo*, taking up the odor of the wine to remedy that disadvantage . . . but it is necessary to take care that the women who hatch the eggs are not at their time of the month, nor having any other infirmity, nor should they be consumptive, because those things would surely kill the worms . . . therefore one wants young women, healthy and clean, of good disposition, and better a bit plump than otherwise, and better if they are virgins, or young girls in their first flower, because they have natural heat and are the most sincere, and most suited for hatching the little worms.[86]

The young women and matrons pictured throughout the book are shown, primarily through costume and gentle attitude, to be of the sort Magino describes. After examination of the second figure, when Magino gets around to discussing how the eggs hatch, Isabella recalls the reader to specifics: "Let me only glance at this third picture, and you get ready to show me how one must take care of the worms after they have come into the light of this world."[87] The third illustration, which is the last of the Roman scenes, shows "how to take care of the seeds to hatch the worms," and is improbably staged in the loggia of a grand palace overlooking the piazza in front of St. Peter's (fig. 108). The carving style of this woodblock print is noticeably different from that of the other figures. All the other illustrations, while not all exactly the same, are consistent with Venetian book illustration of that period. The other scenes are envisioned in terms of a detailed accumulation of pattern, relying on curving contour lines to show tracery and grotesque work, and short parallel hatching lines for areas of shadow. There is a marked emphasis in the second figure, for example, on decorative pattern, from the cloth covering the table at which Magino and a courtly woman work to fill a handkerchief with wine-soaked eggs, to their lace-trimmed and embroidered attire, and in the patterns of the formal garden visible outside the window. These are typical features of Venetian woodblock book illustration and are characteristic of all the pictures in the book, with the exception of Figura Terza. That image, instead, is composed of broad, confident strokes of extremely regular hatching and cross-hatching, with a pronounced concern for legibility through the use of chiaroscuro, and it is almost devoid of decorative surface pattern except for the inventor's breeches, elegantly slashed rather than embroidered as in the other prints. This is also the only signed woodcut in the book, bearing the monogram of the Roman printer and book illustrator Leonardo Parasole in the lower right corner.

This image also flatteringly, and prominently, recorded Domenico Fontana's most important and visible engineering work for Sixtus up to that point: the repositioning of the Vatican obelisk in front of St. Peter's two years before, which would soon be recorded in his own printed and illustrated book.[88] This engineering feat became emblematic of Sixtus's success in performing the impossible through a skilled team of architects and engineers led by Fontana, a world-famous enterprise that the same team was to follow almost immediately with the building of the artisans' housing by the Baths of Diocletian that Magino wanted to occupy. The setting is more urban than the preceding ones, and the newly moved obelisk and the unfinished dome of the basilica are prominently featured through the framing Doric architecture. To the left and right the scene dissolves into generic buildings, domestic and fortified, not only ensuring that attention would not be drawn away from the focal point, but also making it clear that the model for this image was probably one of the Roman guidebooks such as the abundantly illustrated versions of *Le cose maravigliose dell'alma città di Roma*, issued in many editions.

At Isabella's request to take a look at the third illustration, Magino suddenly remembers that they have forgotten to discuss what happens to the eggs at night, though they are heated gently between the women's breasts by day.[89] "Simple me, I did not think of it," says Isabella nonchalantly. Instead of rewarding her honesty, Magino accuses her of heartlessness in endangering the poor worms, but Isabella is on her toes:

I: Don't think, Master Magino, that it is because I am so deprived of reason, nor so little caring, or negligent of my things that I would have failed to take care of the little worms as perhaps you believe and as you say, and that I would let them lie out unprotected on a bench, or too close to the fireplace, or forgotten under some mattress. It was just that I had entered into a kind of reverie, of the sort that you must have done when you neglected to make little signs in your picture using the letters ABC, the way you did in the last two.[90]

Magino concedes her point, but he blames it on the wood-carver since he did not notice the mistake until the block was already carved; by that time, he was in too great a hurry to get the book printed to make another. Besides, he says, in the copies that he will give to important cardinals and princes he will put the missing letters in by hand, which only ruffles Isabella's feathers further: "I see," she sniffs, "that you have not brought one of those to us, the common people!" But in the end, none of the copies of the book I consulted, from the libraries of cardinals, the Venetian republic, and the Medici dukes, contained handwritten letters.[91] The correction to the image was in fact constituted only in the bit of flirtatious repartee that follows it in the text, which again makes reference to the haste with which it was rushed into print.

The problem reoccurred in Figura Ottava (fig. 113), the second image to get past proofreaders without including the little letters in the image that tie it to the key. Isabella, feeling more confident about her knowledge and performance in front of her literary instructor by this time, brings up the problem, but is more generous towards the author than she was before:

I: The drawing is so well done, and sufficiently annotated by you, that I, simple woman, will never know what to add there, nor what about it I should ask you. In this eighth picture, as in the third, the only defect I recognize is that the engraver has forgotten to engrave the letters of the alphabet according to the summary in the declaration of the image. It is true that since few people are represented, and they express what they are doing clearly, one can understand the subject without the letters anyway. Having to print others you should sign them, too, so you do not have to blame the error on negligence, or being in too much of a hurry . . .

M: Having to reprint this, as I believe I may be forced to do because in the kingdom of Naples and in all the other parts of Italy they have already adopted this method, I am resolved to improve it in more than four places, so do not be shocked if in this first impression you do not perceive the excellence that you might wish for, considering that the season had overtaken us before I had wanted to bestow my secrets on anyone (outside of the experiment demonstrated to some princes), since I had not finished sending all the privileges that I finally obtained. This caused a great fury in engraving and printing the figures, shown to you without mature revision, in such a manner that in the morning the drawing and its declaration were made all at once, and that day they were given to the printers – against the wishes of the one who was in charge of them, and against my will, in order not to lose this year as well.[92]

Once again the author uses a printing error to point up the novelty of his silk process and its timely revelation to the world. Is it possible that all the images were made in Rome, and not only the one signed by Parasole, as Magino intimates? While Parasole's image would never be mistaken for anything but Roman, with its allegiances to the drawing style made popular by Antonio Tempesta in the last quarter of the century, nothing else issuing from Roman printmakers resembles Magino's other images, and the one part of the book that is known to have been composed in Rome, the first dialogue, is unillustrated. It seems more likely that both Venetian and Roman printers were guilty of the same lapse of attention, due to the "brevity of time" that was endemic to Magino's project in all its phases, and that the necessity to design and cut the third woodblock in Rome must have arisen either from a lost or damaged block or, much more likely, the presence originally of an out-of-date image of the iconic but ever-changing Piazza S. Pietro that would have offended either Fontana or the pope. If this supposition is correct, then the images (and at least most of the second and third dialogues) were prepared before Magino came to Rome. The *proemio*, as we shall see momentarily, the third block, and those parts of the dialogue that were inserted to correct errors in carving would then have been added when the book was already in press. But it is abundantly clear that Magino was, as he says in the book, adding text in a whirlwind of simultaneous printing and composition as the book was being printed, to correct and explain whatever had issued from the presses the day before.

As impresario of silk, Magino also puts into play another important aspect of Jewish identity in the course of the text, one that associates him with court theater, and is made explicit in the important courtly settings of the illustrations. The actresses performing the silk process in Magino's illustrations are aristocratic women in elaborate costumes that Isabella sometimes remarks on with some envy. In two cases (figs. 110 and 113) the women are attended by a small black page (called *moro*, or *moretto*) who, besides enriching the scenes with the presence of an exotic servant appropriate to the aristocratic company, also seems to be a personification of, and mnemonic device for, the black mulberry tree, the *gelso moro*. In all cases Magino presides over the process in the illustrations just as he does in the text. In the text, he explains the illustrations; in the illustrations, his omnipotent presence presides over the demonstrations of the process. Startlingly for modern viewers of these otherwise decorous images, this means that he is the only adult male present in these otherwise feminine scenes. He is present not only when the women are loading the little packets of wine-drenched seed into their bodices, but also, more incongruously, in Figura Quarta (fig. 109), staged in the bedroom of a Florentine palace across the piazza from a fancifully imagined representation of the Palazzo Vecchio and, presumably, Ammanati's Neptune fountain. Safe from prying eyes below the highly situated open window, a bucket brigade of well-dressed women collect all the packets into long stockings and hand them to a naked woman in a heavily draped bed, who will keep them warm and dry under the covers during the night. Clad only in modest earrings, bracelet, and a single strand of beads around her throat, the night-time caretaker nonetheless reaches out eagerly to accept the unavoidably phallic stockings stuffed full of the handkerchiefs Magino hands her companions. A little brazier on the floor at the foot of the bed promises the warmth her body will provide, as do the glass mullioned windows on either side of the one that is open, giving a view onto the important Florentine setting and advertising Magino's other invention, clear glass windowpanes. There is nothing to suggest that there is anything untoward about Magino's presence in the room; the smiling women are practical and industrious, the inventor stately and composed.

While all the scenes in Magino's *Dialoghi* are self-consciously staged, one of the most overtly theatrical of them takes place in his native Venice (fig. 110). In Rome, Jews inadvertently and unavoidably provided theater as they passed in the street, especially in the wedding and funeral processions that were occasions for public mockery and sometimes violence. They were the unwilling objects of public delight in cruel and grotesque ways, the most famous being when they were stuffed full of food and made to run naked in Roman carnival races.[93] But for Magino, who is described as Venetian throughout the book, the association with theater that he would wish to call to mind was the reputation of Italian Jews for providing opulent comedies and intermezzi at court festivities, as well as the technical skill at rich costume and scenography. All of these are recalled in the detailed illustrations of the *Dialoghi*, playing out in the bedrooms, gardens, and terraces of famous palaces and villas, often taking place between curtains gathered back as if to reveal the players on a proscenium stage (as in figs. 110 and 111).

The most famous Jewish playwright at that time among the northern Italian courts was Leone de' Sommi, called *il cortigiano ebreo*, the author of many plays and intermezzi for the Gonzaga court at Mantua as well as a treatise in dialogue form on theater, or *rappresentazione sceniche*, that dwelt on the importance of rich and elaborate costumes and sets.[94] Sommi himself was exempted by the Gonzaga from wearing the sign on his clothing that set him apart from other Mantuans. Sommi's protector, Ferrante Gonzaga, claimed that such an exemption also separated him from other Jews: by not having to identify oneself as a Jew, one was identified as worthier. Magino the author never seems to try to avoid his religion in the book, although, as we have seen, he did request exemption from wearing identifying signs on his clothing in life, which was denied. On the contrary, he purposefully brings up his religious affiliation and emphasizes it as a positive difference throughout the work. Yet, in these theatrical representations, the author generally takes advantage of the license of literary fiction and the monochrome nature of prints to appear in neutral courtier garb with no identifying sign.

Theatricality seems to be reprised again in the explanation of Figura Ottava (fig. 113), which takes place in a crowded room facing the ducal palace in Turin. We recognize some of the characters and procedures: the small girl spraying trays of seed with wine, the overdressed matron in brocade and lace, Maestro Magino in his usual suit and cape as he gestures towards neat trays of hungry newborn worms, his colorless hat perched on a stool by his knee. Some unusual features catch our eye: a small monkey sitting on the windowsill next to a birdcage seems to be taking notes, inviting us to ape the activities within. An incense burner under the trays of hatchlings sends perfumed smoke into the air, and the small black page from Figura Quinta, in hoop ear-rings and a long cape, stands by Magino's side, tapping on a small drum. The sound of a child playing a drum as well as other musical instruments melodiously played, according to Magino, provide comfort to the little worms at this tender stage when they are weak and tired from all the work they have been doing getting born.[95] The page, the monkey, and the incense also intensify the atmosphere of exotic luxury and trade, appropriate to the great courts that Horatio and Giulio Cesare discourse about in the first dialogue. The theatrical nature of the images and the dialogue format also remind us of the fact that this book was meant to be read out loud as well as to be enjoyed alone. The reader, even if not having heard this particular book declaimed, would be familiar with the many venues in which texts were publicly performed, from charlatans' plays to street cries to religious theater, even to the dramatic proclamation of *bandi*. It also reminds us that among the readers of this book were those who could not actually read written text. Surely this sizeable part of the population was taken into consideration when it was mandated that the book be read out loud at fairs. Magino himself points to this fact in the third dialogue when he is describing the second silk harvest. In the third dialogue he economically reuses some of the woodblock illustrations from the second part of the book, although the process they are illustrating is slightly different from the one they demonstrate in the second dialogue, and they have been given new heading captions (fig. 117). While admiring the third figure of the second harvest (fig. 118) Cesare asks if the reuse of the images might not be misleading, since

> C: . . . in your image [*disegno*] I saw the woman designated with the letter "A" taking the seeds from her breast just as we saw in the first harvest, which is the thing that you said in the declaration underneath the image should not be done in the second. Since not everybody knows how to read, but all those who are not blind can see, perhaps some people acting according to the printed scene will err and it would be safer to change the image.

> M: The lack of time is the cause of this little problem; it is really of no importance other than the impropriety of the scene, because we have to assume that at least one person per house will know how to read, and if not, as could happen in some strange villa, a neighbor will know how to explain it, and one person teaches many.[96]

Magino predictably responds in what has become a leitmotif of the book. This slight exchange brings up many interesting points, and is embellished once again with the theme of time, with the pressure to publish that has weighed so heavily on the author, who has taken various ingenious shortcuts to get the book to press without more expense and without losing another year. As usual, a possible error has provided an opportunity for Magino to display his knowledge of the world: in this case, the reading habits of those agricultural workers who might constitute the primary audience for his book and the method it declaimed beyond the patrician audience to whom it seemed to be addressed. Here Magino relies, as most people probably did, on the presence of one willing and literate individual among a group of able but illiterate workers, through whom knowledge and technology would filter from the printed page to the field. That tertiary

118   Figura Quinta from the first harvest (Dialogo Secondo), and Figura Terza from the second harvest (Dialogo Terzo), Magino Gabrielli, *Dialoghi di Magino Gabrielli, Hebreo Venetiano: sopra l'utili sue inventioni circa la seta*, Rome: Heredi di Giovanni Gigliotti, 1588. Biblioteca Nazionale di Napoli.

reader, the recounter of Magino's invention in a villa so strange or foreign that its inhabitants could not read, would constitute another venue of publication and presentation of the pictures and text. The dialogue, and the use of the pictures in bringing the dramatic text to life, make publication itself a kind of *mise en abyme* as the book and its related actions telescope away from Magino, its first and primary publisher, to the palaces, villas, streets, and fields of Italy, passing on the twin promises of wealth and tolerance discussed among the Venetian inventor, the Neapolitan merchant, and the Roman businessman while we waited with them in the dialogue's vestibule.

   This book, which seemed to partake of so many different aspects of Italian culture, also seemed particularly Catholic when it came to picturing the important steps of the silk process. It was a hybrid pictorial enterprise. The setting in different Italian cities could be keyed to the kinds of views made popular through single-sheet prints marketed to tourists, guidebooks, and illustrated maps. While specifically meant to demonstrate the viability of the silk process at the courts of hoped-for patrons, the settings also register the peripatetic quality of itinerant merchants. The pictured demonstrations themselves, involving Magino and the odd cast of female characters, children, and animals, escape any common genre, though charlatans were sometimes pictured, not in complementary ways, accompanied by monkeys as in Figura Ottava.

The hybrid images in Magino's book, however, seemed to fluidly combine pictorial references to illustrated almanacs and calendars along with the courtly garden scenes. Printed agricultural calendars were cheap and available, providing the kind of knowledge that included information about the position of the sun in the sky, as well as popular information about the best times to plant, to bathe, and to bleed the sick. It was the sort of thing anyone living close to the land would already know about, needing only the precise days on which the moon was waxing or waning in order to plan, so the juxtaposition of this detailed information with activity set in marble palaces seems somehow incongruous to the modern reader. However, such books were popular in the broadest sense, and were widely read and owned. Since the twice-a-year silk process depended on precise timing, and because regulating the heat in the room was important for warming the eggs and cocoons, images of the sun and moon with their requisite facial features are included in almost every figure. Much of the text is given over to warnings about letting the silkworms, in any of the stages of their life cycle, become too warm or too cold, so in one sense the inclusion of the sun in the images also has the effect of making the reader aware of its presence and effects.

In the Prima Figura a blazing sun is labeled "O. Il sole deve illuminar la stanza, mà non offendere il panno coi raggi [the sun must light up the room, but not offend the cloth with its rays]," while a starry moon appearing across the page is labeled "N. Si devono accomodare i bocciuoli à luna mancante di modo che venghino à nascere i vermicelli al principio della crescenze [the cocoons must be set out when the moon is waning, in such a way that the worms will hatch when it first begins to wax]" (fig. 106). The Figura Seconda (fig. 107) shows only the moon in a daylight sky, fatter than before, labeled "P. La Luna vorebbe haver da giorni uno al più trè di crescenza nel nascere de' vermicelli [the moon will want to be one, or at most three days into the waxing phase in the birth of the little worms]." The third picture (fig. 108), the one carved by Parasole, shows a big happy sun nestled in a picturesque bank of clouds over St. Peter's as if to show its approval of the new building, and, if the printer had remembered to label it, would correspond to the key at "E. Il Sole [the sun]." In the nighttime scene in Florence the sun has been replaced by the brazier on the bedroom floor (fig. 109). It reappears in its original spiky form with the moon by its side in Venice in Figura Quinta (fig. 110), where explications of their phases are not to be found among the many important Venetian sites Magino has listed in the key, including the Bucintoro and San Moisè. Figura Sesta (fig. 111) includes the sun and moon with explanations of their phases, while Figura Settima (fig. 112), which takes place in Naples, one of the more important cities in which Magino was to set up the silk business, shows them both through transparent glass windows with no further explanation. However, it is clear that Magino wants us to see how the transparent windows, his important sideline and the business that would in the end provide far more acclaim than the silk process that is the subject of this book, provide a way to protect the room from drafts and also allow the sun to enter: in other words, to light up the room without offending anything with direct rays. Figura Ottava (fig. 113), the very ornate scene in Turin, shows the moon outside the window with nothing about it in the key, but the absence of the sun may lie in Magino's warning that direct sunlight at that stage of the worm's cycle would be more poison than medicine.[97] The last image, Figura Nona set in Genoa (fig. 114), oddly not only neglects to show the sun and moon, but includes an antiquarian close-up of the bottle-glass windows Magino's clear windows would make obsolete.

As a merchant and as a Jew, Magino could be expected to know his way around the important legalisms that related to trade, not only in terms of arranging contracts, but also for leaving a secure paper trail through the use of many notaries, who kept written proof of the many transactions that Magino made in Rome. Notaries at the Vatican drew up the privileges for the silk and crystal enterprises, and notaries on the Capitoline Hill copied them out in clear and sober script into their records, preceding important secondary agreements such as the merchant's promise to give away half his earnings from the silk trade to the pope's sister, Camilla Peretti. Magino used a different notary to write up a further donation to the Count of Olivares, the Spanish ambassador to Rome, in return for being allowed to practice his invention in Naples and Sicily.[98]

Other notaries scattered about the city wrote the contract with the *intagliatore* Francesco da Solaris for providing him with the dozen crystal-studded, gilded chandeliers with figures of cherubs, recorded the lease for a house and workshop in the ghetto, legalized the arrangement for agents to send him copies of his privilege, and prepared agreements to engage others to deal with collecting the taxes on the silk and the wine bottles.[99] A notary registered the contract with the Vatican printer Paolo Blado to print up the astounding number of fifty thousand copies of the announcement of Magino's monopoly on producing those transparent carafes, in the form of a *bando* to be posted all over Rome (this would have meant one *bando* for every two inhabitants of the city).[100] No one has yet found which notary recorded the terms under which Magino hired Leonardo Parasole to carve the Terza Figura for the silk book, or the contract with the heirs of the Gigliotti print shop that would have stated the terms for printing the book.

I have found only one document in Roman archives directly related to the publication of the *Dialoghi*, in which, on 24 May 1588, Magino contracted with Contugo Contughi "to compose and correct at the print shop, and to improve, the three dialogues on the invention of making silk two times a year dedicated to our Lord, Pope Sixtus V" (fig. 119).[101] There were to be fifteen hundred copies made of the book, and they were at that very moment being printed at the shop of the heirs of Giovanni Gigliotti, which supports the rushed and chaotic nature of the printing process as Magino represents it in the book, which really was, at least in part, written on the fly. A note appended to the bottom of the document on 5 July of the same year states that Magino did pay Contughi for the "operas con laborerijs" described above, and that, further, Contughi

119 Agreement between Magino Gabrielli and Contugo Contughi, ASR 30 Not. Cap. uff. 28 Manilius Tondius, vol. 11, c. 1046r. Courtesy of the Ministero per i Beni e le Attività Culturali di Italia.

would contract to revise a declaration in the dialogue from Latin into Italian, and likewise to translate one in Hebrew. This means that at the beginning of July 1588 parts of the *proemio* were still incomplete (the Italian transcriptions of the privilege and the sonnet), and that in all likelihood the "M. Sebastiano Tellarini" mentioned in the text as interpreter of the Hebrew canticle was actually Contugo Contughi. Although Contughi's name does not appear below the canticle, it does come up on page 6 of the first dialogue, when Horatio is discussing with Cesare how difficult it is to come to any consensus about who was the first inventor of most inventions, since some of them seem to have been invented simultaneously in different parts of the world, and the origins of other inventions are shrouded in deepest antiquity: "And in all these matters you might receive better satisfaction from Contugo Contughi, my most dear friend, young as in part you know, as desirous of attaining greatness as he is kept low by fortune, from whom I, too, have taken most of my things."[102] Later on Horatio cites him again, in a discussion about the invention of mirrors (items that Magino was then contracting to supply to Sixtus for his mendicant hospital), as holding the opinion that if anyone could invent a mirror that would show the good or evil in the person reflected, it would be the greatest gift ever given to mankind, and would bring world peace, and would be worthy of remuneration by a magnanimous prelate.[103]

It would seem that Contughi was the author of at least this first dialogue, which is why Horatio could say he took most of his material from that man. Rehearsing the classifications and attributes of known trades and inventions to prove that any particular trade was part of a desirable and recognizable natural order was a humanist preoccupation that necessitated the ability to pull information from various sources and reconfigure it to suit the rhetorical task at hand.[104] Contughi had experience in this sort of library-based work, and at the bottom of the contract the notary specified that he was to undertake the work for Magino with the same care that he used in similar works.[105] In about 1583 or 1584 Contugo Contughi put together a description of the city of Hangzhou, which he called Quinsay using the name given to it by Marco Polo, and dedicated the manuscript to the teenaged Luigi Gonzaga, who was about to take the religious vows that would set him on the path to sainthood. The brief essay was a compilation from already published sources, mostly taken from Marco Polo's account and updated with more modern published information, and is introduced and ended with unusually direct supplications for patronage and payment. Contughi addresses Gonzaga as "my most Illustrious lord" who delights in cosmography and the other sciences, saying that he wants to serve him because of the clarity of his intellect, his goodness, and his name. Rather astonishingly for someone who otherwise seemed to be both literate and familiar with the forms of courtesy, Contughi ends the description sounding inopportunely exasperated and demanding: "Until a more detailed narrative makes it obsolete, this . . . will have to be enough to satisfy your illustrious lordship's noble curiosity and perfect discourse . . . at least I hope that you will compensate me for the obligation in which your desire has placed me, especially considering the brevity of time in which one can say these pages have sketched out difficult and not entirely clear material, at no little inconvenience to me."[106] The reference to writing under time pressure also strikes a familiar note in regard to the authorship of Magino's treatise, as well as the desire for the patronage of a wealthy cleric. Contughi's name does not come up again in literary record, although he is mentioned later in the third dialogue as a person to go to for advice about women.

The descriptions of Hangzhou in Contughi's travel writing include passages about the sensual riches of the Chinese court, dwelling on the variety of pleasurable and exotic tastes, smells, sounds, and colors, including mention of the sexual practices of Chinese women and even of the king. Gonzaga had made it known at an extremely young age that women were of no interest to him, and he would shortly become famous even in the age of reform for the severity of his fasting and asceticism. It will probably never be known whether the fact that Contughi was working as an independent editor in Rome instead of writing letters in the household of a wealthy prince was owed to the tone of his essay or to the fact that Gonzaga took orders with the Jesuits the following year, and ended his very short life nursing plague victims in Rome. Contughi, who was evidently skilled in Latin and Hebrew, might have hoped to put his talents to better use than writing amusing dialogues about the history of inventions for a Jewish merchant charlatan, or so it might have seemed to him when he put into Horatio's mouth a self-deprecating reference to his own store of knowledge and the elusive nature of fame. But his familiarity with Marco Polo did lead him to have Horatio mention that silk had originated in China and, most unusually for the period (or any period), to note that while printing was a recent invention in the West, it had already existed in China for some time.[107]

A decade after the publication of the silk book, Magino shows up in the payment records of Domenico Fontana's Roman workshop when he supplied clear glass windows for the most famous of Fontana's papal building projects, working again in the huge employment machine of the Holy See.[108] Like Camillo Agrippa, he found publication, with its courteous conventions of dedication and epistolary addresses, to be a reasonable method for letting his talents in other areas be made known to those who might be able to stabilize his position on the wheel of fortune closer to the top than to the bottom. Like Agrippa, images were a determining aspect of his finished publication, both adding to its status as a luxury object fit for aristocratic consumption and providing evidence for arcane technical knowledge in a more popular sphere. Agrippa's knowledge shows that he was familiar with the diagrams in which Leonardo da Vinci worked out theories about motion and the human body, and was aware of the role of illustrations in technical and theoretical treatises about science. Magino's familiarity with the print world extended to the visual and performative forms of legal pronouncement, as well as the literary formats of the chapbook, printed play, and illustrated picture books displaying technical information for the pleasure of amateurs, like those by Stradanus and Tempesta. The *Dialoghi* came into being along with an astonishing flurry of manufacturing and printing activity in a very short period of time, with all of the associated legal protocol in manuscript and in print that had to accompany the protection of inventions and the courteous concessions that were all the more necessary for Magino's status as a foreigner and a religious outsider. As in the case of Agrippa, the printed book points to associations with the secretaries of princes and cardinals, learned amateurs and collectors, and men who tried their best to make a living in Rome by working their knowledge of machines, science, and the habits of the papal court. They took part in a public world of letters that used printing and writing imaginatively for the dissemination of practical knowledge through publication in many early modern forms. There is really nothing else quite like the *Dialoghi*, which offers fascinating evidence for what one author (even if a corporate author) felt could be accomplished in the practical world with the publication of a richly illustrated book.

S·PIETRO

6

Talking Pictures:
The Discourse of Images in
Illustrated Dialogues

I began by asking how illustrated books contributed to the formation of the early modern social networks that Roger Chartier described as reading communities. Images enriched books, opened them up to varieties of meaning, and complicated their production exponentially in terms of labor and personnel, bureaucracy and materials. But printed pictures, or even diagrams, continued to be worth the extra planning, work, and expense required by all parties through the middle of the next century.

In preceding chapters we saw that the production, sale, and intellectual consumption of books brought together groups of men, women, and sometimes children (as in the case of Adamo Scultori's heirs, for example) with different concerns that could at some point be satisfied by producing a book. This group includes authors and booksellers, dedicatees and patrons, publishers, typographers, and printers for texts, and artists, draftsmen, and specialized pictorial printers for images. Temporary associations formed to produce particular books created and arose from networks across social lines of all kinds, networks made visible through the act of publication itself. As we have seen, the processes of publication and the relationships worked out in the heat of that battle were referred to in many places in the book: besides being legally noted on the title page, frontispiece, imprimatur, and colophon, the writing, financing, printing, and selling of books were brought up in prefaces, privileges, letters to the reader, dedications, sonnets to the author and sometimes the publisher, and even in the content of the book.

In this context illustrated dialogues, which by definition brought every station of person together around a single subject or proposition, present a special case of the blending of skills and the self-conscious display of the struggle to make ideas public. Illustrated dialogues arose as a genre of illustrated treatise that flourished just through the hundred-year period covered in this book. These dialogues used pictures to bring readers together with the book's interlocutors around

OPPOSITE PAGE   Leonardo Parasole, Figura Terza, *Del governo del seme (Roma)* (detail of fig. 108).

the performance of a craft or technique that had a parallel performance in the skill and daring of crafting a book and bringing it through the press.

The skills described in illustrated treatises like Magni's monologic phlebotomy treatise would have been taught in the presence of other practitioners and through experience, so it seems difficult to explain the considerable amount of trouble spent illustrating them with pictures that often had little to do with the practical work of educating readers about the task at hand. In effect, such a book seems to erase the dimension of conversation that the very formal groups of actors in each picture would have at one point undertaken, a silence that we saw was appropriate to the sickroom and to the comportment of the barber. Artisanal treatises were most often written to elevate the status of a craft by giving it intellectual purchase through theorizing its principles or even simply by making it the subject matter of a book, and Magni's treatise traded on that convention in imaginative ways.[1] Illustrated treatises written in dialogue form, however, were often funny, even to our eyes burlesque, highly theatrical works that do not sit well with the idea of raising the status of a profession to the level of academic gravity that would seem to be required.

Besides the addition of rich and imaginative pictures, the format of written dialogues required extra sophistication with language and a knowledge of literary forms that seems to us to exceed the stated subject matter, which ranged from the motions of the planets, the shape of tornadoes and the density of air, to fencing, culinary techniques, or improvements in the art of sericulture (fig. 120). Crafting a treatise in dialogue form, as Magino carefully pointed out in the first chapter of his treatise on raising silkworms, was an art in itself, bound by certain literary rules for the author and raising certain expectations on the part of the reader. Magino's interlocutors, Cesare and Horatio, let us know what the rules of the game were before we fully entered into it:

> C: I would like to know who the first people were who used dialogues, and, briefly, what they should be, because this discourse that we are having together, will perhaps have something of the form of a dialogue.

> H: In my opinion there must not be more than four people taking part in a dialogue, and perhaps four put to discussion in one dialogue alone is greedy: more than that would make a comedy, and not a dialogue; using two or three is the most common way, and the most praised, this can be used for every kind of material except pastoral, that changes its name to Eclogue; it must be something that can be supposed to be true, and not made up . . . principally one has to observe the decorum and the propriety of the interlocutors.[2]

Magino's amanuensis, presumably Contugo Contughi, has evidently been reading theories about the crafting of dialogues. This was a matter of great interest and strong opinion in the sixteenth century, and there is a fascinating body of literature written about it both then and now. One of the most interesting writers of dialogues who also wrote a treatise on the genre was the Paduan literary figure and rhetorician Sperone Speroni. He was a philosopher of language who, like his friend Annibale Caro, was employed by aristocrats for his skills in oratorical writing; a character based on Caro in fact appears as an interlocutor in his last dialogue, *Dialogo del giuditio di Seno-fante*.[3] Speroni died in 1588, the year that Magino's treatise was published, but before that he wrote dialogues on various subjects important to sixteenth-century literary formation: on the use of the vernacular as a literary language, on love, on usury, on the dignity of women, and others. He lived in Rome from 1560 to 1564 as ambassador of Guidobaldo of Urbino, and was living there again in 1574 when his work was the subject of a denunciation to the Inquisition by an anonymous gentleman.[4] Speroni recounted his interview with the Inquisitor in a book he then wrote in defense of the dialogue form, *Apologia dei Dialoghi*, which he circulated to trusted friends but did not try to publish.[5] At that time Speroni was trying to reprint his early dialogues in Rome, and although he went back into them with the recommendations of the censors and made many changes, he did not receive the necessary *imprimatur* and his books were eventually placed on the Index. The *Apologia* was stimulated in part by an attempt to further efforts to give new life, in Rome, to the dialogues he had written in the earlier part of the century, but it also constituted a thoughtful and fully formed – if contradictory – modern treatise on the literary theory of the dialogue.

Virginia Cox characterizes the first half of the *Apologia* as a vigorous and original defense of the author's liberty to present all sides of an argument without being bound to persuade the reader of a single truth, invoking "the traditional privilege of rhetoric to hold truth conditions in abeyance and [outline] a poetics of the dialogue as 'gioco.'"[6] In the early, freer parts of his *Apologia* Speroni wrote that "the author is not responsible for the ideas expressed in a dialogue: indeed, he may himself be 'ignorant of the truth.' His duty is merely to dramatize the alternative positions which may be held on a given issue, refraining from any simplistic pre-emptive solution."[7] He described the dialogue as an argument made from "probabilities only, so that it never arrives at full certainty, but argues from probable premises to probable conclusions" which may not be conclusively determined by the end.[8] Open dialogues – those that were left inconclusive and provided a space for the free play of a range of ideas – could not by definition be entirely truthful in that they were bound by their dialectical form, and by the mandate to preserve the decorum of each interlocutor, to at least at times present material that was not true. If all interlocutors were performing true positions, there would be no contestation and no point to a dialogue.[9]

The latter half of Speroni's *Apologia* instead points to the author's political duty to act as an upright citizen, which seems to represent a different kind of attention to the Inquisitor's (and the gentleman's) accusations. Cox points out that the first part of the treatise contains an argument for preserving the decorum of the individual interlocutors who must stay in character, to which Magino's Horatio refers above, but by the end of the *Apologia* the larger concern is decorum defined as the literary comportment of the responsible author on the civic stage, in keeping with his actual station in life.[10] The open play of ideas in the dialogue is permissible up to a point, but then the author can take control of the conversation with a declarative injunction to put an end to discussion and resolve the issues to his authorial satisfaction at the end.

120   Frontispieces from illustrated Dialogues:
a. Magino Gabrielli, *Dialoghi di Magino Gabrielli,*
Hebreo Venetiano: Sopra l'utili sue inventioni circa la
seta, Rome Rome: Heredi di Giovanni Gigliotti,
1588. The Lessing J. Rosenwald Collection,
Rare Book and Special Collections Division,
Library of Congress, Washington, DC.

120 b. Camillo Agrippa, *Trattato di scientia d'arme, con*
un dialogo di filosofia di Camillo Agrippa Milanese,
Rome: Antonio Blado, 1553. Cicognara IV 1551,
f. 1r, Biblioteca Apostolica Vaticana.

Like Magino and Agrippa, Speroni placed himself in the (narrated) dialogues in the *Apologia*,
which take place first between himself and the Inquisitor who explains the accusations, and then
between himself and a group of churchmen at the end of Carnival and the beginning of Lent,
a time period that metaphorically connects the changing form of the dialogue permissible
between the Renaissance and the Counter-Reformation to the time period in which he sets his
discussion of the form itself.[11] The open-ended testing of ideas – the appeal to the reader to
judge the facts about any proposition as they appear described with a range of subtlety from the

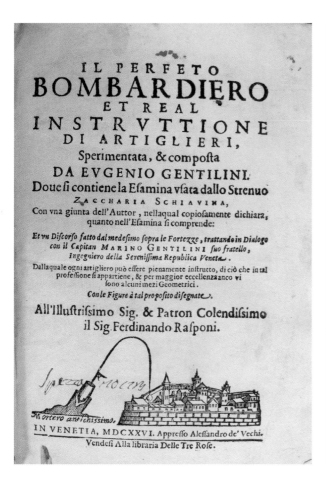

120 c. Eugenio Gentilini, *Il perfeto bombardiero et real instruttione di artiglieri [. . .] Con le figure à tal proposito disegnate*, Venice: Alessandro de' Vechi, 1626. John Hay Library, Brown University Library.

120 d. Angelo Viggiani, *Lo schermo d'Angelo Viggiani dal Montone da Bologna: nel quale per uia di dialogo si discorre intorno all'eccellenza dell'armi, & delle lettere*, Venice: Giorgio Angelieri, 1575. John Hay Library, Brown University Library.

mouths of fully constructed, believable interlocutors who stay in character throughout – was in fact an ideal that seemed to belong to the carnival-friendly era before the institutions of church reform were fully developed, an era that could be seen to correspond to Carnival and end with Lent. This is the period in which illustrated dialogues flourished.

It seems as though Contughi, in patching together material for Horatio and Cesare to use in whiling away the time before Magino appeared in his dialogue, looked to theories of the dialogue form by Carlo Sigonio and also by Speroni. Although it cannot be known how Contughi might

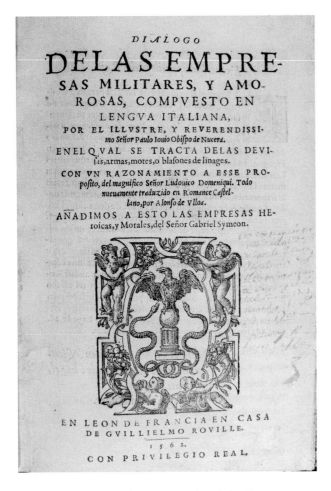

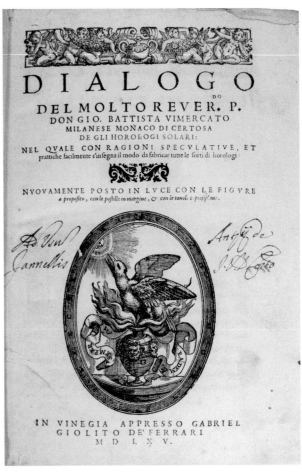

120 e. Paolo Giovio, *Dialogo de las Empresas Militares*, Lyon & France, 1562. John Hay Library, Brown University Library.

120 f. Giovanni Battista Vimercato, *Dialogo del molto reverdo. p. don Gio. Battista Vimercato Milanese monaco di Certosa de gli horologi solari: nel quale con ragioni speculatiue, et prattiche facilmente s'insegna il modo da fabricar tutte le sorti di horologi,* Venice: Gabriel Giolito de'Ferrari, 1565. John Hay Library, Brown University Library.

have had access to the privately circulated copies of this last work of Speroni, a good deal of his ideas about rhetoric were available in his already published dialogues.[12]

Dialogues like the one mined for the life of St. Benedict, in which Gregory the Great enlists the young deacon Peter to draw him out about the lives of the greatest holy men, represent the simplest form of the genre in terms of stylistic ambition and number and intervention of characters – the work is barely a dialogue at all, since Peter is almost entirely reduced to asking short,

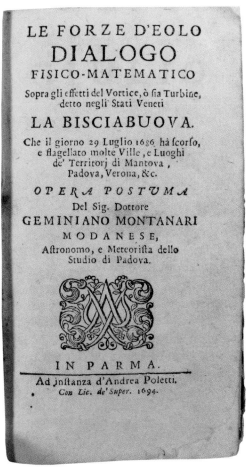

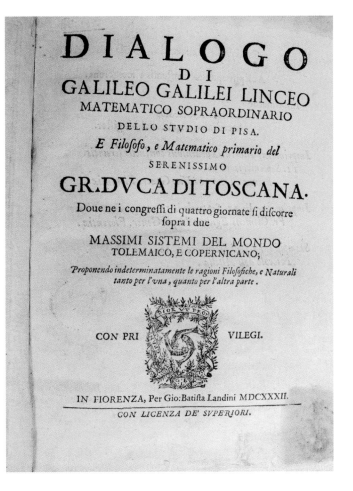

120 g. Geminiano Montanari, *Le forze d'Eolo: dialogo fisico-matematico sopra gli effetti del vortice, ò sia turbine*, Parma: Andrea Poletti, 1694. John Hay Library, Brown University Library.

120 h. *Dialogeo di Galileo Galilei Linceo [. . .] sopra i due massimi sistemi del mondo*, Florence: Giovanni Battista Landini, 1632. John Hay Library, Brown University Library.

leading questions and rarely offers any opinion of his own. The *Dialoghi* of Magino instead involved, at its most crowded, four interlocutors, who were joined pictorially by the women and servants acting out his silk process in the woodcut illustrations. By inserting a warning of the dangers of slipping into eclogue, the reader is reminded that silk production could be considered a subject for pastoral poetry, and had been; Archbishop Vida certainly wrote about it in that style. Unlike Speroni, and in opposition to him, the seventeenth-century Jesuit Sforza Pallavicino insisted that dialogues,

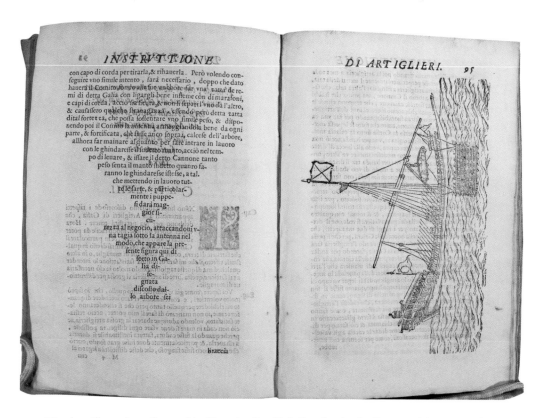

121 Woodcut illustration of a warship, Eugenio Gentilini, *Il perfeto bombardiero et real instruttione di artiglieri . . . Con le figure à tal proposito disegnate*, Venice: Alessandro de' Vechi, 1626, fol. 95r. John Hay Library, Brown University Library.

like rhetoric in general, were for teaching, and that they were primarily tasked with uncovering the truth.[13] With the inclusion of four interlocutors at the grand finale of the book, Magino walked a fine line between comedy and didactic performance, between a philosophical literary form for things supposed to be true and a burlesque comedy about things that were made up.

The presence of so many and such accomplished pictures, carefully engraved, lettered, and keyed, and in most cases themselves central to cuing and guiding the text, is one of the most noticeably excessive properties of these books. Illustrated dialogues not only accommodated their subject matter in the form of a play or a conversation, but also modified their formats to the author's desire to place the picture before the reader's eyes at precise junctures. There were varying degrees of correspondence between the moment of viewing the image under discussion by the people in the book and the moment of reading the text. A Venetian dialogue on bombardments and artillery imitates in typography the shape of a vessel in the woodcut illustration while telling the reader to look over at the picture that "appears below," and a chapter of a Venetian fencing

TERZA
SESTA GVARDIA LARGA, OFFEN-
siua imperfetta; partorita dal rouefcio intiero
difenfiuo, da cui nafcerà il raffettarfi in
guardia alta, offenfiua, perfetta.

P A R T E. 75.
CON. Perche larga? ROD. Per le ragioni mede- *Perche la*
fime, per le quali chiamafsimo la quarta noftra guardia *fefta guar-*
larga, offenfiua per effer nelle parti deftre. CON. *dia fia det-*
Horsù alla fettima guardia. ROD. Volendo uoi, *ta larga,*
Conte di alcuna guardia difenfiua, o ftretta, o larga far *offenfiua.*
nafcere il medefimo rouefcio con quei uolgimenti tut *Come fi*
ti (pur co'l pie deftro innanzi) della uita, delle mani, & *debba far*
de' piedi, come fapete; bifogna che la mano della fpa- *la fettima*
da nel difcendere a baffo; non trafcorra piu giù del gi- *guardia,*
nocchio: ma che di fuori, & dauanti di effo un palmo, *nominata*
fi fermi, & che la punta della fpada guardi al petto inio *ftretta of-*
(uedete come faccio io?) & quefto colpo farà mezo ro *fenfiua,*
uefcio, non hauendo fatto altro che mezo il camino *perfetta.*
dell'intiero rouefcio, & ui formerà una guardia ftretta, offenfiua, che farà la fettima noftra.

SET-

122   *Sesta guardia larga*, etching, Angelo Viggiani, *Lo schermo*, Venice, 1575, fol. 74v. John Hay Library, Brown University Library.

manual in dialogue form, illustrated with etchings, finishes up with a description of a guard that the text promises to show you on the following page (figs. 121, 122).[14] A solution eventually used in many eighteenth-century scientific books with few illustrations can be seen in a late entry to the field, a book by the astronomer Geminiano Montanari, who wrote a dialogue on tornadoes and waterspouts, *Le forza d'eolo*, which used a fold-out at the end of the book to keep the engraved pictures before the reader's eye throughout the experience of reading (fig. 123).[15] Of these three, only the woodcut ship could be printed on the same kind of press as the text, as we saw Principio Fabrizi point out in Chapter 1, when lamenting the time it took to bring an elaborately illustrated book through the press.

As theatrical representations of a conversation, often of a disputation, vernacular dialogues tried to persuade the reader to take sides with speakers in the text, but as we have seen, the reader's pleasure would come in large part from the deferral of that decision, and the invitation to the reader to judge the outcome. Such a personalized judgment by an individual reader would be based in part on the eloquence of the interlocutors, and in part – as in oratory – on their demeanor, the quality of their performances, and their own individual characteristics. Speroni was clear about this in his dialogue on rhetoric, where he talks about how the effect of the presence of the successful orator succeeds or fails in teaching and moving his audience:

> There is an infinite number of accidents that will prevent him from carrying out his function. These include the ugliness of his body, the lack of harmony in his voice. . . . I am talking about moving the audience, not about teaching, for the world knows no greater pain than being made to learn against one's will. . . . Therefore, we will tire ourselves out vainly trying to teach and move people unless we give them delight, for it is by means of delight . . . that we are able to persuade our listeners. . . . Moreover, not content with the delight deriving from words, in order to redouble the pleasure they produce and to sweeten them to perfection, [a speaker] has recourse to gesture and delivery as the seasoning of oratory, the sweetest honey and sugar for our ears and eyes. Indeed, on delivery, thanks to the grace contained in it, depends the efficacy of his oration, for without a good delivery, it is worthless.[16]

Pictures in dialogues amplified the textual delivery, ushering readers into a conversation as into a paper theater: sometimes quite literally a curtained stage, sometimes a charlatan's corner in a familiar piazza. Magino's book, declaimed outdoors at fairs and in public places, meant to be read out loud in farmhouses, and containing images set on a curtained stage, brought multiple theatrical genres into play. The author, along with a cast of his choosing, is made present to the reader not only in the official-looking and generally sober author portrait at the beginning of the book, but also as he takes control of his book from within, delighting and engaging the reader outside with the fullness of his active, speaking self.

Mikhail Bakhtin, focusing on the propensity in Renaissance dialogues for descent into a brawl, characterized the force of the dialogic form of literature as the audience's privilege to participate in "truth in the making." He called attention to the way that early modern dialogues often represented the processes of truth-seeking in a "seriocomic" mode rather than the declarative, monologic form of treatises that presented the reader with a ready-made truth. About the process of dialogue writing, tangible also in its reading, Bakhtin wrote: "Socrates called himself a 'pander': he brought people together and made them collide in a quarrel, and as a result truth was born; with respect to emerging truth Socrates called himself a 'midwife,' since he assisted at the birth. For this reason also he called his method 'obstetric.'"[17] The Socratic dialogue collapsed genres to provoke an enjoyable and dynamic engagement between the author, the reader, and the world evoked in the book. The illustrated dialogues in these books are multi-vocal, contestational compositions that Bakhtin understood to be at the root of the genre of the car-nivalesque in literature, and characteristic of a grain of thinking in exactly the period in which the illustrated books I have discussed here were produced for the pleasure (or delight) and utility – as many of them stated somewhere in their texts – of everybody. Whether dialogues were more properly supposed to be weighted toward the delightful and therefore amusing and

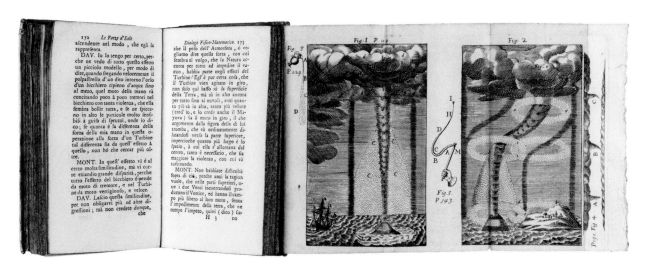

123   Tornadoes and other strong winds, engraving, Geminiano Montanari, *Le forze d'Eolo*, Parma: Andrea Poletti, 1694 (n.p.). Division of Rate and Manuscript Collections, Cornell University Library.

inconclusive, or useful and therefore revealing of certain truths, could be fine-tuned in the presence of illustrations.[18]

Virginia Cox called the oral exchange in dialogues a "fictional shadow to the literary transaction between the reader and the text . . . each argument in a written dialogue is simultaneously part of a fictional conversation and an actual literary exchange."[19] She found that this discursive element of enquiry promoted especially by early sixteenth-century dialogues was stifled a century later by the universal adoption of the conventions of print culture, such as the inclusion of citations, designed to promote purely intellectual engagement in a way that was incompatible with "the fiction of civil conversation."[20] Illustrations and other nonverbal elements in dialogues were conventions of print culture that rather had the opposite effect for a little while. They not only extended the dialogic nature of the work into yet another dimension, but could also render other pictorial treatises, such as Magni's, dialogic. The interpretation of pictures involved readers in the activity of truth-making by further calling attention to the reader's processes of learning, reading, and looking. In this way the overlapping hem of the wandering monk in the *Vita et miracula* of St. Benedict clearly qualifies as dialogic in the connection forged with the reader, but other illustrations do this to a less obvious degree. By bringing to the foreground authorial apprehension about the reader's experience, the sense of reversal of the positions of author and reader is heightened so that they seem to battle for control of meaning in the text.

Revisiting the texts to consider this aspect we can look at the carnival effect on the author's description of publication in the images accompanying the dialogue on cosmography at the end of the fencing manual by Camillo Agrippa of 1553 and Magino's *Dialoghi* for producing silk twice in one year of 1588. An end point for all of this could be Stefano della Bella's famous frontispiece for Galileo's *Dialogue Concerning the Two Chief World Systems* of 1632. Printed in Florence but carefully read and misread by the Inquisition at Rome, the book entered a world no longer prepared to tolerate the carnivalesque, or open dialogue mode in matters of science. One is, in the larger context, firmly in the second half of Speroni, where art must take account of political conditions including the station of the author, and play is curtailed.

The richest moment of carnivalesque reversal in Magino's *Dialoghi*, which is overtly theatrical throughout, comes when he is accosted by Isabella for neglecting to insert the little letters in the woodcut of the third figure that were included in the legend below it (fig. 108). As we have seen, Magino places a full-page illustration in front of every step of his technique, and calls attention to the fact at the beginning of the work: "so that you can see everything better, and share it with whatever simple person or servant you like, I will show you in these illustrations, made from life, everything that I have said up to now, and under each figure there will be a declaration in the letters of the alphabet of what is contained in it."[21] Everything goes well until the interlocutors reach the step in the process that is pictured as taking place in front of St. Peter's, the one that had to be rushed into print and that, perhaps, caused the author to be late to his own book. Isabella, accused by the author of letting her mind wander, points out his lacuna in her own defense: he, not she, had fallen into "some sort of reverie." The reader – whose mind has been taken in multiple pleasurable directions by the textual asides and rich detail of the imagery – is called to awareness of her own attention: had she noticed the lack of letters along with Isabella? Has she been paying full attention to the minutia of the described process? Magino, like Fabrizi, blames the lacuna on the printers – in this case, the wood-carver – and the speed with which he is forced, by circumstances beyond his control (a massive legal system and very important people, along with the cycle of the seasons), to rush the book into print. He refers to the book we are reading as "the first edition," thereby rendering our experience out of date before we have even gotten halfway through the volume, and he calls further attention to the reading process by promising that he will annotate copies for important people by hand. The reader at this point has no choice but to join Isabella in taking offense at the lack of this special handiwork in the volumes destined for important people, and finds herself aligned with Isabella against the author – as not one of the volumes is annotated, of course – and the contours of this offended position for the reader are crafted by the very author.

Digressions like this find a natural space in dialogues: in this case the reader has become involved in scrutinizing the images through the eyes of intelligent, flirtatious Isabella, under the guidance of the wise teacher, whose authority is always reinforced with the designation *inventore*. When he makes a mistake in an image, the carnivalesque is mobilized in the reversal of authorial position; as Isabella points out what the author should have noticed – and the reader joins in the joke at the author's expense – he is denuded, through his own carelessness, of the authorial letterpress designation of *inventore*. But Magino recaptures his position by making his gaffe call attention not only to the error, but also to the difficulties of having seen an illustrated book through the press under extreme time pressure – the book must appear before the next crop of mulberry leaves – and this is a skill of its own that is referred to many times in the text through similar digressions. Galileo, a poet and literary critic himself, wrote that he liked to use the dialogue form precisely because it would allow him to stray from the "rigorous observation of mathematical laws, and . . . allow digressions, which are sometimes no less interesting than the main topic."[22]

Both images and digressions are rhetorical tools that can be used in dialogues to guide the reader's passage from one idea to another. Jon Snyder called digression in a dialogue "a provisional point of passage in discourse on the way toward another, still concealed meaning, lying somewhere between the next message and the next response."[23] Images in technical and scientific dialogues are either pictorial or diagrammatic, and as such they might be expected either to digress or to focus – but they are most interesting when they are used productively in both ways. This is true of Agrippa's fencing treatise with its dialogue at the end about the cosmos, dedicated to Cosimo de' Medici, the first Grand Duke of Tuscany. It is also true of the *Dialogue Concerning the Two Chief World Systems*, dedicated to Ferdinando II, the fifth Grand Duke of Tuscany, by Galileo, a university professor who used all kinds of strategies and analogies to reach a wider audience than the universities where he taught.[24] These books appeared in 1553 and 1632, at about either end of this proposed century of illustrated dialogues, and show two more ways that images in dialogues deployed the easily reversible situations that characterized the carnivalesque, cued the reader's knowledge of reading pictures as well as of reading text, and also demonstrated how authors dealt with their heightened self-consciousness about the role of publication in their conceptions of the book.

Agrippa's dialogue on the cosmos that followed his geometrically inspired fencing treatise was an opportunity for the author to address his most serious concerns about his right to publish his theories. He framed these doubts with pictures, showing as much of the courtier's nonchalance as he could muster about going public with his ideas. The question his dialogue was supposed to put to rest was that of his qualification to discourse publicly about natural laws. In his book, he brings the question to an authority in the world of letters, but also one who is quite at home in the realm of the carnivalesque: Annibale Caro, the author of scenes of drunken disorientation and men riding monuments through the streets of Rome in his comedy, *Gli straccioni*.

We saw that the images at the beginning and end of Agrippa's fencing treatise show two very different pictures of contestation that flank a work about fighting; then comes the dialogue advertised prominently in the book's title: *Treatise on the Science of Arms with a Philosophical Dialogue*. These paired pictures (figs. 55 and 63) flank and frame the treatise on dueling – linking the physical form of contestation pictured in the treatise to the dialogue, which is a textual, and therefore a verbal, contestation. At the outset we see an image of an academic disputation of words and gestures, at the end of the treatise the image of the mugging. Neither image illustrates the dialogue taking place between Agrippa and Caro in Caro's apartments in the Palazzo Farnese, but it is in that dialogue that we learn what the images are meant to show as Agrippa identifies his enemies as students of Euclid and Aristotle.

In the dialogue, Agrippa has come to the home of Caro to ask the question, "shall I publish this book?" meaning the treatise, but in effect it is both the treatise and the legitimating dialogue we are in the act of reading. When Caro voices doubts, the dialogue proceeds among reasonable and articulate gentlemen in which the character named Agrippa convinces the character named Caro – each name indexical to a known, living individual in Roman literary circles – of the breadth and depth of his learning. Their conversation is introduced by, but also stands in stark contrast to, the comedic image of the mugging in the forum that precedes it.

But there is another academic disputation that lurks behind the picture. The presentation of academic orthodoxies alongside the prominent codpieces so important to the silhouette of the gentlemen in Agrippa's book calls to mind the disputation, graphically described by Rabelais, that took place between humanists and scholastics he published about twenty years earlier. This described the trip to Paris made by the English cleric Thaumaste when he engaged in a public dispute with Pantagruel.[25] In challenging Pantagruel to an intellectual duel "in the manner of the Academics," Thaumaste proposes a disputation to be conducted wholly without words: "I want to debate by signs alone, without speaking; for the matters are so arduously difficult that human words would not suffice to explain them to my satisfaction."[26] This too is a story of friendship and alliances, for Pantagruel's devoted friend Panurge begs to stand in for his master as if a second at a duel, and accessorizes his codpiece with "a lovely lock of silk, red, white, green, and blue, and inside it had put a fine orange." Panurge makes thunderous use of the codpiece during the debate, shaking it at the increasingly agitated Thaumaste until his opponent lets loose "a grand baker's fart," simultaneously relinquishing control of his bowels and of the disputation.[27]

Besides distinguishing visually between the powerful and the powerless brawlers' bodies, Agrippa's illustrator has further underscored the fruitfulness of the studies of the corps of gentlemen and the barren nature of the academics' investigations through the articulation of the *mise-en-scène*. The academics appear against a half-buried landscape of barely sketched-in ruins without visible order or detail, and are removed from actual experience of them by a well-fortified medieval wall. The gentlemen are instead framed by a fully excavated and restored Doric colonnade guarded by heroic sculpted figures; Agrippa is emphasized by the strategic placement of a magnificently erect obelisk bearing legible hieroglyphics.

Here and in its pendant, the engraving of the university disputation, the scholastics are cast as foolish, argumentative quibblers about nothing, hair-splitters, obfuscators of truth whose access to natural knowledge and Aristotelian precepts was owed solely to corrupted books, whose acquaintance with nature was so far removed from practice or observation that it was based on "commentaries on commentaries."[28] These men were famous among sixteenth-century humanists for their self-imposed distance from the experience of nature as well as the means to express anything about it clearly with their awful distortions of the Latin language. In the words of Gerard Lister

(Listrius), the Dutch humanist friend of Erasmus who wrote a commentary on that other bastion of anti-scholasticism, *In Praise of Folly*, the scholastics were good for "nothing else than second intentions, common natures, quiddities, relations, ecceities, and countless other questions even more trifling than these trifles. And because they dream of those monstrosities, they appear subtle in their own eyes and condemn with stern eyebrows persons who spurn these questions and penetrate to the real things themselves."[29]

Agrippa, whose profession and self-representation made him the expositor of "the real things themselves," visualizes the showdown in the forum in a way that might have had its roots in any of the anti-academic attitudes published and made popular in the sixteenth century by men like Thomas More, John Colet, Juan Luis Vives, and especially Rabelais and Erasmus, who all had fun at the expense of the schoolmen. The airless room of the public disputation in the frontispiece and the raunchy brawl in the forum, as well as the critical presence of Annibale Caro, would seem to point Agrippa's quarrel with the academics back to Rabelais, but without quite the unleashed literary spirit of the carnivalesque that both Caro (in the fantastic assault on Rome in *Gli Straccioni* and the bawdy earlier *La Nasea*) and Rabelais took pleasure in evoking.[30]

As Agrippa will demonstrate in future books, gentlemen have the most lively interest in science, fired by questions that stem from natural curiosity, and fed with answers gained from experience. Joseph Connors mentions a letter from the Benedictine mathematician Fra Benedetto Castelli, who in 1631 wrote to his friend and teacher, Galileo, that he was "begged by a group of charming, literate gentlemen to explain the principles of geometry to them."[31] But Agrippa was a gentleman among gentlemen, not a scholar among gentlemen. Furthermore, as a practical man, his practice of record was engineering, not intellectual work like Caro, nor even, as far as we know, actual military work like Alfonso Soderini. In his dialogue Agrippa makes his representative character demonstrate to Caro in Caro's private rooms that he is not putting himself in harm's way by falsely representing himself as a university educated man. He describes the subject matter of the image to the left of the text as a scary dream in which he was accosted in the forum by students of Euclid and Aristotle who were "calling me presumptuous in wanting to discuss similar things, I not having studied." His own knowledge, we already know, is gained through practice.

At the end of the last day of discussion an overexcited Caro commands the author to make any necessary clarifications to the pictures that are needed, and to get the book to the printer's as quickly as he can, even though it is a holiday. The dialogue, with Caro's stamp of approval, allays any reservations that the reader might hold about Agrippa's right to be an author. But the reader enters the book in holiday mode, prepared to lay aside the author's defensive self-presentation of possible failings, having already encountered the image of the disputation placed at the beginning: a confident scene of an already triumphant battle made clear through its pictorial ridicule of those who would hold opposing views.

By the mid-seventeenth century lavishly illustrated dialogues were already very much on the wane in Italy; the few I have found are illustrated solely with diagrams. Galileo's *Dialogue Concerning the Two Chief World Systems* is in fact wholly illustrated with schematic diagrams used in unusually imaginative ways, fully integrated with the formation of the dialogue's argument and shaping its rhetoric.[32] The single pictorial image is Stefano della Bella's magisterial frontispiece (fig. 124), which presents the book's subject matter and grain of argumentation more persuasively and actively than the usual architectural frontispiece, declaring author, title, dedicatee, and publisher in pictorial form on an entirely integrated simultaneous visual field.[33] This print, which then greatly influenced succeeding seventeenth-century scientific frontispieces, seems to show a philosopher's conversation taking place in an outdoors version of Agrippa's academic contestation. Simplicio, the Aristotelian pedant in the Galileo frontispiece, is an updated version of the schoolman cast as Agrippa's opponent in the academic disputation that appeared at the beginning of his book. Both images deploy that stereotype of the ridiculous scholastic, almost a *commedia dell'arte* figure himself, popular since Erasmus and well aired in print. Some scholars have found that the rhetorical form of the dialogue, as a genre, contributed to this judgment by endowing Galileo's argument with a persuasive strength beyond the force of his words. Others have pointed to the fact that Galileo never engaged in anything much like a dialogue with the Aristotelians, and that the part of Simplicio was written as such an obvious stereotype that the *Dialogue* took place in an echo chamber where the notoriously combative Galileo addressed like-minded readers.[34] Both the dialogue format and aspects of the frontispiece attracted the notice of the Inquisitors who found Galileo in violation of his promise not to teach or publish Copernican ideas.

The frontispiece, as frontispieces should, organized the *materia* of the book as well as setting the tone, establishing personalities for the interlocutors and demarcating their positions, while pointing to the demonstrative analogies that Galileo would use to argue his points.[35] The characters are absorbed in the discussion of a small armillary sphere held by the central figure, standing before a harbor where a ship and fortifications provide the hazy backdrop to their makeshift open-air stage. Galileo himself had been a gentleman member of the Florentine Accademia del Disegno since 1613, and cared enough about art and drawing to have had a determining voice in the creation of his frontispiece.[36] The young Stefano della Bella grew up around the court and knew Galileo. Most of della Bella's theatrical work as a Medici set designer, which made him well known as a master of illusion in matters of costume and theater, dated to the period following this. Although he left Florence to work in Rome the year after Galileo's book came out, he had already etched prints of Medici celebrations and his teacher, Jacques Callot, had been working in this vein from the beginning of the century.[37] In the frontispiece he makes putti hold aloft the grand ducal crown while they part a fringed curtain announcing that the work is a dialogue dedicated to the current Medici duke, whose coat of arms has become a ring of planets with Galileo's name at the center. The open curtain both reveals the conversation and frames it as a theatrical entertainment under ducal patronage.

In the text, Galileo refers to the discourse as a performance, calling it a "rappresentazione," and the reader has already encountered this picture and felt as if seated at a theater before continuing on to read Galileo's announcement in the preface that he is bringing these truths openly to the

124   Stefano della Bella, frontispiece, etching, Galileo Galilei, *Dialogo dei massimi sistemi . . .* ,
Florence: Landini, 1632. Hay Lownes QB721. G3x, John Hay Library, Brown University Library.

"theater of the world."[38] The figure playing Copernicus, who is called Salviati in the text and represents Galileo's point of view, stands confidently with belly relaxed and shoulders back in approved early modern academic posture. Holding a paddle-shaped model of the heliocentric universe in one hand, he extends the other toward the men, palm upright, fingers curled gently, demonstrating the obvious with delicacy and tact.[39]

On the left Simplicio, representing Aristotelian academic orthodoxies, stabs at the air with a crooked finger, the dark, close-fisted hand on the reader's left opposing the light, receptive one on the reader's right. Hatless, he rests his weight on a cane, infirm and off-balance, leading with an over-sized chin sprouting a scraggly beard, displaying the characteristic physiognomy of a philosopher and the posture of an old fool. The stick he leans on and the shadow he leans into illustrate Galileo's scornful definition of the Aristotelians as "Peripatetics in name only, satisfied with worshipping shadows without walking around."[40] A man named Sagredo, costumed in a turban to play the part of the Alexandrian Ptolemy, stands in the middle, displaying the armillary sphere. He is grouped more closely with the geocentric Aristotelian, while Salviati, whose role is to represent the Copernican heliocentric idea, stands slightly detached from the group. In the dialogue, Salviati asks Sagredo several times to declare to which side of the argument he feels pulled, an ambivalence made visible here in his position as mediator of the little group.[41]

The two characters playing ancient philosophers wear togas from their eras, while Salviati, playing Copernicus, is portrayed in the academic robe any professor of Galileo's own time had to wear. Also referred to as a toga, it is the item of clothing that was made fun of in Agrippa's illustrations visually and by Galileo in a famous mocking poem about the teaching garment he hated, describing it as the identifying accoutrement of academic pomposity and the legitimizing cover-up for untruths.[42] Anton Francesco Doni also did send-ups of this garment – it was considered the distinguishing mark of the pedant but also a symbol of the deceitful practices of the clergy. In the dialogue Galileo uses the metaphor of changing costume to refer to changing one's position in an argument, and has Salviati say, self-protectively: "Before going further I must tell Sagredo that I act the part of Copernicus in our arguments and wear his mask. As to the internal effects upon me of the arguments which I produce in his favor, I want you to be guided not by what I say when we are in the heat of acting out our play (*la rappresentazione della favola*), but after I have put off the costume, for perhaps then you shall find me different from what you saw of me on the stage."[43] For that reason it is not just out of convenience that each man is labeled on the hem of his robe – the robes are themselves costumes of authority, symbols of the difference between the physical person trying on a discussable position on the stage of the "theater of the world" and the declarative proposition of certain truth itself.

This is also a theme of Speroni's *Apologia*, where the author writes that true science may only be followed through Aristotelian scholastic method; dialogue is something else. Cox explains: "The alternative *via* of dialogue is less arduous, but also less profitable, leading to the vineyards and gardens of 'gioco' rather than the sober farmlands of truth. A dialogue may *sound* as though it is saying something, but this appearance of substance is illusory. In fact, the dialogue can provide no more than a beguiling simulacrum of knowledge, as persuasive and as deceptive as a parrot's imitation of speech."[44] The dialogue form as Speroni wanted it to be understood leaves the judgment of truth to the reader; the dialogue's author only presents a situation, and the reader comes to a conclusion that may be completely apart from any presented in the text.

In successive editions of Galileo's *Dialogue* from 1635 on, the frontispiece was copied in such a way as to erase the idea that this was a *rappresentazione* of any kind by giving the figure of Salviati the features of Copernicus instead of just annotating his garments, setting the scene between the pillars of Hercules, and turning the Medici *impresa* back into its inert and monologic self (fig. 125). The figures were made into avatars of the ideas they espouse, labeling them on the ground at their feet instead of leaving them as costumed representatives of intellectual positions that were as changeable as their togas.[45] The reworked frontispiece also had Copernicus make eye contact with the viewer, pictorially enlivening his voice over that of the others, and by using a dry and hard engraving style, it made the distance appear as legible as the foreground. Della Bella instead took advantage of the indistinct quality that the technique of etching can achieve to illustrate the limits of the mind in understanding what is plainly available to it. The mists of ignorance, both Salviati and Sagredo say at different times, may be dispelled by the light of reason. The human mind "is clouded with deep and thick mists, which become partly dispersed and clarified when we master some conclusions and get them so firmly established and so readily in our possession that we can run over them very rapidly."[46] Reasoning, moving through an argument step by step, is laborious, but that is how the human mind works, as opposed to the divine mind, which perceives truth in bright revelatory illumination: "like light in an instant."[47]

Like most awkward translations, the copyist of the later frontispiece, in clarifying what were perceived to be the main points of the picture, entirely erased its dialogic nuance in della Bella's representation of Galileo's characters as searching and thinking – of the presence, manifest in the very form of the text, of minds working through inchoate ideas that were swirling around the speakers and the readers. Della Bella's use of step etching, in which lines are selectively bitten deeper or shallower to make them appear darker or paler, made many of Galileo's triumphs and concerns visual as dim images in a mind struggling to understand. In the text, Galileo makes a point of using scholastic method as he laboriously moves the reader through the book's arguments bit by bit. Even Simplicio hopes out loud that this activity will result in dissolving the mists that cloud certainty: "I hope that a time will come when the mind will be freed . . . and all the mists which keep it darkened will be swept away."[48]

Towards that end, a seemingly straightforward diagram that illustrates a passage in the text was, in fact, drawn by Simplicio with the encouragement of Salviati, who inverts Simplicio's plaintive request for clarification as he begs for a diagram in order to make their discussion easier to understand (fig. 126). Salviati agrees that he can have a diagram, but makes Simplicio draw it himself before the reader's eyes to show that he does indeed actually understand the propositions that he thinks are obscure. Simplicio follows his step-by-step instructions to diagram what he himself thinks about the disposition of the sun and the planets, unhesitatingly mapping out a

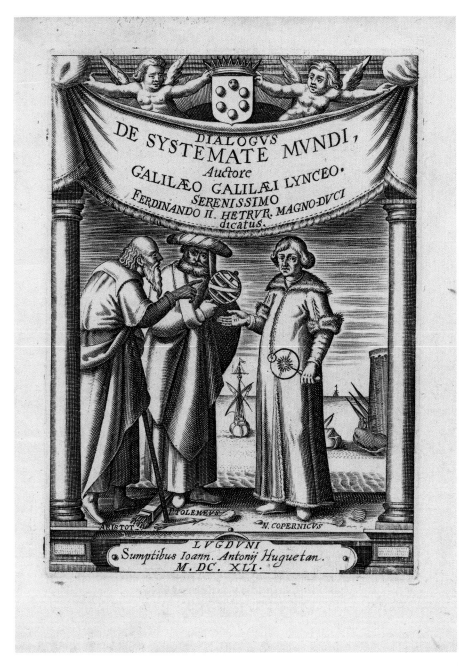

125   Frontispiece, engraving, Galileo Galilei, *Dialogus de systemate mundi – Systema cosmicum,
in quo quatuor dialogis, de duobus maximis mundi systematibus, Ptolemaico & Copernicano . . .
disseritur*, Lyon: Jean-Antoine Huguetan, 1641. Houghton ★IC6 G1333 Ef635sb, Houghton
Library, Harvard University.

SIMP. Sia questo segnato A. il luogo del globo terrestre.
SALV. Bene sià. So secondariamente, che voi sapete benissimo,
che essa terra non è dentro al corpo solare, nè meno a quello
contigua, ma per certo spazio distante, e però assegnate at Sole qual'altro luogo più vi piace remoto dalla terra a vostro be-
neplacito, e questo ancora contrassegnate.
SIMP. Ecco fatto: Sia il luogo del corpo solare questo segnato O.

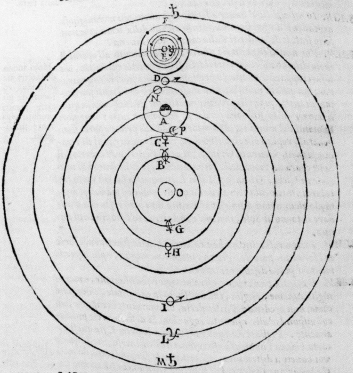

SALV. Stabiliti questi due, voglio, che pensiamo di accomodar'il
corpo di Venere in tal maniera, che lo stato, e mouimento suo
possa sodisfar'a ciò, che di essi ci mostrano le sensate apparen-
ze,

126 Woodcut diagram of the sun and the planets, Galileo Galilei, *Dialogo dei massimi sistemi* . . . , Florence: Landini, 1632, fol. 320v. Hay Lownes QB721. G3x, John Hay Library, Brown University Library.

perfectly heliocentric view of the heavens with his own hand, and yet still doubting the truth of what he himself has convinced the reader is the case. Salviati points this out: "See also what great simplicity is to be found in this rough sketch."[49] As Lorraine Daston has discussed, Galileo's overriding disagreement with Aristotelians was that they let plausibility take the place of observable, demonstrative certainties. We see this critique put into action through the reader's persuasive inclusion in the act of observing Simplicio's persistent blindness to what he has clearly demonstrated to everyone who can read or see, a blindness underscored in the frontispiece by the man's cane that now does double duty as a blind man's stick.[50] Rather than presenting the diagram as a fait accompli, Galileo activates it dialogically, turning its creation into a rhetorical tool, capitalizing on the reader having shared in the experience of the image in the making under Salviati's guidance.[51]

There is much to say about this frontispiece that seems to capture the sense of Galileo's scientific *commedia* pictorially in ways that not only future translators, but also the inquisitors, failed to do. Galileo was immersed in the worlds of courts in which visual language was tremendously important; he himself drew and was a critic of art in his own right.[52] He would certainly have paid attention to the crafting of a frontispiece, the first encounter a reader would have with his dialogue. Mists are important throughout the text, and so are the ship, the cannons, and the tide rolling in at the feet of the men standing on the sandy beach. All of these come up either in the dialogue or in previous treatises Galileo wrote about the trajectory of bodies in motion — weights dropped from the masts of moving ships, or cannonballs projected from their guns — and the motions of the tide, which was the original title for this dialogue. In spite of the open air setting for the frontispiece, this dialogue actually takes place inside a Venetian palace. However, a previous unpublished dialogue on motion had been situated on a beach, with the interlocutors drawing their diagrams with sticks in the sand.[53]

In Galileo's text, ships provide a closed system of movement in which to demonstrate, contrary to Aristotelian arguments that the earth has to be immobile, that the earth can indeed move without moving away from you. Salviati describes an experiment involving a bowl of fish swimming in water below deck on a ship, their movement remaining constant whether the ship is speeding or standing still:

> Shut yourself up with some friend in the main cabin below decks on some large ship, and have with you there some flies, butterflies, and other small flying animals. Have a large bowl of water with some fish in it; hang up a bottle that empties drop by drop into a wide vessel beneath it. With the ship standing still, observe carefully how the little animals fly with equal speed to all sides of the cabin. The fish swim indifferently in all directions; the drops fall into the vessel beneath; and, in throwing something to your friend, you need throw it no more strongly in one direction than another, the distances being equal; jumping with your feet together, you pass equal spaces in every direction. When you have observed all these things carefully (though there is no doubt that when the ship is standing still everything must happen in this way), have the ship

proceed with any speed you like, so long as the motion is uniform and not fluctuating this way and that. You will discover not the least change in all the effects named, nor could you tell from any of them whether the ship was moving or standing still. . . . The fish in their water will swim toward the front of their bowl with no more effort than toward the back, and will go with equal ease to bait placed anywhere around the edges of the bowl.[54]

Della Bella seems to have incorporated Giovanni Battista Landini's printer's mark, with its three interlocking fish in a circle, into the design of the frontispiece to provide a neat preview of the imperturbability of the fish in their bowl below the deck of the ship. The publisher, through his sign – we could say, through publication itself – thereby becomes a fourth force in Galileo's demonstration.[55] In spite of the inclusion of Landini's initials and motto, "Grandior ut proles," at the top of the circle, the inquisitors at the court of the Barberini pope, Urban VIII, hypersensitive to the language of emblems and blind to the pictorial nature of the frontispiece, were suspicious of the fish. They understood them as a mockery of the Barberini commanders of the papal troops by punning on the Barberini triad of bees, present on another of Galileo's frontispieces, in uncomplementary ways (figs. 127, 128).[56] Like Simplicio, they knew that meaning was intended, but were unable to parse what they felt was a mysterious and possibly seditious emblem rather than the pictorial acknowledgment of the importance of publication itself. The three fish unperturbed by movement in their bowl emblematized instead the space of open discourse for the three interlocutors dispersing the mists of confusion through an informal disputation at the edge of the sea.

The costumes of authority, removable and exchangeable though they are, were not here being tried on at a dress rehearsal. Publication constituted a level of authority and responsibility for one's positions, as Speroni would come to believe, that Agrippa and Magino understood even before they were likely to come up before the Inquisition for unpopular, not to mention heretical, ideas. The idea of what Speroni thought a dialogue could accomplish in the first part of his *Apologia*, a truth that would only be understood to be true as "the final meaning of one particular dialogic event," may have been what Galileo was hoping to achieve, putting the case to the reading public for a final decision.[57] Magino (or Contugo) understood that at its most basic a dialogue was supposed to treat of things that were not imaginary but in entertaining ways, and he kept his literary musings well within the area of Jewish mastery: the amusing intermezzo in the ongoing economic melodrama of papal finances. Galileo's literary and artistic formation in Florentine academies, and among Venetian and Florentine nobility, would have made him familiar with that distinguishing aspect of images that is also one of the most distinguishing aspects of written dialogues before the Counter-Reformation: the refusal for any one aspect of the text "to congeal into one-sided seriousness."[58] Landini's fishbowl, one might say, here emblematized a belief in the ability of texts to hold their own under the various and changing types of scrutiny that attended early modern publication: a bubble that would soon be burst by the frustrated churchmen who did not find this humorous.

127 Barberini bees on frontispiece, Galileo
Galilei, *Il saggiatore: nel quale con bilancia
esquisuta e giusta siponderano le cose contenute
nella Libra astronomica e filosofica di Lotario
Sarsi Sigensano / scritto in forma di lettera all'ill.
mo . . . Virginio Cesarini . . . dal Sig. Galileo
Galilei*, Rome: Giacomo Mascardi, 1623. John
Hay Library, Brown University Library.

128 Landini's printer's mark, title page,
Galileo Galilei, *Dialogo dei massimi
sistemi . . .* , Florence: Landini, 1632. Hay
Lownes QB721. G3x, John Hay Library,
Brown University Library.

Illustrated dialogues, and illustrations in books in general, staged a subject in ways that distanced the author equally from certain truths and possible falsehoods by presenting a situation to the reader that would only be resolved in the act of reading. In this way the reader, too, became an author. This was true of the carnivalesque cinquecento dialogue form, and was a property of dialogues strengthened by the active reading that images could prompt. Such works made the author present to the reader in far more local and knowable ways than we usually think of printed books as doing. They also called attention to the processes of scanning, and looking, activating the engagement between readers and printed pictures with a sense of novelty and self-awareness so characteristically beloved in this period. The illustrations were often puzzlingly opaque to the explication of any particular process or invention, but in combination with the text they eloquently evoked the social conditions in the realm of the reader where the possessor of expertise became a knowledgeable author. They allowed, maybe at times even forced, the reader's engagement with the world in the book in a way that text alone could not do, providing another way to understand communities of reading, learning, and looking as they were formed through the use of pictures outside the university setting.

Abbreviations

ASR. Archivio di Stato di Roma.

BAV. Biblioteca Apostolica Vaticana.

Jacques-Charles Brunet, *Manuel du libraire et de l'amateur de livres, contenant 1: Un nouveau dictionnaire bibliographique* . . . , 5th ed., [6 vols.] Paris: G.-P. Maisonneuve & Larose, 1965–66.

Dialogues. Grégoire le Grand, *Dialogues*, ed. Adalbert de Vogüé, trans. Paul Antin, Paris: Les Éditions du Cerf, 1979, vol. 2, 120–249.

DBI. Dizionario Biografico degli Italiani, Rome: Istituto della Enciclopedia Italiana, 1960– (and http://www.treccani.it/biografie/).

JSAH. Journal for the Society of Architectural Historians.

JWCI. Journal of the Warburg and Courtauld Institutes.

Charles LeBlanc, *Le Graveur en Taille Douce*, Leipzig, R. Weigel, 1847.

Masetti Zannini. Gian Ludovico Masetti Zannini, *Stampatori e librai a Roma nella seconda metà del Cinquecento*, Rome: Fratelli Palombi, 1980.

Mortimer. *Catalogue of Books and Manuscripts, Part 2: Italian 16th Century Books*, compiled by Ruth Mortimer under the supervision of Philip Hofer and William A. Jackson, Cambridge, Mass.: Belknap Press of Harvard University Press, 1974.

PMLA. Publications of the Modern Language Association of America.

RB. Terrence G. Kardong, *Benedict's Rule: A Translation and Commentary*, Collegeville, Minn.: Liturgical Press, 1996.

Vita. Vita et Miracula Sanctissimi Patris Benedicti Ex Libro ii Dialogorum Beati/ Gregorii Papae et Monachi collecta/ Et ad instantiam Devotorum Monachorum Congregationis eius: idem Sancti Benedicti Hispania: rum aeneis typis accuratissimè/delineata . . . con licentia Superiorum, 1579.

OPPOSITE PAGE   Attributed to Giovanni Ambrogio de' Predis, Prince Massimiliano Sforza Choosing Between Vice and Virtue, Donatus, *Ars grammatica* of Massimiliano Sforza (detail of fig. 27).

Notes

The sixteenth- and seventeenth-century books discussed here are rarely paginated in a standardized manner. Citations are not always consistent between books but will nevertheless direct readers to the proper place in each case. Where a proemio is unpaginated, it has been given a Roman numeral and either r (for recto) or v (for verso): i r, i v, ii r, ii v, etc. Where all pages are numbered in the modern fashion, they are referred to as "pages." Where they are numbered on one side of the sheet only, or in a book with a variety of pagination styles, they are referred to as "folios." In giving publication information, modern spelling is used although the books themselves often used Latin place names (Lugduni for Lyon, for example, or Romae for Rome). Where more than one press was responsible for the publishing of a book, this information is included here with the original wording retained. Several of the books discussed are available online in digitized versions, but as urls often change, they are not given here and readers are advised to search for these to consult whilst reading this text.

1 Pictures and Readers in Early Modern Rome

1 See Marco Ruffini, *Le imprese del drago: Politica, emblematica e scienze naturali alla corte di Gregorio XIII (1572–1585)*, Rome: Bulzoni, 2005, 45–52.

2 *Missale fratrum Carmelitarum Ordinis beatae Dei genitricis Mariae*, Rome: ex typographia Iacobi Tornerij, 1587 (excudebant Alexander Gardanus, & Franciscus Coattinus). See ASR, 30 Not. Cap. uff. 9 Garganus, vol. 5 (1586–7) fols. 122r–v, for the contract of 1586; Notari del A. C. Fabritius uff. 5, vol. 2467 (May–June 1588), fol. 623, for the sales; and Notari del A. C. Fabritius uff. 5, vol. 2468 (July–Aug. 1588), fol. 684, 21 August 1588, for the negative outcome. The Thomas de Victo-

ria mentioned here must be the composer Tomás Luis de Victoria, although he moved back to Spain in 1587, returning to Rome for an extended period only in 1590. The document in a mixture of Spanish, Latin, and Italian quotes Victoria as saying: "Io non voglio pagare perche non ha ordine bastante."

3 The book, by Benito Perera, is *Benedicti Pererii Valentini e Societate Iesu Commentariorum in Danielem prophetam libri sexdecim. Adiecti sunt quatuor indices, unus quaestionum alter eorum quae pertinent ad doctrinam moralem, & usum concionantium; tertius locorum Sacrae Scripturae; quartus generalis, & alphabeticus*, Rome: in aedibus Populi Romani; apud Georgium Ferrarium, 1587.

4 The Tornieri documents are ASR, 30 Not. Cap. uff. 9 Garganus, vol. 5 (1586–7), fol. 187r: "Domenicus Basa, Bartholomeus de Grassis et Jacobus Tornierij omnes librarij ad Peregrinum . . . promisserunt et . . . pro eius rata promisset D. Antonio Cardinale Caraphe noncupate, et S. Maria in Via Lata di Urbe presenti per imprimere seu imprimi facere opus noncupatij commentarij Danielis RP Benedicti perrerij Societas Jesù . . . Actum Romae in Reg. Sct Eustachij in Palatio di habitione di Illmi d. Cardinalis presentibus." The note inserted in the volume in a different hand from the notary's reads: "cose da patteggiar col Basa et co gli altri che fanno stampar l'opra sopra Daniele/Della bonta della carta, & del inchiostro, del tirar bene, e secondo che gli sara dato corretto la copia, e poi la prima stampa, cosi habbino a stampare. Se, qualche foglio fossi di carta bruta, o scorettamente tirato, e, che al giudiccio del Illmo. Cardinale si dovessi retornar a stampare promettano di farlo. Che promettano di far ogni giorno un foglio, e di mandarlo a correggere al collegio del Giesu a hora commode si come fra loro saranno d'accordo. Che promettano di dar gratis al auctor del opera fin'a sessanta volumi, e per mandar a varij collegij della religion, vender al auctor fin a cinquanta o sessanta volumi, tre o quatro giuli manco di quello che publicamente si venderanno qui a Roma. Le lettere che l'hanno usar nello stampar Daniel saranno queste: per il testo, il sopra silvio, per il commento, il silvio, per le citationi dentro il commento, il corsivo, per le annotationi marginali, lettera tonda bella." The *Silvio* was a chancery lettering developed by the Venetian typographer Francesco Marcolini da Forli. Scipione Casali, *Annali della tipografia veneziana di Francesco Marcolini da Forli*, Forli: M. Casali, 1861, xii.

5 See Adrian Johns, *The Nature of the Book*, Chicago: University of Chicago Press, 1998, 99, and 310 and chapter 8, for the roles of booksellers as "undertakers" of entire printing projects.

6 For this contract and more on the contracts and associations of Roman booksellers at this time, see Masetti Zannini, 204–5 and 285–6.

7 See Horatio R. F. Brown, *The Venetian Printing Press*, London, J. Nimmo, 1891, 81–95; See also Martin Lowry, *The World of Aldus Manutius: Business and Scholarship in Renaissance Venice*, Ithaca: Cornell University Press, 1979; and Jane A. Bernstein, *Print Culture and Music in Sixteenth-Century Venice*, New York and Oxford: Oxford University Press, 2001. See Michael Bury, *The Print in Italy, 1550–1620*, London: British Museum, 2001, 172–7, for many of the same concerns in the field of Venetian pictorial printing.

8 Antonio Martini, *Arti, mestieri e fede nella Roma dei papi*, Bologna: Cappelli, 1965.

9 For reading communities see Roger Chartier, *The Order of Books*, trans. L. G. Cochrane, Stanford: Stanford University Press, 1994. See also Adrian Johns's restated definition of print culture in *Nature*, 28–31; also Pamela O. Long, *Openness, Secrecy and Authorship: Technical Arts and the Culture of Knowledge from Antiquity to the Renaissance*, Baltimore: Johns Hopkins University Press, 2001, for an interest in the ideals of publication among different groups for effecting a parity between classes and professions in the sixteenth century in the case of practical, skill-based knowledge.

10 David Gentilcore, *Medical Charlatanism in Early Modern Italy*, Oxford and New York: Oxford University Press, 2006, 336–60.

11 Magino di Gabrielli, *Dialoghi . . .*, Rome: Gli Heredi di Gioranni Gigliotti, 1588, fol. 72: C: "ma tutti quegli che non sono ciechi veggono, potrebbe esser che alcuno reggendosi secondo il ritratto delle figure fallisse, di modo che più sicuro era il mutare il disegno." M: "La brevità del tempo è causa di questo poco disordine, che oltre l'improprietà della vista non è di momento nessuno, perche deviamo presupporre che almeno uno per casa sappia leggere, e se non; come potrà succedere in qualche strana villa; lo sapra intendere il vicino, & uno l'insegna a molti."

12 Gabrielli, *Dialoghi . . .*, fol. 41: M: "Havete veduto ancor bene à vostro modo questo disegno?" I: "Signor si, dite pure à vostro piacere quello che ho à fare mano in mano, che tengo le orecchie attente à voi, se ben qualche volta rivolgo gl'occhi alla figura."

13 Peter Stallybrass and Allon White, *The Poetics and Politics of Transgression*, Ithaca: Cornell University

Press, 1986; William Eamon, *Science and the Secrets of Nature*, Princeton: Princeton University Press, 1994. On Roman *bandi* see Rose Marie San Juan, *Rome: A City Out of Print*, Minneapolis: University of Minnesota Press, 2001, 23–55.

14 *Gazzetta universale*, vol. 5, 16 January 1778: "Io Domenico Zito, Regio Trombetto della Gran Corte della Vicaria, dico di aver pubblicato il sopra scritto Bando e quanto in esso si contiene, per tutti i luoghi soliti e consueti di questa Fedelissima Città, alta et intelligibili voce, more praeconis, ut moris est; e averne affisse le copie. Domenico Zito."

15 Michel de Certeau, *The Practice of Everyday Life*, trans. S. Rendell, Berkeley: University of California Press, 1984, 175–6.

16 Chartier, *Order*, 23. For the autonomy of the reader in Chartier and the historiography of print, see Adrian Johns, "How to Acknowledge a Revolution," in *American Historical Review* 107, no. 1 (February 2002): 106–25.

17 Mary Carruthers, *The Craft of Thought*, Cambridge: Cambridge University Press, 1998, 68–9, discussing the distinction between medieval *memoria* and imagination, quotes Mary Warnock, *Memory*, London: Faber, 1987, 34: "what distinguishes memory from the imagination is not some particular feature of the [mental] image but the fact that memory is, while imagination is not, concerned with the real." See also M. Warnock, *Imagination*, Berkeley and Los Angeles, University of California Press, 1976, 16, 24. The role of memory in imagination is also described at length in Murray Wright Bundy, *The Theory of Imagination in Classical and Mediaeval Thought*, Chicago: University of Illinois, 1927, 249–69.

18 For authorial teams see Evelyn Lincoln, "Invention, Origin, and Dedication: Republishing Women's Prints in Early Modern Italy," in M. Biagioli, P. Jaszi, and M. Woodmansee (eds.), *Making and Unmaking Intellectual Property: Creative Production in Legal and Cultural Perspective*, Chicago: University of Chicago Press, 2011, 339–57.

19 On the historiography of the book see Robert Darnton, "What is the History of Books?" *Daedalus* 3, no. 3 (Summer 1982): 65–83; and Adrian Johns's beautifully lucid "Science and the Book

in Modern Cultural Historiography," *Studies in the History of Science* 29, no. 2 (1997): 167–94.

20 Mikhail Bakhtin, *The Problem of Dostoevsky's Poetics*, trans. Caryl Emerson, Minneapolis: University of Minnesota Press, 1984, 106–60.

21 See the introduction by Gigliola Fragnito in G. Fragnito (ed.), *Church, Censorship and Culture in Early Modern Italy*, Cambridge: Cambridge University Press, 2001, 1–13; and of course Carlo Ginzburg's seminal *Formaggio e i vermi*, Turin: Einaudi, 1976, translated as *The Cheese and the Worms*, Baltimore: Johns Hopkins University Press, 1980.

22 J. L. Heilbron, *Galileo*, Oxford and New York: Oxford University Press, 2010, 106.

23 Principio Fabrizi, *Delle allusioni, imprese, et emblemi del sig. Principio Fabricij da Teramo sopra la vita, opere, et attioni di Gregorio XIII pontefice massimo libri VI nei quali sotto l'allegoria del drago, arme del detto pontefice, si descriue anco la uera forma d'un principe christiano; et altre cose, la somma delle quali si legge doppo la dedicatione dell'opera all'ill.mo et ecc.mo s. duca di Sora*, in Rome: Bartolomeo Grassi, 1588 (Romae: apud Iacobum Ruffinellum), 385–6. For an excellent discussion of his letter, and for Grassi's share in the treatise, see Ruffini, *Le imprese del drago*, 45–52.

24 *Comento di Cristophoro Landino fiorentino sopra la comedia di Danthe Alighieri poeta fiorentino*, Florence: Nicholo di Lorenzo della Magna, 1481.

25 See Ruffini, *Le imprese del drago*, 48, for Grassi's preference for such illustrated books structured around pictures.

26 See Johns, "Science and the Book," for the formation of the history of the book as a field, and the trajectory of its motive questions.

27 The first known Italian printed illustrated book is the *Meditations* of Cardinal Torquemada, Rome: Ulrich Han, 1467. For the beginnings of printing in Italy see Lamberto Donati in Francesco Barberi, *Mostra del libro illustrato romano del Cinquecento*, Rome: Biblioteca Angelica, 1950, 9–10; Frederick John Norton, *Italian Printers, 1501–1520: An Annotated List*, London: Bowes and Bowes, 1958; Luigi Balsamo, "The Origins of Printing in Italy and England," *Journal of the Printing Historical Society* 11 (1976–7): 48–63; S. H.

Steinberg, *Five Hundred Years of Printing*, rev. J. Trevitt, London: British Library, 1996; Brian Richardson, *Printing, Writers and Readers in Renaissance Italy*, Cambridge and New York: Cambridge University Press, 1999, 134; Andrew Pettegree, *The Book in the Renaissance*, New Haven and London: Yale University Press, 2010, 49.

28 See Stephen Orgel, "Textual Icons: Reading Early Modern Illustrations," in Neil Rhodes and Jonathan Sawday (eds.), *The Renaissance Computer*, London and New York: Routledge, 2000, 60.

29 Michael Lynch and John Law, "Pictures, Texts and Objects: The Literary Language Game of Bird-Watching," in Mario Biagioli (ed.), *The Science Studies Reader*, New York: Routledge, 1999, 317–41.

30 See Anna Modigliani, "Tipografi a Roma (1467–1477)," in M. Miglio and O. Rossini (eds.), *Gutenberg e Roma*, Naples: Electa Napoli, 1997; and the introduction by Massimo Miglio in Paola Farenga (ed.), *Editori ed edizioni a Roma nel Rinascimento*, Rome: Roma nel Rinascimento, 2005, vi–xiii, and in the same volume, Anna Modigliani, "Printing in Rome in the XVth Century: Economics and the Circulation of Books," 65–76.

31 San Juan, *Rome*, 129–60.

32 D. S. Chambers, "The Economic Predicament of Renaissance Cardinals," in William M. Bowsky (ed.), *Studies in Medieval and Renaissance History*, vol. 3, Lincoln: University of Nebraska Press, 1966, 289–313; and Mary Hollingsworth and Carol M. Richardson (eds.), *The Possessions of a Cardinal: Politics, Piety and Art, 1450–1700*, University Park: Pennsylvania State University Press, 2010.

33 For printing and the past in Rome see Anthony Grafton, *Bring Out Your Dead: The Past as Revelation*, Cambridge, Mass.: Harvard University Press, 2001, especially 31–62.

34 The title page is inscribed: SPECULUM ROMANAE MAGNIFICENTIAE. OMNIA FERE QUAECUNQUE IN URBE MONUMENTA EXTANT, PARTIM IUXTA ANTIQUAM, PARTIM IUXTA HODIERNAM FORMAM ACCURATISS. DELINEATA REPRAESENTANS. Accesserunt non paucae, tum antiquarum, tum modernarum rerum. Urbis figurae nunquam antehac aeditae / Roma tenet propriis monumenta sepulta ruinis plurima, quae profert hic rediviva liber. Hunc igitur lector scrutare benigni, docebit urbis maiestas pristina quanta fuit. ("A Mirror of Roman Magnificence – Representing nearly all the monuments in the city, however small, some ancient in form, some modern, most accurately sketched – The images of the city, never displayed before now, constitute not a small number of ancient and modern subjects / Rome holds many monuments immersed in her own ruins, which this book brings forward. Therefore most benevolent reader, examine closely this book, it will show how great was the former grandeur of the city.") Transcription and translation from the website "The Speculum Romanae Magnificentiae, Digital Collection," accessed 10 August 2011: http://speculum.lib.uchicago.edu/search.php?search[0]=Title+page&searchnode[0]=all&result=1.

35 See Peter Parshall, "Antonio Lafreri's *Speculum Romanae Magnificentiae*," *Print Quarterly* 23, no. 1 (2006): 3–27; and Rebecca Zorach, *The Virtual Tourist in Renaissance Rome: Printing and Collecting the Speculum Romanae Magnificentiae*, Chicago: University of Chicago Press, 2008, with a bibliography on the *Speculum*.

36 Pasha Ahmet, *Antiquités Romaines expliquées dans les memoires du Comte de B★★★. Contenant ses avantures, un grand nombre d'histoires & anecdotes du tems très-curieuses, ses recherches & ses découvertes sur les antiquités de la ville de Rome & autres curiosités de l'Italie*, The Hague: Jean Neaulme, 1750, 193.

37 For the financial obligations and arrangements of publishers, besides individual articles in the *DBI*, see Masetti Zannini; the many books of published Roman archival documents by Achille Bertolotti; and the dictionary still in process: Marco Menato, E. Sandal, and G. Zappella (eds.), *Dizionario dei tipografi e degli editori italiani: Il Cinquecento*, Milan: Editrice Bibliografica, 1997. Most usefully the internet database *EDIT16: Censimento nazionale delle edizioni italiane del XVI secolo*, http://edit16.iccu.sbn.it/web_iccu/ihome.htm, allows researchers to search by title, publisher's name, date, and other rubrics, along with short biographical entries and bibliography, to find patterns inacces-

38   See the entry "Bartolomeo Bonfadino," in Menato, Sandal, and Zappella, *Dizionario*, 173–4. For Parasole and Bonfadino, see ASR, 30 Not. Cap. uff. 21 Franciscus Grillus, vol. 74 (1609), fols. 614r–v, 615r, 17 August 1609.

39   See F. Barberi, "Blado, Antonio," in *DBI*, vol. 10, 753–7. The Blado print shop was still at the same location in 1572.

40   A. Modigliani, " Massimo, Pietro," in *DBI*, vol. 72, 15–16.

41   On the *Mirabilia* and other guidebooks, see Eunice D. Howe (trans. and ed.), *The Churches of Rome*, Binghamton, NY: Medieval and Renaissance Texts and Studies, 1991. See also Ludwig Schudt, *Le guide di Roma: Materialien zu einer geschichte der römischen topographie*, Vienna: B. Filzer, 1930; San Juan, *Rome*, 57–93; and Amy Marshall, *Mirabilia urbis Romae: Five Centuries of Guidebooks and Views*, Toronto: University of Toronto Library, 2002. See also Sergio Rossetti, *Rome: A Bibliography from the invention of printing through 1899. Vol. I. The Guide Books*. Florence: Olschki, 2000.

42   For Gigliotti, see the online database of the Instituto Centrale per il catalagto Unico delle biblioteche italiane e per le informazion bibliografiche (ICCU), EDIT 16.

43   *Le cose maravigliose dell'alma citta di Roma, dove si tratta delle chiese, stationi, indulgenze, & reliquie dei corpi santi, che sono in essa . . .*, Rome: appresso Giouanni Osmarino, alla chiavica di Santa Lucia, 1571, 1572, 1573, 1574, 1575 (in Spanish and Italian editions), 1579, 1580, 1581, 1584, and 1585, each time with improvements and additions such as how to obtain indulgences or the names of all the popes, emperors, and Christian princes; and *L'antichità di Roma di m. Andrea Palladio, raccolta brevemente da gli auttori antichi, et moderni. Aggiuntovi un discorso sopra li fochi de gli amichi*, Rome: Giovanni Osmarino Gilioto, alla chiavica di Santa Locia.

44   *Ludovici Demontiosii Gallus Romae hospes: Ubi multa antiquorum monimenta explicantur, pars pristinae formae restituuntur; Opus in quinque partes tributum*, Rome: Ioannem Osmarinum, 1585.

45   *Il piacevole viaggio di Cuccagna. Di novo ritrovato & stampato a commodità di tutti i buon compagni, che*

desiderano andare in quel paese, Rome: per Giovanni Osmarino Giliotto, *c.*1586, was typical of the genre.

46   Carla Casetti Brach, "Bonfadino, Bartolomeo," in Menato, Sandal, and Zappella, *Dizionario*, 173–4. For a list of the sixteenth-century publications bearing the printer's mark of Bartolomeo Bonfadino (1583–1600), alone and in partnership with Tito Diani (1583–5), see the online database *EDIT16*. His imprint appears in books printed until about 1607; see also C. Casetti Brach in *DBI* under "Diani, Tito," vol. 39, 650–2. See also Massimo Ceresa, *Annali typografici di Guglielmo Facciotti ed eredi (1592–1640)*. Rome: Bulzoni, 2000, 1–52; for Bonfadino 19, 44.

47   Barberi, "Blado, Antonio," 753–7.

48   *Icones operum misericordiae cum Iulii Roscii Hortini sententiis et explicationibus pars prior eorum quae ad corpus pertinent*, Romae: impensis Bartholomaei Grassii Rom. bibliopolae; incidebat Marius Cartarius, 1586 (Romae: ex typographia Bartholomaei Bonfadini in via Peregrini, 1585).

49   *Herbario nuovo di Castore Durante medico, & cittadino romano. Con figure, che rappresentano le vive piante, che nascono in tutta Europa, & nell'Indie orientali, & occidentali*, Rome: Per Iacomo Bericchia & Iacomo Tornierij, 1585 (in Roma: nella stamperia di Bartholomeo Bonfadino, and Tito Diani, 1585).

50   *Speculum, & exemplar christicolarum. Vita beatissimi patris Benedicti, monachorum patriarcae sanctissimi. Per r.p.d. Angelum Sangrinum abbatem Congregationis Casinensis carmine conscripta*, Romae, 1587 (Romae: ex typographia Bartholomaei Bonfadini, in via Pelegrina, 1587).

51   *Petri Angelii Bargaei Commentarius de obelisco ad sanctiss. et beatiss. d.n.d. Xystum V pont. max. Huc accesserunt aliquot poetarum carmina, quorum, partim ad idem argumentum, partim ad eiusdem summi Pont. laudem pertinent*, Rome: Bartholomaei Grassij, 1586 (Rome: ex typographia Bartholomaei Bonfadini, 1587).

52   *Discorso di Pietro Paolo Magni piacentino sopra il modo di fare i cauterij ò rottorij à corpi humani, nel quale si tratta de siti, oue si hanno da fare, de ferri che usar vi si debbono, del modo di tenergli aperti, delle legature, & delle palline, & dell'utilità che da essi ne vengono, cose utilissime non solo à barbieri, ma à tutte*

le persone, che n'hanno bisogno, Rome: appresso Bartolomeo Bonfadino, 1588 (Rome: appresso Bartolomeo Bonfadino, 1588); Giovanni Battista Cavalieri, *Effigies pontificum Romanorum cum eorum vitis in compendium redactis*, Rome: Bartholomaei Bonfadini, 1595.

53 Bartolomeo Crescenzio, *Nautica Mediterranea*, Rome: appresso Bartolomeo Bonfadino, 1602/1607.

54 ASR, 30 Not. Cap. uff. 21 Franciscus Grillus, vol. 74 (1609), fols. 614r–v, 615r, 17 August 1609: locatio domus Giovanni Battista Raimondi Cremonensii, witness is Bartolomeo Bonfadino brescianensis.

55 ASR, 30 Not. Cap. uff. 28, vol.13, (1589) fols. 610r–629r, 637r–638r. See Valeria Pagani, "The Dispersal of Lafreri's Inheritance 1581–89 – III: The de' Nobili–Arbotti–Clodio Partnership," *Print Quarterly*, 28: 2 (2011): 119–35.

56 The inventory of Luigi Zannetti's shop made in September 1607, appears in ASR, 30 Not. Cap. uff. 21, vol. 69, (1507) fols. 151r–152v. The details of the arrangements with his wife and children appear on the preceding pages, and another division of the property after Faustina's marriage to a Genoese goldsmith, Anibale de' Fabiani, can be found in ASR, 30 Not. Cap. uff. 21, vol. 80 (1611), fols. 228r–239r, 15 June 1611. For Curzio Lorenzini, see Achille Bertolotti, "Le tipografie orientali e gli orientalisti a Roma nei secoli XVI e XVII," in *Rivista europea* 9:2 (1878): 247.

57 For Rosato Parasole, see Masetti Zannini, 124; from the low settlement price Masetti Zannini concurs that the work was decorative.

58 See, among many other things, Paul Saenger, *The Space Between Words*, Stanford: Stanford University Press, 1997; Thomas N. Corns, "The Early Modern Search Engine: Indices, Title Pages, Marginalia and Contents," in Neil Rhodes and Jonathan Sawday (eds.), *The Renaissance Computer*, London and New York: Routledge, 2000, 95–105; Peter Stallybrass, "Books and Scrolls: Navigating the Bible," in J. Anderson and E. Sauer (eds.), *Books and Readers in Early Modern England*, Philadelphia: University of Pennsylvania Press, 2002, 42–79; Ann M. Blair, *Too Much to Know: Managing Scholarly Information Before the Modern Age*, New Haven and London: Yale University Press, 2010.

59 See Robert K. Barnhart (ed.), *The Barnhart Dictionary of Etymology*, New York: H. W. Wilson Co., 1988, 508. "A spiritual illumination," *c.*1375, from O.Fr. *illustration*, from L. *illustrationem* (nom. *illustratio*) "vivid representation" (in writing), lit. "an enlightening," from *illustrare* "light up, embellish, distinguish," from "in" + *lustrare* "make bright, illuminate." Mental sense of "act of making clear in the mind" is from 1581. Meaning "an illustrative picture" is from 1816. *Illustrate* "educate by means of examples," first recorded 1612. The sense of "provide pictures to explain or decorate" originated in 1638.

60 *Passio domini nostri Iesu christi secundum seriem quattuor evangelistarum: per quendam fratrem ordinis Minorum de observantia: accuratissima opera devotissimaque expositione illustrata: magnorumque virorum sententijs compte adornata*, Basel impressa, 1513.

61 Giovanni Antonio Panteo (trans.), *Ars et theoria transmutationis metallicae cum Voarchadumia, proportionibus, numeris, & iconibus rei accommodis illustrata, Ioanne Augustino Pantheo Veneto authore*, Paris, 1550.

62 *Tabula in grammaticen Hebraeam, authore Nicolao Clenardo. A Iohanne Quinquarboreo Aurilacensi a mendis quibus scatebat repurgata, & annotationibus illustrata*, Parisiis, 1550; Caspar Bauhin, *Anatomica corporis virilis et muliebris historia; Caspari Bauhini d. anato. botanic. Basileens. ord. Hippocrat. Aristotel. Galeni auctoritae illustrata & novis inventis plurimis aucta. Cum indice locupletissimo*, Lyon: Joannem Le Preux, 1597; or Fulvio Cardulo, *Passio sanctorum Martyrum, Getulij, Amantij, Cerealis, Primitiui, Symphorosae, ac septem filiorum, ntis [sic] & digressionibus illustrata Fulvij Carduli presbyteri e Societate Iesu*, Rome: Franciscum Zannettum, 1588.

63 The Apianus of 1551 announced new figures: *Cosmographia Petri Apiani, per gemmam Frisium apud Lovanienses medicum & mathematicum insignem, iam demum ab omnibus vindicata mendis, ac nonnullis quoque locis aucta, figurisque novis illustrata: additis eiusdem argumenti libellis ipsius Gemae Frisiim*, Paris: Vaeneunt apud Viuantium Gaultherot, via Iacobea; sub intersignio D. Martini, 1551, while the 1588 Marliani included figures among the improvements in this edition: *Bartholomaei Marliani Urbis*

Romae topographia accurate, tum ex veterum, tum etiam recentiorum auctorum fontibus hausta, nunc denuo mendis omnibus sublatis, & figuris illustrata, castigatissime in lucem edita. Cui accessere Hieronymi Ferrutij Romani quamplures additiones ex his primum quae celebria in vrbe erecta & aucta sunt. Cum indice rerum, & locorum locupletissimo ad studiosorum omnium communem vtilitatem, Venice: 1588. Carlo Ruini's anatomy of the horse: *Anatomia del cauallo, infermita et suoi rimedii. Opera nuoua, degna di qualsiuoglia Prencipe, & caualiere, & molto necessaria a Filosofi, Medici, Cauallerizzi, & Marescalchi. Del Sig. Carlo Ruini senator bolognese. Adornata di bellissime figure, le quali dimostrano tutta l'Anatomia del cauallo . . .* , Venice: Gasparo Bindoni, 1599.

64 Some Venetian examples of this would be *Prattica utilissima et necessaria di cirugia dello eccell.mo m. Giouanni Di Vico, con nuoue figure adornata. Con il compendio dell'eccellente m. Marian Santo da Barletta suo discepolo. Tradotta nuouamente di latino in lingua volgare per lo eccellente fisico m. Pietro Rostino, a beneficio & utilita universale. Con tre tauole copiosissime*, Venice: Vincenzo Valgrisio, & Baldessar Costantini, 1557; *Le devote, & pie meditationi di S. Bonauentura cardinale. Sopra il misterio dell'humana Redentione. Nuouamente con somma diligentia ristampata, & ricoretta, & di varie figure adornata*, Venice: Domenico Cavalcalupo, 1581; and *Aminta: favola boschereccia / del sig. Torquato Tasso; di novo corretta & di bellissime, & vaghe figure adornata*, Venice: presso Aldo, 1590. The title of the Ripa book is *Iconologia di Cesare Ripa perugino . . . Di nuovo revista, & dal medesimo ampliata di 400 & piu imagini, e di figure d'intaglio adornata . . .* , Rome: appresso Lepido Facij, 1603.

2   Pictures for the Ear of the Heart

1   *Vita.* See Mortimer, cat. 362; Brunet, vol. 4, 418 [21740] mentions only the Arnolfini edition of 1596. See also E. Lincoln, "The Devil's Hem: Allegorical Reading in a Sixteenth-century illustrated Life of St. Benedict," in C. Baskins and L. Rosenthal (eds), *Early Modern Visual Allegory: Embodying Meanings*. Aldershot: Ashgate, 2007, 135–53.

2   See *Dialogues.* The second book of the *Dialogues, De vita et miraculis venerabilis Benedicti*, is the earliest account of the life of Saint Benedict.

3   For theories of monastic reading, *memoria*, and the imagination discussed in this chapter I have relied consistently on the work of Mary Carruthers: *The Book of Memory*, Cambridge and New York: Cambridge University Press, 1990, and *The Craft of Thought*, Cambridge: Cambridge University Press, 1998; Lina Bolzoni, *La stanza della memoria*, Turin: Einaudi, 1995; Jean Claude Schmitt, "L'imagination efficace," in Klaus Krüger and Alessandro Nova (eds.), *Imagination und Wirklichkeit*, Mainz: Philipp von Zabern, 2000, 13–20; and the seminal work by Murray Wright Bundy, *The Theory of Imagination in Classical and Mediaeval Thought*, Urbano: University of Illinois, 1927.

4   Francis Clark, *The "Gregorian" Dialogues and the Origins of Benedictine Monasticism*, Leiden and Boston: Brill, 2003. Dante, *Paradiso*, XXXII: 34–6; see Patrizia Castelli (ed.), *Iconografia di San Benedetto nella pittura della Toscana: Immagini e aspetti culturali fino al XVI secolo*, Florence: Centro d'incontro della Certosa di Firenze, 1982, 15–19.

5   There is little information about the publication history of the first edition of this book, and none mentioning the name of the first publisher other than the Spanish Benedictines. See n. 54, below.

6   For Bernardino Passeri, see the entry by Christopher L. C. E. Witcombe in Jane Turner (ed.), *The Dictionary of Art*, vol. 24, New York: Grove, 1996, 234.

7   For Aliprando Capriolo (Caprioli), see B. Passamani, "Caprioli, Aliprando" in *DBI*, vol. 19, 209–10; and Guido Suster, "Dell'incisore Trentino Aliprando Caprioli," *Archivio Trentino* 18 (1903): 144–206. For the Madruzzo family, see Rotraud Becker, "Madruzzo, Cristoforo," and "Madruzzo, Giovanni Ludovico" in *DBI* vol. 67 (2007): 175–80, 181–6.

8   Capriolo signs with his monogram, a conjoined "ac"; only Plate 25 contains his full name: "Aliprando Capriolo fe." Passerus's signature is a version of "Bernardinus Passerus invenit." A list of the monograms and signatures in the book appears in Iolanda Olivieri, "Il corvo e il pane: Iconografia degli episodi legati al cibo nei cicli

figurativi," in Angela Adriana Cavarna *et al.*, *Il cibo e la regola*, Rome: Biblioteca Casanatense, 1996, 129–42. For monograms as artisan's signatures, see Pamela O. Long, *Openness, Secrecy, Authorship: Technical Arts and the Culture of Knowledge from Antiquity to the Renaissance* Baltimore: Johns Hopkins University Press, 2001, 216; and Evelyn Lincoln, "Invention and Authorship in Early Modern Italian Visual Culture," *DePaul Law Review* 52, no. 4, (Summer 2003): 1112.

9 *Dialogues*, 128.

10 Carruthers, *Craft*, 103, 108–9.

11 Huldreich Coccius, *Opera D. Gregorii Papae huius nominis primi*, Basel, 1564, see Clark, *"Gregorian" Dialogues*, 9–10 and 220–32, for the issue of Gregory's authorship.

12 Clark, *"Gregorian" Dialogues*, 10, with a history of arguments for and against Gregory's authorship of the *Dialogues*: 7–24. Clark is decisive about the work's having been composed in the seventh century by a Roman notary.

13 Clark, *"Gregorian" Dialogues*, 10 cites Martin Chemnitz, a Centuriator of Magdeburg, in a discussion of the origins of purgatory in his *Examen Decretorum Concilii Tridentini* (1565–73 and republished in Frankfurt in 1596), who wrote that the *Dialogues* could not have been written by Gregory.

14 *The Dialogues of Saint Gregory, surnamed the Greate: Pope of Rome & the first of that name. Divided into Four Books, wherein he intreateth of the Lives and Miracles of the Saintes in Italie . . .*, trans. P. W., Paris: Charles Boscard, 1608, 5. Early English Books Online: http://gateway.proquest.com/openurl?ctx_ver=Z39.882003&res_id=xri:eebo&rft_id=xri:eebo:image:21746:28, accessed 9 May 2011.

15 The plate references the *Dialogues*, chapter 14, whereas the event pictured occurs in chapter 12.

16 *RB* 48, "De opera manuum cotidiana," 381–4, 399–401. Kardong defines *lectio divina* as reading of the bible in a "prayerful" way that involved meditation and interpretation as well as memorization, but emphasizes that *how* medieval monks practiced it is not clear today. De Vogüé adds reading as a third term to "ora et labora;" see Adalbert de Vogüé, *The Rule of Saint Bendedict*, trans. J. B. Hasbrouck, Kalamazoo, Mich.: Cister-

cian Publications, 1983, 242. See also Jean Leclercq, "Lectio divina," *Worship* 58, no. 3 (1984): 239–48; and Brian Stock, *After Augustine: The Meditative Reader and the Text*, Philadelphia: University of Pennsylvania Press, 2001, 101–14.

17 Carruthers, *Book* and *Craft*, and Bolzoni, *La stanza*, propose historical contexts for this kind of reading.

18 For the Monteoliveto frescoes, see Carlo Enzo, *Le storie di San Benedetto a Monteoliveto Maggiore*, Milan: Silvana, 1980.

19 *Dialogues*, 135; Carruthers, *Craft*, 4–5 and 110–11.

20 *Dialogues*, 132.

21 Carruthers, *Craft*, 110–11, 113.

22 For the picture within a picture, see Sixten Ringbom, "Action and Report: The Problem of Indirect Narration in the Academic Theory of Painting", *JWCI* 52 (1989): 34–51.

23 *Vita*, Plate 26: "Sanctus ait, Non barbarico delebitur ense; / Tempore sed multo fulminibusque cadet. / Quae cecinit vates, quis non ita diruta cernit, / Moenia, cum thermis, templa, theatra, domos?"

24 Jerome Nadal, *Evangelicae historiae imagines: ex ordine Evangeliorum, quae toto anno in Missae sacrificio recitantur, in ordinem temporis vitae Christi digestae*, Antwerp: Martin Nutius, 1593. Jerome Nadal, *Adnotationes et meditationes in Evangelia quae in sacrosancto Missae sacrificio toto anno leguntur cum Evangeliorum concordantia, historiae integritati sufficienti*, Antwerp: Martinus Nutius, 1594; second edition 1595. See the introduction by Walter Melion in Jerome Nadal, *Annotations and Meditations on the Gospels*, vol. 1, trans. and ed. Frederick A. Homann, Philadelphia: Saint Joseph's University Press, 2003; and Paul Debuchy, "Spiritual Exercises of Saint Ignatius," *The Catholic Encyclopedia*, vol. 14, New York: Robert Appleton Company, 1912; http://www.newadvent.org/cathen/14224b.htm, accessed 9 December 2012; Thomas Joseph Campbell, SJ, *The Jesuits, 1534–1921: A History of the Society of Jesus from its Foundation to the Present Time*, New York: The Encyclopedia Press, 1921, 17; John W. O'Malley, *The First Jesuits*, Cambridge, Mass.: Harvard University Press, 1993, 37–50. See also Danilo Zardin, "Le *Adnotationes et meditationes* illustrate di Nadal

sui Vangeli del ciclo liturgico: il modello e li riuso," in Erminia Ardissino and Elisabetta Selmi, (eds.), *Visible Teologia. Il libro sacro figurato tra Cinquecento e Seicento in Italia*, Rome: 2012, 3–23.

25 Dom Francisco García de Cisneros, *Exercitatorio de la vida espiritual*, Montserrat: Juan Luschner, 1500.

26 Diego Jiménez, in the *proemio* of the *Adnotationes*, quoted in Nadal, *Annotations*, vol. 1, 101.

27 *RB* 42, 344–6. On abstention from speaking, see *RB* 6, "De taciturnitate", 118–19.

28 Bundy, *Theory of Imagination*, 123.

29 For prolepsis and its role in effecting unity in narrative, see Gérard Gennette, *Narrative Discourse: An Essay on Method*, Ithaca: Cornell University Press, 1980, 67–79.

30 Carruthers, *Craft*, 69.

31 *Vita*, Plate 10: "Ad tesqua interea plures trahit aemula virtus, / Bis sex claustra quibus construit ampla Pater. Bissenos illis vitae morumque magistros / Proeficiens, paucos aggregat ipse sibi; / Eutitius, Maurum, Placidum Tertullus eisdem / Addunt, Romana nobilitate satos."

32 Jon Whitman, *Allegory: the Dynamics of an Ancient and Medieval Technique*, Cambridge, Mass.: Harvard University Press, 1987, 266. For the importance and then distrust of allegory as permitting a too-generous expansion of religious meaning, see Elizabeth A. Clark, *Reading Renunciation: Asceticism and Scripture in Early Christianity*, Princeton: Princeton University Press, 1999, 70–103.

33 "Cumque vir Dei venisset in eodem monasterio, et constituta hora, expleta psalmodia, sese fratres in orationem dedissent, aspexit quod eundem monachum, qui manere in oratione non poterat, quidam niger puerulus per vestimenti fimbriam foras trahebat": *Dialogues*, 152. English translation in St. Gregory, *The Second Book of the Dialogues, 1638*, ed. D. M. Rogers, English Recusant Literature 1558–1640, vol. 294, Ilkley and London: Scolar Press, 1976, 22–3.

34 *Vita*, Plate 11: "Orandi, monachus, ne templa reliqueret hora, / Voce Patris monitus, deserit illa tamen / Vidit ab aethiopis spectro ludique trahique / Hunc Pater, et Mauro posse videre dedit / Quem simul ut virgae percussit verbere, nunquam / Orandi passus taedia prima fuit."

35 *Vita*, Plate 4: "At puer impatiens laudis, nutrice relicta, / Sublaci rupes, et loca vasta petit. / Inventum in rigida monachus Romanus eremo, / Quis sit, et unde venit, quo velit ire rogat / Ille refert sua vota, habitus velamina sacri / Poscit, et à Sancto suscipit illa sene."

36 De Vogüé, 275.

37 See David Brakke, "Ethiopian Demons: Male Sexuality, the Black-Skinned Other, and the Monastic Self," *Journal of the History of Sexuality* 10, no. 3/4 (July–Oct. 2001): 501–35 (for temptation regarding labor: 508).

38 Katherine Park, "The Organic Soul," in Charles Schmitt, Q. Skinner, and E. Kessler (eds.), *The Cambridge History of Renaissance Philosophy*, Cambridge and New York: Cambridge University Press, 1988, 464–84 (466).

39 *Dialogues*, 137–41, English translation in *St. Gregory*, 9–11.

40 *St. Gregory* (1638), 10.

41 Rufinus of Aquileia, *Historia monachorum sive de vita sanctorum partum*, ed. Eva Schulz-Flügel, Berlin: W. de Gruyter, 1990.

42 John Cassian, *Collationes partum XXIV*, available in English as *John Cassian: The Conferences*, trans. and ed. Boniface Ramsay, New York: Paulist Press, 1997. See also David Brakke, *Demons and the Making of the Monk*, Cambridge, Mass.: Harvard University Press, 2006, 242–3, on Cassian, demons, and the Benedictines; and Elizabeth A. Clark, "Foucault, the Fathers and Sex," in James W. Bernauer and Jeremy R. Carrett (eds.), *Michel Foucault and Theology: The Politics of Religious Experience*, Burlington, Vt.: Ashgate, 2004, 39–56 (46), and more broadly, Clark, *Reading Renunciation*.

43 Brakke, "Ethiopian Demons," 508–10.

44 *RB* 48, 381–402; and Brakke, *Demons*, 242–5.

45 Clark, "Foucault"; and Brakke, "Ethiopian Demons," 506.

46 *RB* 20, 206–9, "It is not in many words, but in the purity of the heart and tears of compunction, that we are heard." The notion, *compunctione lacrimarum*, is from Cassian (*RB*, 208). See also Catherine Vincent, "Discipline du corps et de l'esprit chez les Flagellants au Moyen Âge," in *Revue Historique* 302: 615 (July/September 2000): (593–614).

47 Massimiliano Sforza was the son of Lodovico il Moro and Beatrice d'Este (1493–1530, ruled 1500–12). I am grateful to Isabel Fiorentini for providing information about these images in the Biblioteca Trivulziana.

48 The property was ceded in 1472. Arnaldo Bruschi, *Bramante architetto*, Bari: Laterza, 1969: 987.

49 Clark, *Reading Renunciation*, p. 61, referring to John Chrysostom, *Contra Marcionista et Manichaeos*, 1: "reading does not soffice for a holy life unless knowledge is added to it . . .".

50 An inscription in the church of Santa Barbara records the patronage of Arnolfini in restoring the church in 1601. The Spanish edition of the *Vite* is *Vita et miracula sanctissimi Patris Benedicti ex libro ii Dialogorum Beati Gregorii Papa, et Monachi collecta per Thomam Triterum, et e Latina in Hispanicum lingua conversa per D. Franciscum Cabrera. Adiuncta vita, et effigie eiusdem. / S. Benedicti / Ad Philippum Hispanioarum et Indiarum principem / Sumptu Paullini Arnolfini Lucen(sis)* / Rome, 1597 (Biblioteca Casanatense, K II 21 CCC). The *libraro* Paolino di Giovanni Arnolfini, canon of the cathedral of Lucca, also worked in Lyon, and by 1603 is documented as having a bookshop in Parione. See Masetti Zannini, 211. The dedicatory page reads: "Ser.mo et (. . . mo) Principe [Philip II]. Essendomi risoluto di far stampare la vita, et li miracoli di S. Benedetto uno de maggiori Santi di Dio, ho voluto farla tradurre di Latinio in lingua spagnuola per beneficio publico, et per commodità de la Spagna, nella quale ha fiorito, et fiorisce la Religione di questo Santo Padre. Et non havendo a fin hora havuto altra occasione di mostrare la reverenza, et divota osservanza che devo alla persona di VA Serenissima ho voluto dedicarle la presente opera arrichita de la vita del Santo et del ritratto di esso non mai più stampato: perche trattandosi di uno de maggiori Santi de la Chiesa con veniva che la mandassi sotto la protettione di VA Serenissima sendo il maggior Principe del mondo, et figliuolo del maggior difensore de la Religion nostra. Degni si dunque d'accetarla con un minimo cenno de la grandezza de la'animo suo; che con maggiore occasione cercarò di mostrarle la infinita devotione che le porto. Dio la conservi, et le conceda lunghissimi anni di vita,

et vera felicita. Di Roma il di primo febraro – 1597. D.S.A. Serenissima / Humilissimo et devotissimo servitore / Paulino Arnolfini Luchese."

51 Angelo Sangrino, *Speculum Exemplar Christicolarum Vita / Beatissimi Patris Benedicti Monachorum Patriarchae Sanctissimi*, Rome: Bartolomeo Bonfadino, 1586 and 1587. Brunet, vol., col. V 126. On Sangrino (Angelo de Faggiis de Castro Sangri), see Barry Collett, *Italian Benedictine Scholars and the Reformation: The Congregation of Santa Giustina of Padua*, Oxford and New York: Clarendon Press, 1985, 90 and 253. The Sangrino edition is discussed and partially illustrated in Olivieri, "Il corvo e il pane," 137–8. It measures 170 × 115 mm, as compared with the original edition, which is 332 × 210 mm. It seems to have been printed both with and without the extended textual commentary, although Sangrino's name is not always attached to the version without the commentary. The caption on the wandering monk plate reads: "Daemon ab aede sacra puerum trahit, hunc pater ictu / Percutieris virgae, cogit adesse choro. / Ist chorus arx dilecta Deo, protecta supernis / Coelitibus, sathane valde odiosa domus." See also Guido Arbizzoni, "Immagini per le vite dei Santi," in Ardissino, 83–113, especially 89–92.

52 Loren Partridge, "Discourse on Asceticism in Bertoja's Room of Penitence in the Villa Farnese at Caprarola," *Memoirs of the American Academy in Rome* 40 (1995): 145–74 (159). For Odoardo Farnese, see Clare Robertson, *The Invention of Annibale Carracci*, Milan: Silvana, 2008, 113–41; and C. Robertson, "Farnese, Odoardo," in *DBI*, vol. 45, 112–19. Also, the *Vita Beatissimi Patris Benedicti Monachorum Patriarchae Sanctissimi* ab Andrea Vaccario / Successor, Roma 1611, is a reduced copy of the Passeri images, either by Folcari or Philippe Thomassin.

53 Collett, *Italian Benedictine Scholars*, 157. See 157–212 for the Benedictine arguments at the Council of Trent.

54 Collett, *Italian Benedictine Scholars*, 208 and elsewhere. Christopher L. C. E. Witcombe, Copyright in the Renaissance, Leiden: Brill, 2004, 188, says that this was printed in Rome for the Spanish Benedictine congregation at S. Benedetto di Val-

ladolid under Cristobal de Aguero but does not give the source of that information. The 1993 English translation (*Saint Benedict, Life and Miracles*, ed. Rev. Bede Peay, trans. Mary Jean Lutz-Bujdos, Subiaco: Errebigrafica, 1993) attributes the Latin captions of the 1579 edition to the Spanish Benedictine Antonio Suarez. Jesús María González de Zarate, "Aportaciones del Coro Alto de San Benito de Valladolid a la Iconografia de San Benito," *Boletín del Seminario de Estudios de Arte y Arqueología* 52 (1986): 357–68, attributes the publication of the Roman volume to Fray Juan de Guzmán, the Procurator General in Rome of the Spanish congregation of Benedictines in Valladolid (citing Ernesto Zaragoza, Introduction to *Vita et Miracula Sanctiis Patris Benedicti*, Zamora, Ed. Monte Casino 1980). In the absence of citations to any further source for these attributions, they have to be treated as suppositions at this point. The Spanish Cardinal Pedro de Deza (1520–1600) received his cardinal's hat in 1578, the year before the first edition of the *Vita* was printed. He had served as chancery judge at Valladolid and in 1578 became the chancery president; Richard Kagan, *Lawsuits and Litigants in Castile 1500–1700*, Chapel Hill: University of North Carolina Press, 1981, 173 (chapter 5: The Chancillería of Valladolid). Deza would be a possible candidate for helping to bring this volume to press. See also David Freedberg, *The Power of Images: Studies in the History and Theory of Response*, Chicago: University of Chicago Press, 1989, 178–88 for Nadal and meditative response to printed images.

55   See Elizabeth A. Clark and David Brakke, following and enlarging on Foucault's notion of commentary as a way of representing texts to support the culture and concerns of their own communities, in which renunciation was "exalted as the pinnacle of Christian achievement" (Clark, *Reading Renunciation*, 8–10). Sexual renunciation as a technology of the self was designed to control the self-knowledge that is sexually produced (Brakke, "Ethiopian Demons," 506). See also Clark, *Reading Renunciation*, 59–61, on the usefulness of hagiography in teaching saintly (monastic) practice.

56   RBI, 2–3: "a Obsculta, o fili, praecepta magistri, et inclina aurem cordis tui, et admonitionem pii patris libenter excipe et efficaciter comple, ut ad eum per oboedientiae laborem redeas, a quo per inoboedientiae desidiam recesseras. Ad te ergo nunc mihi sermo dirigitur, quisquis abrenuntians propriis voluntatibus, Domino Christo vero regi militaturus, oboedientiae fortissima atque praeclara arma sumis."

3   Camillo Agrippa's Cosmology of Knowledge

1   Camillo Agrippa, *Trattato di scientia d'arme, con un dialogo di folosofia di Camillo Agrippa Milanese*, Rome: Antonio Blado, 1553. For Agrippa (Milan *c.*1520 – Rome 1600) see G. L. Barni, "Agrippa, Camillo," in *DBI*, vol. 1, 503; Elio Nenci, "Camillo Agrippa: Un ingegnere rinascimentale di fronte ai problemi della filosofia naturale," *Physis* 29 (1992): 71–119. There is a longer bibliography on the fencing treatise, much of it looking at it in the context of the development of the sport, but see Carlo Bascetta, *Sporti e giuochi: Trattati e scritti dal xv al xvi secolo*, Milan: Edizioni il Polifilo, 1978, 185–208, and most recently: Camillo Agrippa, *Fencing: A Renaissance Treatise*, trans. and ed. Ken Mondschein, rev. ed., New York: Italica Press, 2013.

2   On the semiotic qualities of an author portrait, see Michael Camille, review of Hans Beltung, *Bild und Kulte: Eine Geschichte des Bildes vor dem Zeitalter der Kunst*, *The Art Bulletin* 74, no. 3 (September 1992): 517; and for a continuing tradition from antiquity to the late Middle Ages of author portraits as visual proof of an author's intelligence and erudition: Eve R. Hoffman, "The Author Portrait in Thirteenth-Century Arabic Manuscripts: A New Islamic Context for a Late-Antique Tradition," *Muqumas* 10 (1993): 6–20, especially 14.

3   Giovanni Paolo Lomazzo, *Rime . . . divise in sette libri*, Milan: Paolo Gottardo Pontio, 1587, book IV (*Dove si contengono varie dimostrationi, essempi, historie, riprensioni & altre fantasie dichiarate sotto metafora . . .*), fol. 230: "Et le figure, che già Carlo

Urbino / Fè nel trattato di Camillo Agrippa / Mostrarono il schermir à una fantasma." See Giulio Bora, "Note Cremonesi, II: L'eredità di Camillo e i Campi," *Paragone Arte* 28, no. 327 (May 1997): 55–88.

4 See Pamela O. Long, *Openness, Secrecy, Authorship: Technical Arts and the Culture of Knowledge from Antiquity to the Renaissance*, Baltimore and London: Johns Hopkins University Press, 2001, especially 245–50.

5 I have only used the word "science" where Agrippa used "scientia" when he seems to have meant the spectrum of mathematically based knowledge with which he was familiar, including mathematics, geometry, optics, astronomy, and engineering. While it is anachronistic to use the word science here as it is used now, Agrippa used the word "scientia" quite purposefully. I like Deborah Harkness's explanation of her use of the word in the Introduction to her book, *The Jewel House*, New Haven and London: Yale University Press, 2007, xv–xvii, and have tried to stay close to that use of "science" and "scientific" in this book when speaking of early modern practice.

6 For Agrippa's practice engineering see Nenci, "Camillo Agrippa," 71–5; and Alessandro Birali and Paolo Morachiello, *Immagini dell'ingegnere tra Quattro e Settecento: Filosofo, soldato, politecnico*, Milan: Franco Angeli, 1985.

7 Camillo Agrippa, *Trattato di Camillo Agrippa Milanese di trasportar la guglia in su la piazza di San Pietro*, Rome: Francesco Zannetti, 1583, fol. 5r; see note 27 below. The letter by Claudio Tolomei attesting to his fame of 26 July 1543 is quoted in Elisabeth MacDougall, *Fons Sapientiae: Renaissance Garden Fountains*, Washington, DC: Dumbarton Oaks, 1978, 12: "Ma molto più lodai M. Agrippa, il quale oltra a tante altri benefizj fatti al Popolo Romano, e dopo gli aquedotti rifatti e riedificati dell'Appia, dell'Aniene, e della Morziá, già guasta e caduti, egli ancora condusse questa aqua vergine in Roma: la qual sola di tutte l'altre acque è rimasa ancor viva, e viene a Roma e sovviene a molti bisogni, e fa nobili que' giardini, che le son d'appresso, benche ancora ella sente i morsi della vecchiezza, e del tempo; e buona parte se n'è già perduta, la qual potrebbe con l'industria, e dili-

genza degli uomini agevolmente riguardarsi."

8 For the mechanics of Agrippa's hydraulic system as well as its history, see Leonardo Lombardi, "Camillo Agrippa's Hydraulic Inventions on the Pincian Hill (1574–1578)," *Waters of Rome Occasional Journal* 5 (2008): 1–10. See also Paula Hoffmann, *Il Monte Pincio e la Casina Valadier*, Rome: Edizioni del Mondo, 1967, 473, who transcribes the verses, now lost, documented during the papacy of Sixtus V as inscribed on a fountain at the villa, preserved in the Vatican Codice Barberiniano XXX, 89, fol. 537v: "Loco de Medici, prima di Montepulciano cardinale. In facciata d'una fontana, in capo del giardino, che ne cadeva in sontuosa pila di marmo bianco, piena di figure più che di mezzo rilievo, ancora dov'era sacrificio d'un toro et altre attioni diversi: 'Vergine acquam duxit tantum Ma(uo)rtis in agrum / (Marcus) Agrippa, et opus dicitur egregium. / At collis in pincii verticem Camillus Agrippa / Extulit, ingenium cernitur eximium."

9 Giovanni Pietro Rossini, *Il Mercurio Errante* . . . , Rome: Zenobj, 1715, 406. See also David R. Coffin, *The Villa in the Life of Renaissance Rome*, Princeton: Princeton University Press, 1997, 229–30.

10 Achille Bertolotti, *Artisti lombardi a Roma nei secoli XV, XVI, e XVII*, vol. 2, Milan: U. Hoepli, 1881, 68, citing R. Signaturarum Pio V, fol. 29.

11 Camillo Agrippa, *Nuove invenzioni di Camillo Agrippa Milanese sopra il modo di navigare*, Rome: Domenico Gigliotti, 1595. A copy of this book at the John Carter Brown Library, Providence, R.I. (H595.A279n), bears an inscription in pencil noting that it was purchased in Rome in 1599 for "baiochi dodici." One baiocco was equal to 1/100 of a silver scudo.

12 Agrippa, *Nuove inventioni*, fol. 40: "C: Voi sapere ch'io ho composto dodici libri, ciascuno de'quali ho intitolato col nome de'mesi, havendo intentione, che tutta l'opera insieme si chiami l'anno dell'Agrippa, si che il primo sia intitolato Gennaro, il secondo Febraro, il terzo Marzo, & cosi gli altri di mano in mano secondo l'ordine de'Mesi. Hora di questo dodici ne ho solo mandati tre in luce, cioè, Gennaro, Febraro, & Marzo, si che questo viene ad esser il quarto chiamato Aprile, & cosi credo habbiate inteso l'ordine."

13   Mario Biagioli, "The Social Status of Italian Mathematicians 1450–1600," *History of Science* 27, no. 1 (1989): 41–95. Galileo was expected to teach the casting of genitures, and made extra money casting them for others while he was at Pisa; see J. H. Heilbron, *Galileo*, Oxford and New York: Oxford University Press, 2010, 46 and 90.

14   Brian Richardson, *Printing, Writing and Readers in Renaissance Italy*, Cambridge and New York: Cambridge University Press, 1999, 133, dates the concept and first use of the word to the publisher Gabriel Giolito in Venice.

15   On the *libro unitario* in antiquity see Armando Petrucci, *Writers and Readers in Medieval Italy*, New Haven and London, Yale University Press, 1995, p. 1.

16   LeBlanc 112, Johan C. J. Bierens de Haan, *L'oeuvre gravé de Cornelis Cort, graveur hollandais, 1533–1578*, La Haye: M. Nijhoff, 1948, cat. 59, 79–80 59 after Muziano or Federico Zuccaro. State II is inscribed C. Cort. fe. above the date 1568. The image appeared printed in reverse published by Sadeler (1550–1600) with the caption: "Domine salvum me fac. Modicae fidei, quare dubitasti. Math.14," pictured here (New Hollstein 52, copy d: F. W. H. Hollstein, *The New Hollstein: Dutch and Flemish Etchings, Engravings and Woodcuts 1450–1700*, Amsterdam: Rijksmuseum, 1993). The Graffico copy of the central figures bears the line from Matthew 14 in a caption below, and is inscribed "Camillo grafico fe. Roma," without mention of Cort or any other inventor. A copy of Agrippa's book in the Vatican library (BAV Cicognara IV, 1552) does not contain the Graffico copy common to the rest of the edition, but instead the Cort print itself is tipped in and folded to the size of the other pages. For Graffico and his relation to Agrippa and his family, see Evelyn Lincoln, "Camillo Graffico, Printmaker and Fountaineer at the Villa Farnesina," *Print Quarterly*, 29, no. 3 (September 2012): 259–80.

17   Agrippa, *Nuove invenzioni*, fol. 3r: Ma dirò solo, benche in questa opera mia si tratti della Navigatione per Mare, esserne un'altra per Terra, nella quale non si truova manco fortuna, tempesta, naufragij nel praticar con gli huomini di mala mente, che sia nel Mare tra Scilla, & Cariddi a mezza notte il verno. Et però m'è parso di dedicarla a V.S. Illustrissima, accioche vegga quel che vi sarà a proposito per pelegrinar questo tempestoso mare di tribulatione, & per compir questa navigatione vitale, quale soggiace a tanti pericoli. Contra si fatti pericoli nelle navigationi ordinarie è assai profittevola la Calamita. Ma nella navigatione terrestre tanto turbulenta, quanto si sà, è assai più profittevole rimirar [3v] sempre Iddio, qual non ha bisogno di aiuto, si come ha la Calamita. Perche da lui procede ogni bene, & per lui si può entrar nel vero porto della salute. Si che sarà dunque bene, che questa Opera vada in luce per dar alli professori della navigatione queste inventioni mie, quali saranno, come io spero, molto giovevole, & piaceranno a quelli, che si dilettano di tali intelligentie. Di V. S. Ill[ustrissi]ma. & Rev[erendissi]ma / Humi[llissi]mo. Ser[vito]re / Camillo Agrippa."

18   On the role of an expert, see Edward Said, "Opponents, Audiences, Constituencies, and Community," in Evan Selinger and Robert P. Crease (eds.), *The Philosophy of Expertise*, New York: Columbia University Press, 2006, 370–94 (379). For noble and gentleman practitioners in the early modern period and further bibliography, see Simon Pepper, "Artisans, Architects and Aristocrats: Professionalism and Renaissance Military Engineering," in David J. B. Trim (ed.), *The Chivalric Ethos and the Development of Military Professionalism*, Leiden and Boston: Brill, 2003, 117–48 (especially 121–3).

19   Eric H. Ash, *Power, Knowledge and Expertise in Elizabethan England*, Baltimore and London: Johns Hopkins University Press, 2004, 8–17.

20   Biagioli, "Social Status," 44–51. See also R. Ago, *Gusto for Things*, tr. B. Bouley, C. Tazzara and P. Findlen, Chicago: University of Chicago Press, 2013, 174–8.

21   Pepper, "Artisans," 123–4, on the soldier's attitude to military architecture in the early sixteenth century: "Architects, together with other civilian *dottori*, [. . .] were to keep out of a subject of which they knew nothing and could learn nothing from their classical libraries because 'Books don't fight.'"

22   Nenci, "Camillo Agrippa."

23 Frangipane is Agrippa's interlocutor in *Dialogo . . . del modo di mettere in battaglia . . .* (1585).

24 See Josephine von Henneberg, *L'oratorio dell'Arciconfraternita del Santissimo Crocifisso di San Marcello*, Rome: Bulzoni, 1974, 107. For the bust of Frangipane, see Rudolph Wittkower, *Art and Architecture in Italy, 1600 to 1750*, Harmondsworth: Penguin, 1958, 173–7.

25 The fencing manual is still known to fencers, and thanks to Rachel Hooper for pointing out that Agrippa's methods were invoked by William Goldman in *The Princess Bride*, New York: Ballantine, 1973, 110, 117, as well as in the movie version of the same book. For an explanation of the book in the context of contemporary fencing, see Mondschein in Agrippa, *Fencing*.

26 Agrippa, *Trattato di trasportar*. See most recently Brian A. Curran, A. Grafton, P. O. Long, and B. Weiss, *Obelisk: A History*, Cambridge, Mass.: Burndy Library, 2009, 107–9, with further bibliography.

27 Agrippa, *Trattato di trasportar*, fol. 5: "Alla venuta mia in Roma, che fu all 26 d'Ottobre nel 1535, io sentiva raggionare di portar la guglia sicuramente in su la piazza di S. Pietro, et erano all'hora in predicamento per conto di questa impresa Antonio Sangallo degnissimo huomo, et il gran Michel Angelo Bonarota, & infiniti altri."

28 Curran *et al.*, *Obelisk*, is a thorough study of the history and meaning of the Roman obelisks from antiquity through the early modern period; for the state of the Roman obelisks in the Renaissance, see 61–7.

29 Michele Mercati's treatise is *De gli obelischi di Roma*, Rome: Domenico Basa, 1589. For Mercati, see E. Andretta, "Mercati, Michele," in *DBI*, vol. 73, 606–11.

30 See Claudio de Dominicus, *Membri del Senato della Roma Pontificia: Senatori, Conservatori, Caporioni e loro Priori e Lista d'oro delle famiglie dirigenti (secc. X–XIX)*, Rome: Fondazione Marco Besso, 2009. Both men were from the *rione* Colonna; Fabrizio is listed as serving in 1580 (139), Agapito in 1584 (141), where he is noted as *priore* of the *caporioni*. The senators, conservators, *caporioni*, and *priori* of the Campidoglio represented the *ceti dirigenti* of Rome at the time and established families in this class in the future. The name seems to have been Milanese; there was a painter Ambrogio di Stefano da Fossano (Ambrogio Bergognone, d. Milan, 1524) active in the circle of Leonardo da Vinci and the Milanese court at the end of the fifteenth century. See Janice Shell, "Bergognone, Ambrogio," Grove Art Online. Oxford Art Online, http://www.oxfordartonline.com/subscriber/article/grove/art/T008121, accessed 20 August 2011. The address of two brothers, Pier Antonio and Serafino da Fossano, from a noble Milanese family, appears inserted among the collected miscellaneous notes left by Leonardo da Vinci, on casting a bronze sculpture of a horse; see Jean Paul Richter, *The Notebooks of Leonardo da Vinci*, New York: Dover, 1970, vol. 2, 427, n. 1413.

31 Domenico Fontana, *Della trasportatione dell'obelisco vaticano*, Rome: Domenico Basa, 1590, with engravings by Natale Bonifacio after images by Fontana. See Curran *et al.*, *Obelisk*, esp. 109–38; Paolo Portoghesi and Adriano Carugo, *Della trasportatione dell'obelisco vaticano, 1590*, Milan: Il Polifilo, 1978; Bern Dibner, *Moving the Obelisks*, Cambridge, Mass.: Burndy Library, 1991.

32 Camillo Agrippa, *Dialogo . . . sopra la generatione de venti, baleni, tuoni, fiumi, laghi, valli, & montagne*, Rome: Bartolomeo Bonfadino and Tito Diani, 1584.

33 David C. Lindberg, "Science and the Early Church," in *God and Nature: Historical Essays on the Encounter between Christianity and Science*, Berkeley: University of California Press, 1986, 32–58 (18–48).

34 Ann Blair, "Mosaic Physics and the Search for a Pious Natural Philosophy in the Late Renaissance," *Isis* 91, no. 1 (March 2000): 33–4; and John Monfasani, "Aristotelians, Platonists and the Missing Ockhamists: Philosophical Liberty in Pre-Reformation Italy," *Renaissance Quarterly* 46, no. 2 (1993): 247–76.

35 See P. Portone, "Este, Luigi d'," in *DBI*, vol. 43, 383–90.

36 By 1603, Neroni was the court cosmographer for Cosimo II de' Medici in Florence; see M. Biagioli, "Social Status," 49; and "Galileo the Emblem Maker," *Isis* 81, no. 2 (June 1990): 231 n. 3. Biagioli

finds him listed under "Architettori, pittori e altri manifattori" rather than grouped with engineers or mathematicians in court rosters; see his *Galileo, Courtier*, Chicago: University of Chicago Press, 1993, 159. For Neroni at Tivoli, see David Coffin, *The Villa d'Este at Tivoli*, Princeton: Princeton University Press, 1960; for Neroni as a cartographer of fortresses, see Daniela Lamberini, "Collezionismo e patronato dei Medici a Firenze nell'opera di Matteo Neroni, 'cosmografo di granduca'," in P. Carpeggiani and L. Patetta (eds.), *Il disegno di architettura*, Milan: Guerrini e Associati, 1989, 33–8. Amelio Fara, *Il sistema e la città: Architettura fortificata dell'Europa moderna dai trattai alle realizzazioni 1464–1794*, Genoa: SAGEP, 1989, says that it was in Rome that he first made maps for the then Cardinal Ferdinando. Lamberini's *Il mondo di Matteo Neroni, cosmografico medico*, Florence: EDIFIR, 2013, had not yet appeared when this book went to press.

37 For the Tipografia Orientale Medicea, see Sara Fani and Margherita Farina eds., *Le vie delle lettere: la Tipografia Medicea tra Roma e l'Oriente*, Florence: Mondragon, 2012; Guglielmo Enrico Saltini, "Della Stamperia Orientale Medicea e Giovan Battista Raimondi," *Giornale Storico degli Archivi Toscani* 4 (Oct.–Dec. 1860): 237–308; Alberto Tinto, *La Tipografia Orientale Medicea*, Lucca: Maria Pacini Fazzi, 1987; and Robert Jones, "The Medici Oriental Press (Rome 1584–1614) and the Impact of its Arabic Publications on Northern Europe," in G. A. Russell, (ed.), *The "Arabick" Interest of the Natural Philosophers in Seventeenth-Century England*, Leiden: Brill, 1994, 88–108.

38 For Raimondi, see G. J. Toomer, *Eastern Wisedome and Learning: The Study of Arabic in Seventeenth-Century England*, Oxford: Clarendon Press, 1996, especially 22–5; Jones, "Medici Oriental Press," 88–108; Tinto, *La Tipografia*.

39 For the literary style of Baldi's *Vite*, see G. Federici Vescovini, "Les *Vite di matematici Arabi* de Bernardino Baldi" in L. Nauta and A. Vanderjagt (eds.), *Between Demonstration and Imagination*, Leiden: Brill, 1999, 395–408. See also Alfredo Serrai, *Bernardino Baldi: La vita, le opere, la biblioteca*, Milan: Edizioni Sylvestre Bonnard, 2002; and Elio Nenci, *Bernardino Baldi (1553–1617), studioso rina-*

scimentale: Poesia, storia, linguistica, meccanica, architettura, Milan: Franco Angeli, 2005.

40 The trial is reported from the archival records in Achille Bertolotti, "Le tipografie orientali e gli orientalisti a Roma," *Rivista evropea* 9:2 (1878), 19–20. The Avicenna was the abridgment of the canon of medicine: *Kutub 'al-qānūn fī 'al-ṭibb quibus addidi sunt in fine eiusdem libri Logicæ, Physicæ, & Metaphysicæ, Arabice nunc primum impressi*, Rome: Tipographia medicea, 1593. The first Arabic grammar was *Kāfiyah li-Ibn al-Ḥājib. Grammatica Arabica dicta Caphiah, autore filio Alhagiabi* (1592); the second was probably the *Alphabetum Arabicum* (1593). The last was *De geographia universali hortus cultissimus, mire orbis regiones, provincias . . . , earumque dimensiones et orisonta describens* (1592). See Toomer, *Eastern Wisedome and Learning*, 23.

41 Bertolotti, *Le tipografie*, 9; deposition of Nicandro Filippini di Fondi: "detto Matteo soleva usare il tocca lapis che io gli l'ho visto in man in una penna di ottone che con esso ho visto ci disegnava le pitture, perchè si diletta anco dipingere et fare li globi."

42 Lombardi, "Camillo Agrippa's Hydraulic Inventions," 2–7.

43 Camillo Agrippa, *Dialogo di Camillo Agrippa Milanese del modo di mettere in battaglia presto & con facilità il popolo di qual si voglia luogo con ordinanze & battaglie diverse*, Rome: Bartolomeo Bonfadino, 1585. Bonfadino's press brought out a second edition in 1635 dedicated to Muzio Frangipane.

44 Camillo Agrippa, *La virtù, dialogo di Camillo Agrippa Milanese sopra la dichiarazione de la causa de' Moti, tolti de la parole scritte nel Dialogo de' Venti*, Rome: Stefano Paolini, 1598.

45 Stefano Paolini cast type for Oriental languages at the Medici press until he founded his own collegio working for the Propaganda Fide in the seventeenth century. See Bertolotti, *Le tipografie*; and Toomer, *Eastern Wisdome and Learning*, 24. See also Willi Henkel, *Die Druckerei der Propaganda Fide: Eine Dokumentation*, Munich: Schöningh, 1977; and Barbara Tellini Santoni and Alberto Mondadori, *Libri e cultura nella Roma di Borromini*, Rome: Retablo, 2000, 80.

46 For hydraulic projects regarding the Po, see

Cesare Maffioli, *La Via delle Acque (1500–1700): Appropriazione delle arti e trasformazione delle matematiche*, Florence: Olschki, 2010; and A. Fiocca, D. Lamberini, and C. Maffioli, *Arte e scienza delle Acque nel Rinascimento*, Venice: Marsilio, 2006.

47 Pamela Long noticed that Fontana's obelisk design actually did include some components of Agrippa's idea, most notably the sheathing of the obelisk. See Curran et al, *Obelisk*, 209.

48 Cornelis Meyer, *L'arte di restituire a Roma la tralasciata navigatione del suo Tevere*, Rome: Lazzari Varese, 1685.

49 See Luisa M. Dolza, "Reframing the Language of Inventions: The First Theatre of Machines," in Wolfgang Lefèvre, J. Renn, and U. Schoepflin (eds.), *The Power of Images in Early Modern Science*, Basel and Boston: Birkhäuser, 2003, 89–207; and Marcus Popplow, "Why Draw Pictures of Machines? The Social Contexts of Early Modern Machine Drawings," in Wolfgang Lefevre (ed.), *Picturing Machines*, Cambridge, Mass.: MIT Press, 2004, 17–50.

50 Agrippa, *Nuove invenzioni*, fol. 40: "A: Perche non ci havete messo il trattato dell'arme, & quello della sfera? C: Perche, quando io feci quelli, io non haveva questo pensiero. A: Havetene altri delli libri? C: Sig. si, parte fatti, & parte nell'intelletto. Sopra il che non mi bisogna tentar piu oltre; perche col tempo si diranno."

51 After a very brief discussion of justice, Agrippa continues: "Ma per non esser' questo il proposito mio vengo al secondo capo, cioé all'intelligentia del'arme: ne la quale consiste la vita e la vittoria di chi l'usa." Agrippa, *Trattato di scientia d'arme* (1553), proemio, Prima Parte, fol. 1r.

52 See Steven Shapin, "'A Scholar and a Gentleman': The Problematic Identity of the Scientific Practitioner in Early Modern England," *History of Science* 29 (1991): 279–327, and his *A Social History of Truth: Civility and Science in Seventeenth-Century England*, Chicago: University of Chicago Press, 1994, especially the Introduction, and chapters 2 and 8.

53 Shapin, *Social History of Truth*, 107–14.

54 Giorgio Agrippa, bombardier, died in Florence in 1596, his son, Giacomo Agrippa, claiming his inheritance in Rome (ASR, 30 Not. Cap. Petrus

Paulus Stella, vol. 70, c. 158, 27 June 1615). The scant archival records are complicated by the existence of a nephew of Camillo with the same name who lived with him towards the end of his life, and who may have been a son of Giorgio Agrippa. Records in the Archivio del Vicariato, *Stati d'anime, S. Lorenzo in Damaso*, 1599, fol. 49r, register Camillo Agrippa, age eighty, living with M. Vincenzo Aiutoli, a priest; records of the same office from 1595, fol. 42r, show him living with his nephew Camillo, at which time he is seventy-seven years old. See also Nenci, "Camillo Agrippa," 80, for a house in Via Pellegrini being ceded to Agrippa for his lifetime by the Maffei family in 1584.

55 Agrippa, *Trattato di scientia d'arme* (1553), [fols. iir–v].

56 Paul Grendler, *The Universities of the Italian Renaissance*, Baltimore: Johns Hopkins University Press, 2002, 73–82; Antonio Ricci, "Lorenzo Torrentino and the Cultural Programme of Cosimo I de' Medici," in K. Eisenbichler (ed.), *The Cultural Politics of Duke Cosimo I de' Medici*, Aldershot: Ashgate, 2001, 103–19; and Michel Plaisance, "Culture et politique à Florence de 1542 à 1551," in A. Rochon (ed.), *Les écrivains et le pouvoir en Italie à l'époque de la Renaissance*, Paris: Université de la Sorbonne Nouvelle, 1973, 148–242.

57 Matthew Landrus, *Leonardo da Vinci's Giant Crossbow*, Berlin: Springer-Verlag, 2010, 17.

58 Camillo Agrippa, *Trattato di scienza d'arme et un dialogo in detta materia*, Venice: Antonio Pinargenti, 1568, dedicated to Don Giovanni Manriche (Camerieri di S. M. Cesarea) and with a preface and illustrations, copied from Agrippa's book in reduced form and printed several to a page, by Giulio Fontana. The book also appeared in an edition of 1604 (Venice: Roberto Meglietti).

59 Achille Marozzo, *Opera nuova chiamato duello, o vero, fiore dell'armi de singulari abattimento, offensivi, & diffensivi, composta per Achilee Marozzo gladiatore Bolognese . . .* , Modena: Antonio Bergola, 1536. See Agrippa, *Fencing*, xvi–xvii.

60 See the illustration of *Vein Man* fig. 85.

61 Herbert S. Matsen, "Students' 'Arts' Disputations at Bologna around 1500," *Renaissance Quarterly* 47, no. 3 (Autumn 1994): 533–55.

62 For an examination of the culture of mathematics in Renaissance Milan and its practical applications through the case of one practitioner, see Alexander Marr, *Between Raphael and Galileo: Mutio Oddi and the Mathematical Culture of Late Renaissance Italy*, Chicago: University of Chicago Press, 2011.

63 Cesare Cesariano, *Di Lucio Vitruuio Pollione De architectura libri dece, traducti de latino in vulgare affigurati, comentati, & con virando ordine insigniti: per il quale facilmente potrai trouare la multitudine de li abstrusi & reconditi vocabuli a li soi loci & in epsa tabula con summo studio expositi & enucleati ad immensa utilitate de ciascuno studioso & beniuolo di epsa opera . . .*, Como: Gottardo da Ponte for Agostino Gallo and Aloisio Pirovano, 15 July 1521. The book contained pictures of waterwheels and other machines as well as figures and shapes. See Samuel Edgerton, *The Heritage of Giotto's Geometry: Art and Science on the Eve of the Scientific Revolution*, Ithaca: Cornell University Press, 1991, 170.

64 Luca Pacioli, *De divina proportione*, Venice: Paganius, 1509. See also J. V. Field, "Piero della Francesca's Mathematics," in Jeryldene M. Wood (ed.), *The Cambridge Companion to Piero della Francesca*, Cambridge and NY: Cambridge University Press 2003, 155–7, for Piero's illustrations of Archimedean polyhedrons in spheres.

65 For the circulation of Leonardo's manuscripts, see Carlo Pedretti, Introduction, in Carlo Pedretti (ed.), *Libro di pittura: Codice Urbinate lat. 1270 nella Biblioteca Apostolica Vaticana*, 2 vols., Florence: Giunti, 1995, 11–82.

66 Carlo Pedretti and Sergio Marinelli, "The Author of the Codex Huygens," *JWCI*, 44 (1981): 214–20; and Ugo Ruggeri, "Carlo Urbini e il Codice Huygens," in *Critica d'arte* 11, (1978): 167–76. For the artist's work and known biography, see Giuseppe Crillo, *Carlo Urbino da Crema, disegni e dipinti*, Parma: Grafiche STEP Editrice, 2005, 16–19 for work with Agrippa, p. 18 for sojourn in Rome, pp. 80–4 for the Codex Huygens.

67 Claire Farago, "The Defense of Art and the Art of Defense," *Achademia Leonardi Vinci* 10 (1997): 13–22 (14–15).

68 Pamela O. Long, "Power, Patronage and the Authorship of *ars*: From Mechanical Know-How to Mechanical Knowledge in the Last Scribal Age," *Isis* 88, no. 1 (1997): 1–41.

69 Agrippa, *Trattato di scientia d'arme* (1553), cfol.Iv, emphasis mine.

70 Agrippa, Prima Parte: De la seconda guardia signata per E. cap. IX, fols. XVIv–XVIIr: ". . . Ritrovandosi uno pur nella Terza larga di passo tanto luntano dal nemico che non potesse esser toccato da lui, caso che tentasse premere per forza la sua spada, ritirarebbe la mano à dietro in Seconda come in questa figura, che pur tiene la spada in mano in Scurcio, benche non appaia per esser'in prospettiva, & sta in passo largo, come di sopra."

71 Agrippa, *Trattato di scientia d'arme* (1553), Prima Parte, fol. XIX: "come si vede ne la sequente figura, con le tante linee tirate in schena, da li doi punti de li occhi, segnata cosi, per dar' à conoscere, che li occhi benche siano doi, non però ponno vedere piu d'un punto per volta, non potendo, naturalmente andar le'linee loro, à Paralela, ma à Piramide, à finire in un punto solo."

72 Martin Kemp, "Leonardo and the Visual Pyramid," *JWCI* 40 (1977): 128–49. On Renaissance optical theory, see also Martin Kemp, *The Science of Art: Optical Themes in Western Art from Brunelleschi to Seurat*, New Haven and London: Yale University Press, 1990; David Lindbergh, *Theories of Vision from al-Kindi to Kepler*, Chicago: University of Chicago Press, 1976.

73 Agrippa, *Trattato di scientia d'arme* (1553), Parte Seconda, Cap. 1, fol. XXXIXr.

74 Giorgio Vasari, *Le vite de' più eccellenti pittori scultori e architettori: nelle redazioni del 1550 e 1568*. Ed. R. Bettarini and P. Barocchi. Vol. 6, Florence, Sansoni, 1966–, 389–90. Vasari also sent a portion of his *Lives* to Caro for a response while they were in progress. Caro responded approvingly with a letter of 11 December 1547. See Annibale Caro, *Lettere famigliari*, ed. Aulo Greco, Florence: Olschki, 1957–61, vol. 2, 50–1; and Marco Ruffini, *Art Without an Author: Vasari's Lives and Michelangelo's Death*, New York: Fordham University Press, 2011, 141.

75 See Katie Scott, "Authorship, the Académie, and the Market in Early Modern France," *Oxford Art Journal* 21, no. 1 (1998): 27–41, for the idea that

in the ancien régime the idea of a published or publishing author was considered vulgar and pushy, "someone who does not speak naturally; who is very full of himself, in short, a person who is both singular and pretentious," 34.

76 Agrippa, *Trattato di scientia d'arme* (1553), fols. LXIIIr–v: "Di poi mi pareva con l'aiuto di molti genti'homini amici mei, et col mio che mi diffendevo: il che non penso voglia predire altro, se non che forse alcuni allevi di Euclide, o di Aristotile, vorranno imputarmi, di quel ch'io dico, & io col mio aiuto, & altri miei Patroni mi diffenderò: Si che in ogni modo voglio dicchiararle, per le var' via ogni mala impressione che potesse havere ogn'uno, chi vedesse quelle figure, & per mostrare al mondo se ben non ho studiato, che naturalmente posso parlar ancor'io di qualche cosa con ragione. & se volete vederne voi la prova, pigliate quel mio libro in mano, & ritrovate le figure, ch'adesso vi darò à conoscere per termini di littere, come si fanno: se però non vi annoia questa Theorica."

77 On anti-Aristotelianism in the sixteenth century, see Cesare Vasoli, "The Renaissance Concept of Philosophy," in C. B. Schmitt (gen. ed.), *The Cambridge History of Renaissance Philosophy*, Cambridge and New York: Cambridge University Press, 1988, 71.

78 Pamela O. Long, *Artisan/Practitioners and the Rise of the New Sciences 1400–1600*, Corvallis: Oregon State University Press, 2011, 30–2 and 127–9; Long, *Openness, Secrecy, Authorship*, 16–45 and 210–50. Also on craft and knowledge, see Pamela H. Smith, *The Body of the Artisan: Art and Experience in the Scientific Revolution*, Chicago: University of Chicago Press, 2006, especially the Introduction, 3–24; Richard Sennet, *The Craftsman*, New Haven and London: Yale University Press, 2008.

79 Long, *Openness, Secrecy, Authorship*, 16. See also Martin Jay, *Songs of Experience*, Berkeley: University of California Press, 2005, 13; and William A. Wallace, "The Culture of Science in Renaissance Thought," *History of Philosophy Quarterly* 3, no. 3 (July 1986): 281–91.

80 Shapin, *Social History of Truth*, 359–64.

81 Julia Annas, "Moral Knowledge as Practical Knowledge," in Evan Selinger and Robert P. Crease (eds.), *The Philosophy of Expertise*, New York: Columbia University Press, 2006, p. 286.

82 Jay, *Songs*, 17–18. See also Ash, *Power*, for the changing nature of the role of experience in the formation of expertise, especially 14.

83 Long, *Openness, Secrecy, Authorship*, 22 and 38–42.

84 Long, *Openness, Secrecy, Authorship*, 211.

85 Kenneth Keele, *Leonardo da Vinci's Elements of the Science of Man*, New York and London: Academic Press, 1983, 63, and 302–3.

86 Pedretti, *Libro di pittura*, vol. 2, cap. 750, 438; trans. from Richter, *Notebooks*, 843, 1213–15.

87 Nenci, "Camillo Agrippa," 82–9, discusses Agrippa's seemingly Copernican ideas about the earth having an independent center; the center of the earth in this dialogue is also mentioned in Agrippa, *Fencing*, 116–17.

88 For Barbagrigia, see Aulo Greco, *Annibal Caro, cultura e poesia*, Rome: Edizioni di storia e letteratura, 1950, 71–3.

89 Davide Stimilli called my attention to the medal, and translated and edited a comment on it by Aby Warburg: Davide Stimilli, "Le forze del destino riflesse nel simbolismo all'antica, 1924," in D. Stimilli (ed.), *Aby Warburg: La dialettica dell'immagine, Aut aut* 321/322 (May–August 2004), Milan: Il Saggiatore, 2004, 16–17. The medal is described in Lore Börner, *Die italienische Medallien der Renaissance und des Barock (1450 bis 1750)*, Berlin: Gebr. Mann Verlag, 1997, 116, cat. 464; J. Graham Pollard, *Italian Renaissance Medals in the Museo Nazionale of Bargello, vol. 3, 1513–1640*, Florence: Associazione Amici del Bargello, 1984–5, 1371, cat. 799. Philip Atwood, *Italian Medals c.1530–1600 in British Public Collections*, London: British Museum, 2003, 394, cat. 964, also provides a brief biography of Bonini, noting that he is documented in Rome from 1557 until 1585. For Caro and iconography, see Clare Robertson, "Annibale Caro as Iconographer," *JWCI* 45 (1982): 160–81.

90 Stimilli, "Le forze," 17.

91 Fritz Saxl, "Macrocosm and Microcosm in Mediaeval Pictures," *Lectures*, vol. 1, London: The Warburg Institute, 1957, 58–72 (68).

92 Saxl, "Macrocosm," 72.

93   Castelvetro wrote his response by 1555; it was printed in Parma in 1558. For the Caro–Castelvetro controversy, see Elisabetta Arcari, "Polemica Caro Castelvetro: Ragione del Castelvetro contro il Caro con postile autografe dell'autore," in Gino Belloni and Riccardo Drusi (eds.), *Vincenzo Borghini: Filologia e invenzione nella Firenze di Cosimo I*, Florence: Olschki, 2002, 315–18; and Karen Pinkus, *Picturing Silence*, Ann Arbor: University of Michigan Press, 1996, 123–28. Thanks to David Stimilli for pointing out the relevance of this to Agrippa's story.

94   The murdered poet was Alberigo Longo, a friend of Caro's, a deed for which Castelvetro was tried and acquitted. See John Addington Symonds, *Renaissance in Italy, Vol. 4: Italian Literature*, New York: Scribner, 1941, 249, and Valentina Gallo, "Alberigo Longo," in *DBI*, vol. 65: 686–87.

95   Gigliola Fragnito (ed.), *Church, Censorship and Culture in Early Modern Italy*, Cambridge: Cambridge University Press, 2001, 41: the expurgation was undertaken in 1600 "with such manifest flouting of the rules that the vicar decided that no further expurgations should be assigned to laymen."

96   Ettore Camasasca (ed.), *Lettere sull'arte di Pietro Aretino*, vol. 2, 1543–55, Milan: Edizione di Milione, 1957, 42–3. Caro, *Lettere famigliari*, Vol. 1, 149–50, Vol. 3, 259–63. Erica Tietze-Conrat, "Neglected Contemporary Sources Relating to Michelangelo and Titian," *The Art Bulletin* 25, no. 2 (June 1943): 155, characterizes Corvino as "a busybody always willing to offer his services, although conscious of never getting anywhere," (155). He was also the *maestro di casa* of Cardinal Giulio Ascanio Sforza, a Farnese nephew. See Charles Davis, "Ammanati, Michelangelo and the Tomb of Francesco del Nero," *Burlington Magazine* 118, no. 80 (1976): 472–84, who mentions Corvino as a collector on p. 479.

97   Filippo Titi, *Descrizione delle Pitture, Sculture e Architetture esposte in Roma*, Rome: Marco Pagliarini, 1763 (reprint, Florence, 1966): 159, also mentioned in Giuseppe Vasi, *Itinerario istruttivo di Roma*, Rome: Marco Pagliarini, 1763.

98   Agrippa, *Trattato di scientia d'arme* (1553), fol. LXX. Nenci, "Camillo Agrippa," 79, and Agrippa, *Fencing*, xx–xxi, also offer brief characterizations of this group of men.

99   For the collection of antique statues, see J. F. Orbaan, *Documenti sul barocco in Roma: Società romana di storia patria*, Rome, 1920, 356: and Jean Jacques Boissard, *Topographiae Urbis Romae*, Frankfurt, 1597, 76. For Ruffino as an advisor on sculpture, see Loren Partridge, "The Sala d'Ercole in the Villa Farnese at Caprarola, Part I," *The Art Bulletin* 53, no. 4 (December 1971): 467–86 (for Ruffino, 482, for Garimberto, 482 and 484); and L. Partridge, "The Farnese Circular Courtyard at Caprarola: God, Geopolitics, Geneology, and Gender," *The Art Bulletin* 83, no. 2 (June 2001): 259–93. For the association between Ruffino and Rafael Bombelli: S. A. Jayawardene, "Rafael Bombelli, Engineer-Architect: Some Unpublished Documents of the Apostolic Camera," *Isis* 56, no. 3 (Autumn 1965): 298–306; and S. A. Jayawardene, "The Influence of Practical Arithmetics on the Algebra of Rafael Bombelli," *Isis* 64, no. 4 (December 1973): 510–23.

100   Palazzo Gaddi was demolished in 1694 along with the parish church of San Biagio to make way for Carlo Fontana's additions to Bernini's Palazzo Montecitorio.

101   Clifford M. Brown and Anna Maria Lorenzoni, "Major and Minor Collections of Antiquities in Documents of the Later Sixteenth Century," *The Art Bulletin* 66, no. 3 (September 1984): 496–507; and Clifford M. Brown, *Our Accustomed Discourse on the Antique: Cesare Gonzaga and Gerolamo Garimberto, Two Renaissance Collectors of Greco-Roman Art*, New York: Garland, 1993.

102   Girolamo Garimberto, *Il capitano generale*, Venice: Giordano Zietti, 1556; G. Garimberto, *Della fortuna libri sei di Girolamo Garimberto*, Venice: Michele Tramezzino, 1547.

103   Alessandro Nova and Alessandro Cecchi, "Francesco Salviati e gli editori," in Catherine M. Goguel (ed.), *Francesco Salviati (1510–1563) o la Bella Maniera*, Milan: Electa, 1998, 66–74.

104   Philippe Costamagni, "Il ritrattista," in Goguel, *Salviati*, 47–52; and for Caro as intermediary for Salviati's commissions, see Patricia Rubin, "The Private Chapel of Cardinal Alessandro Farnese in the Cancelleria, Rome," *JWCI* 50 (1987): 82–112.

105 For Simone Verovio: Thomas W. Bridges, "Simone Verovio," in *The New Grove Dictionary of Music and Musicians,* 2nd ed., ed. Stanley Sadie, London and New York: Macmillan, 2001, vol. 26, 489–90. For van Buyten's work on handwriting manuals, see Christopher L. C. E. Witcombe, *Copyright in the Renaissance*, Leiden: Brill, 2004, 289–90; and A. F. Johnson, "A Catalogue of Italian Writing Books of the Sixteenth Century," *Signature*, new ser. 10 (1950): 22–48; Stanley Morison, *Early Italian Writing Books: Renaissance to Baroque*, Boston: Godine, 1990, 128–29.

106 Agrippa's name appears as an active voting member of the Compagnia di San Giuseppe between 1582 and 1586, Archivio della Pontefice Insigne Accademica Artistica dei Virtuosi al Pantheon, *Verbali del le Congregazioni* 1543–87. For Muziano's academy, see Melchior Missini, *Memorie per servire alla storia della Romana Accademia di S. Luca*, Rome: de Romanis, 1823, 19. See also Peter M. Lukehart (ed.), *The Accademia Seminars: The Accademia di San Luca in Rome, c. 1590–1635*, Washington, DC: National Gallery of Art, 2009; and documents and commentary at the website "The History of the Accademia di San Luca, 1590–1635: Documents from the Archivio di Stato di Roma," http://www.nga.gov/casva/accademia/intro.shtm. Accessed 31 March 2013.

107 Agrippa, *Nuove invenzioni*, fol. 4v: "Benigni Lettori, l'Authore di queste navigationi ha per intentione, che voi troviate alcune cose, che vi piaceranno: se però ve ne delettarete. Perche il diletto fa l'huomo piu diligente, & accorto nelle scientie, & nelle arti, le quali mettendo in prattica diventarà eccellente piu di fatti, che di parole. Si che Spiriti gentili vederete questa mia Opera con amore, perche l'Authore ve la da con desiderio, & speranza grande di sodisfarvi, sperando anchora, che voi l'augumentrate: perche non si da fine ne a scientiia, ne ad arte." The engraving is printed on a finer paper than that used for the text. Comparison with the images from the *Trattato di scientia d'arme* suggest that the same artist may have been responsible for the image; if so, then the plate may have been made for an earlier book planned by Agrippa but finally used for this one. The treatise on navigation may also have been written earlier than the publication date, as it contains an epigram by Petri Mallardi Santonis, who had died *c.*1573, as reported in the posthumous collection of his epigrams, *Petri Maillardi Santonis Epigrammatum libri duo . . .* , Turin: apud haeredes Nicolai Bevilaquae, 1576.

108 Henri Broise and Vincent Jolivet, "Pincio (Jardins de Lucullus)," *Mélanges de l'École française de Rome – Antiquité*, 110, no. 1 (1998): 492–5, where Agrippa is referred to as the architect of Ferdinand de' Medici.

109 For personifications of Nature in the sixteenth century, see Katharine Park, "Nature in the Person: Medieval and Renaissance Allegories and Emblems," in L. Daston and F. Vidal (eds.), *The Moral Authority of Nature*, Chicago: University of Chicago Press, 2004, 50–73, especially 57–9. For the popularity of the image in the sixteenth century, see Kathleen Wren Christian, "The De' Rossi Collection of Ancient Sculptures, Leo X, and Raphael," *JWCI* 65 (2002): 132–200, especially 167–76, who notes that one such statue, unknown today, was in the collection of Agrippa's friend Gerolamo Garimberto (p. 176; see Brown, *Our Accustomed Discourse*, 179).

110 The sculptures were made *c.*1534 by Raphael da Montelupo and Giovanni Angelo Montorsoli; see Estelle Lingo, "The Evolution of Michelangelo's Magnifici Tomb: Program vs. Process in the Iconography of the Medici Chapel," *Artibus et Historia* 16, 32 (1995): 91–100. The figures of the saints were installed by 1559; see Till Verellen, "Cosmas and Damian in the New Sacristy," *JWCI* 42 (1979): 274–7. Michelangelo's design for the Magnifici tomb contains plans for allegorical figures for that group that correspond closely to those in Agrippa's image, particularly to the figure of the *vita contemplativa*. The print by Cornelis Cort (New Hollstein 219.1) is dated to 1570.

111 "Intendi nobil donna le querele delle Muse, e delle regolate scientie, e le differentie che sono tra noi, e secondo la sapientia tua giudica le nostre tante contrarietà accio che viviamo in pace." For this translation and for her acute interpretation of the allegory itself I am indebted to Cristelle Baskins, who also noted the similarity between the pose of the figure of the *vita contem-*

plativa and Michelangelo's *capitano*/Lorenzo in the New Sacristy. The figural grouping is also similar to the image of Sixtus V and the Muses painted *c.*1586 on the vault of the staircase next to the Last Judgment wall of the Sistine Chapel by artists working under Giovanni Guerra and Cesare Nebbia: see Corinne Mandel, "Felix Culpa and Felix Roma: On the Program of the Sixtine Staircase at the Vatican," *The Art Bulletin* 75, no. 1 (1993): 65–90.

112   Archivio Vicariato, Santa Maria del Popolo, Libro di Morti, 1595–1620, 70: "Anno mese et die predictus [1 January 1600] Magnificus Dominus excellens peritus ac sapiens Architectus dominus Camillus Agrippa Medioleanensis sub cura Santi Laurentij in Damaso in Via Pelegrini in domo sua proprio habitans. Sacrementis receptis inscrizione Sta. Mariae Ecclesiae anno etatis sue 50 die. . . . Suum clausit euius corpus ad nostram ecclesiam delatum entra l'anuam maorem hora 24 sepultum fuit." The full notice of his age is illegible, but an entry in the Archivio Vicariato, *Stati d'anime, S. Lorenzo in Damaso*, 1599, fol. 49r, registers Camillo Agrippa, age eighty, living with M. Vincenzo Aiutoli, a priest. For Agrippa's practice as engineering, see Nenci, "Camillo Agrippa," 71–5; and Birali and Morachiello, *Immagini*.

4   The Care of the Body in Pietro Paolo Magni's Manual for Barber-Surgeons

1   ASR, Not. AC, vol. 2446, fol. 738r. The original agreement is recorded here as notarized by "Ascanium Mazziottum notarium R.mi. D. Vicarii" on 6 July 1584, but I was unable to find the document.

2   The first edition is *Discorsi di Pietro Paolo Magni piacentino intorno al sanguinar i corpi humani, il modo di ataccare le sanguisuche e ventose è far frittioni è vescicatorii: con buoni et utili avertimenti*, Rome: Bartolomeo Bonfadino, and Tito Diani, 1584. The 1586 edition, Pietro Paolo Magni, *Discorsi . . . sopra il modo di sanguinare . . .* , Rome: Bonfadino, 1586, was issued by Bonfadino alone. See the appendix to this chapter for all the editions of the book. Brunet, vol. 3, 1298–9 [7502].

3   For phlebotomy and the humors: Noga Arikha, *Passions and Tempers: A History of the Humours*, New York: HarperCollins, 2007, especially 89–92. She notes the reluctance of physicians to engage with the open bodies of patients as a factor in the lack of attention to updating old phlebotomy manuals. See also Gianna Pomata, "Practicing between Earth and Heaven: Women Healers in Seventeenth-Century Bologna," *Dynamis* 19 (1999): 119–43. Arikha, Pomata, and Monica Green, "Women's Medical Practice and Health Care in Medieval Europe," *Signs* 14, no. 2 (Winter 1989): 434–73, all discuss the perceived threat to the institution of medicine from women; see below. A doctor writing in 1895 about the disappearance of venesection from the medical scene talks about the suddenness of the phenomenon, saying that thirty years before it would still have been in good use, and that even then he was not sure it would not come back into fashion: "the lancet is carried idly in its silver case; no one bleeds; and yet . . . my friends retain their lancets, and keep them from rusting": W. Mitchell Clarke, "The History of Bleeding, and its Disuse in Modern Practice, *British Medical Journal* 2, no. 759 (17 July 1875): (67–70) 68.

4   "Cesaris et Felicis eius ex d. q. Adam filiorum infantium": ASR, Not. AC, 2446 vol. 5, fol. 738v. Adamo's death is recorded by the Congregazione dei Virtuosi al Pantheon and in a letter to the duke of Mantua from Attilio Malegnani; see Valeria Pagani, "Adamo Scultori and Diana Mantovana," *Print Quarterly* 9, no. 1 (1992): 72–87 (85).

5   The only other noticeable change in the illustrations is the addition of shading on the floor of the "Figura Universale" in the 1586 edition.

6   Palumbo, noted as *d. Petrum Paulum Palumbum peritum electus*, was a bookseller and publisher from Novara, active in Rome from 1563; see Michael Bury, *The Print in Italy, 1550–1620*, London: British Museum, 2001, 230. His testament was made in 1599; see Masetti Zannini, 91.

7   One scudo, which could be of gold or silver, was equal to 100 baiocchi; the giulio, a silver coin created by Julius II, was equal to 0.1 of a scudo or 10 baiocchi. U. Benigni, "Mint," *The Catholic Encyclopedia*, vol. 10, New York: Robert Appleton

Company, 1911, 335. The document states that "inter ipsos socios controversia super illius computis introducta," a disagreement that had resulted in naming Bartolomeo Grassi as an expert valuator during Adamo's lifetime. The books therefore came to 62 scudi, and various other adjustments, including 6 giulii for a piece of copper in Adamo's possession, resulted in Magni giving up another eighty books. For Grassi, who seems to have made something of a sideline in this kind of evaluation, see Bury, *Print in Italy*, 227; and Marco Ruffini, *Le Imprese del drago: Politica, emblematica e scienze naturali alla corte di Gregorio XIII (1572–1585)*, Rome: Bulzoni, 2005, 45–50.

8 The plate for the frontispiece engraved by Cherubino Alberti does not appear among those that reverted to the possession of Alberti, for which see Christopher L. C. E. Witcombe, "Cherubino Alberti and the Ownership of Engraved Plates," *Print Quarterly* 6, no. 2 (1989): 160–9.

9 Among others, see Nancy Siraisi, *Medieval and Early Renaissance Medicine*, Chicago: University of Chicago Press, 1990, 18–365, and especially 20, on literacy as a measure of intellectual and social demarcation in the field. Michael R. McVaugh, "Bedside Manners in the Middle Ages," *Bulletin of the History of Medicine* 71, no. 2 (1997): 201–23: "[For medieval readers] the urine flask had the iconic significance of a modern physician's white coat: it represented the expertise, the specialized knowledge possessed by the physician, and the respect that that expertise deserved" (203).

10 On the use, content, and significance of barbersurgeons' belt-books, see Peter Murray Jones, "Image, Word and Medicine in the Middle Ages," in J. A. Givens, K. M. Reeds, and A. Touwaide (eds.), *Visualizing Medieval Medicine and Natural History 1200–1550*, Aldershot: Ashgate, 2006, 1–11 (9–11). One of these books is illustrated in Siraisi, *Medicine*, 33, fig. 3.

11 *I Statuti, ordini e costitutioni della Università de Barbieri, & Stufaroli dell'Alma Citta di Roma*, Rome: Antonio Blado, 1559. The *proemio* to the statutes begins by offering praise to the Virgin Mary and the patron saints and ends with the injunction not to work on feast days, particularly that of Cosmas and Damian "nostri advocati."

12 Giuseppe Borghini, "Salasso e l'opera di Pietro Paolo Magni Piacentino," *Piacenza sanitaria* 4, no. 9 (1956): 3–30, says that the bowls are the sign of the barbers, already in use in images of Cosmas and Damian in the thirteenth century (19).

13 Manfredo Tafuri, "Capriani, Francesco, detto Francesco da Volterra," in *DBI* vol. 19, 190.

14 Federico Zuccaro, *Scritti d'arte di Federico Zuccaro*, ed. D. Heikamp, Florence: Olschki, 1961, 65 and 70; See also Janis Callen Bell, "Alberti," in *Grove Art Online. Oxford Art Online*, http://www.oxfordartonline.com/subscriber/article/grove/art/T001520pg2 (accessed 1 October 2011).

15 The dedication remained essentially unchanged except for tidying up the form of title to reflect the change in his position after the death of the pope the year before. Bianchetti had been the Bolognese *maestro di camera* for Gregory XIII 1572–85: Ludwig von Pastor, *The History of the Popes, from the Close of the Middle Ages*, London: J. Hodges, 1891, 26. He had also been a canon of St. Peter's Basilica and acted as sacristan in 1575, when he was responsible for embellishments to Michelangelo's statue of the Pietà: Kathleen Weil-Garris Brandt, "Michelangelo's Pietà for the Cappella del Re di Francia," in *Michelangelo: Selected Scholarship in English*, ed. William E. Wallace, Hamden, Conn.: Garland, 1995, 217–60 (228). For his position as *maestro di camera* of Farnese, see Borghini, "Salasso."

16 The sonnets are by Celso Cittadini, who contributed many sonnets to books printed in Rome in these years, and Gioan Filippo Montagnesio, DM; the 1586 edition bears an additional short epigram by Petri Maillardi Santonis. The privilege is dated 1 October 1583.

17 See, for example: Hans von Gersdorff, *Feldtbuch der Wundartzney*, Strasbourg, 1540; Stefan Falimirz, *O ziolach i o moczy ich . . .*, Crakow, 1534. An exception is the engraving by Giorgio Ghisi showing a scene of bloodletting using cups, for which see R. and M. Lewis and S. Boorsch, *The Engravings of Giorgio Ghisi*, New York: Metropolitan Museum of Art, 1985, cat. 1.

18 Katharine Park, *Secrets of Women. Gender, Generation, and the Origins of Human Dissection*, New York: Zone Books, 2006, 127–9, for dissection

taking place in a domestic setting in a French fifteenth-century illustration.

19   John C. Burnham, "How the Concept of Profession Evolved in the Work of Historians of Medicine," *Bulletin of the History of Medicine* 70, no. 1 (1996): 1–24 (23), on the function of the traditional priestly role of the physician in regard to the patient–healthcare giver relationship, in contributing to the professionalization of the field of medicine.

20   Borghini, "Salasso," 18

21   Claudine Herzlich, "Modern Medicine and the Quest for Meaning: Illness as a Social Signifier," in Marc Augé and Claudine Herzlich (eds.), *The Meaning of Illness: Anthropology, History and Society*, London: Harwood Academic Publishers, 1995, 151–74 (154–5), discusses illness as a deviant social state in that it renders the ill person inactive, while participation in a cure that meets societal expectations constitutes rational behavior (154–5). For the status of the doctor and distance from bodily care, see Pomata, "Practicing," 129.

22   Herzlich, "Modern Medicine," 161: "the language used to express health and illness is not a language of the body, of organic facts, it is a language of the individual's relationship with the socialized exterior, with society."

23   Antonio Martini, *Arti, mestieri e fede nella Roma dei papi*, Bologna: Cappelli, 1965, 229.

24   Martini, *Arti*, 229.

25   *Discorso di Pietro Paolo Magni sopra il modo di fare cauterij* . . . , Roma: Bartolomeo Bonfadino, 1588. In his preface Magni writes: "in my discourse on bleeding, which I published in years past, I discussed (although very briefly) blistering, and that being almost the same operation as cauterization, researching it with the same diligence, I believed this to be more like a continuation of the first book, and a perfection of it, than a new thing." He makes a point of saying that this book is not intended to discuss everything known about cauterization, but written to help barbers, which is why it is in Italian, since few in Italy could read Latin ("percioche qual utile possoni i Barbieri trarne da simili libri composti latinamente, se non ve ne niuno, ò molto pochi, e massimamente in Italia, che habbia cognitione della lingua Latina").

26   Tomaso Garzoni, *La piazza universale di tutte le professioni del mundo*, vol. 2, ed. G. B. Bronzini, Florence: Olschki, 1996, 1057, *Discorso CXL*, "De'barbieri": "Sono anco di molti scandali cagione di questo, che acconciano in modo certi vecchi ganimedi, radendo loro sotto il mento, e nelle guancie i peli sottili, che i tavanoni tratti dal lichetto del mele, volano al scuro sopra di loro, né mai si fornisce di lascivire come si deve." Sandra Cavallo, *Artisans of the Body in Early Modern Italy*, Manchester: Manchester University Press, 2007, 54–7, instead sees the barber in the seventeenth century as practitioner of a noble art, and discusses the importance of barbers' cosmetic work in relation to fashion and hygiene.

27   Martini, *Arti*, 229–32, and Garzoni, *La piazza universale*, 1010, on "Stufaruoli", after a short discussion of the glories of the bathing practices and establishments of the ancients: "Ma a proposito nostro i stufaruoli attendono a lavare, a far sudare, a metter cornetti, a cacciare i peli, e mondar tutta la vita dell'huomo nelle stuffe loro. . . . E i lor difetti sono intorno alle spurcitie della carne, perché son pochi stufaruoli che non siano ruffiani, et che non tengano camera a nolo, meschiando la munditia esteriore con l'immunditia interna in quelle stufe, che son ricetto di mille vergognose, e dishoneste libidini carnali." See also David Gentilcore, "Charlatans: Regulation of the Marketplace and the Treatment of Venereal Disease in Italy," in Kevin Patrick Siena (ed.), *Sins of the Flesh: Responding to Sexual Disease in Early Modern Europe*, Toronto: Centre for Renaissance and Reformation Studies, 2005, 62–3. See also Anna Esposito, "Stufe e bagni pubblici a Roma nel Rinascimento," in Massimo Miglio *et al.* (eds.), *Taverne, locande e stufe nel rinascimento*, Rome: Roma nel Rinascimento, 1999, 77–91, who notes the generally low repute in which *stufe* were held in cinquecento Rome.

28   Bernardino Ramazzini, *De morbis artificum*, Modena, 1700, translated as *Diseases of Workers*, trans. W. C. Wright, Ontario: OH&S Press, 1993, 174–5. Martini, *Arti*, 230; J. B. de C. M. Saunders and C. D. O'Malley, *The Illustrations from the Works of Andreas Vesalius of Brussels*, New York: Dover, 1950, 25, on bath attendants and barber-

surgeons alike forming the market for single-sheet anatomical illustrations showing the venous system.

29 *Statuti, ordini e costitutioni della Venerabil Compagnia & Università de Barbieri di Roma*, Rome: Stamperia della Rev. Camera Apostolica, 1641. The 1614 decision was recorded in this edition. It states that they decided "per pace & quiete di tutta la Università, di riformare li statuti della loro arte de Barbieri, & da essi affato levare, e cassare detti Stufaroli" (29). Martini, *Arti*, 230–1, reports the date as 1613. In a similar way, Jonathan Sawday comments on the connection between the surgeon and the executioner, and the importance for the anatomist of distancing himself from that role in representations of the performance of dissection: Jonathan Sawday, "The Fate of Marsyas: Dissecting the Renaissance Body," in *Renaissance Bodies: The Human Figure in English Culture c. 1540–1660*, London: Reaktion Books, 1990, 111–35 (117, and 122).

30 Magni, *Discorsi*, 1586, 1–2: "Ilche considerando io, & havendo consumato hormai tutta l'età mia in quest'Alma Città di Roma, in cavar sangue à coloro, che di ciò hanno havuto bisogno; & in mettere le sanguisughe, ventose, vessicatorij, & in far le debite fregagioni ordinate da Medici; & vedendo, che in queste cose molti errano in diversi modi, con grandissimo pericolo, & danno de gli infermi: perciò ho composto la presente opera, dalla quale ne potranno i Barbieri (di quelli parlo, che n'hanno bisogno) trarre, senza dubbio, non picciolo giovamento: perche oltre gli avvertimenti, & molti secreti contenuti in essa, da me industriosamente trovati, hò eletto uno stile chiaro & facile, il più che ho potuto, ricordandomi molto bene, che scrivo à i Barbieri, & per insegnar loro di cavar sangue, non di ben parlare: le quali cose se havessero i miei censori considerate, non sarebbe loro per aventura paruto strano, perche io habbia alcune cose scritto meno che toscanamente." The first page of the book of anatomy of Mondino, written in Italian in the famous illustrated edition of 1493, also purports to demonstrate a physical procedure "not by observing an elevated style [*non observando stile alto*], but I will give you knowledge of it according to the manual operation [*secondo la manuale operatione*]".

31 Magni, *Discorsi*, 1586, 2: "Non mi hanno già mosso le persuasioni di coloro, che molto teneri mostrandosi dell'honor mio, mi consigliavono (essendo io Chirurgo, & amesso dal Collegio dei Medici di Roma) ad honorar l'opera col nome di Chirurgo, invece di Barbiere; si perche l'intention mia in questa opera non fù mai altra, che di giovare altrui, sì anco, perche non tratto dell'arte del sanguinare, se non quanto s'appartiene alla pratica, & alla professione del Barbiere; allegando appresso molte belle isperienze fatte da me medesimo, la qual cosa haverebbe senza dubbio offeso gli animi de Chirurgi d'hoggi, che (qual si sia la cagione) abboriscono grandemente tal mestiero."

32 Andreas Vesalius, *De humani corporis fabrica*, Basel: Iohannis Oporinus, 1543, fol. ii, partial English translation available as *On the Fabric of the Human Body*, vol. 1, trans. W. F. Richardson and J. B. Carman, San Francisco: Norman Publishing, 1998, xlix. Also see Katherine Park, *Doctors and Medicine in Early Renaissance Florence*, Princeton: Princeton University Press, 1985, 8, for surgery being an academic discipline in Italy while it was not in France or England. University trained surgeons in Italy, therefore, felt themselves certifiably distanced from barbers. Jerome J. Bylebyl, "Interpreting the *Fasciculo* Anatomy Scene," *Journal of the History of Medicine* 45 (1990): 285–316, brings up the continuing prejudice in Vesalius's time against surgeons in spite of their university training and the preference among university trained doctors for the profession of physician rather than surgeon (312, n. 75). See also Pomata, "Practicing," 122.

33 Siraisi, *Medicine*, 35, discusses the scorn heaped on "illiterates, rustics, old women, empirics and Jews" by physicians and surgeons who were literate in Latin. This is taken up further in Alison Klairmont Lingo, "Empirics and Charlatans in Early Modern France: The Genesis of the Classification of the 'Other' in Medical Practice" *Journal of Social History* 19 (Summer 1986): 583–603.

34 Vesalius, *De humani corporis*, fols. ii–iii; Saunders and O'Malley, *Illustrations*, 34.

35   Gianna Pomata, "Barbari e comari," in G. Adani and G. Tamagnini (eds.), *Medicina, erbe e magia*, Milan: Silvana editore, 1981, 161–83 (168); and Pomata, "Practicing," 123: "From the perspective of the medical elite, a huge gap divided the academically trained physicians from apothecaries and surgeons: as put by Paolo Zacchia, the author of the most important medico-legal treatise of the seventeenth century, the physician 'treats the body by using his intellect, not his body;' apothecaries and surgeons, by contrast, use their hands rather than their minds: they 'cure the body with the body,' and as such their work is on a par with that of servants. Medical rank, in this perspective, was crucially related to distance from the body."

36   Sachiko Kusukawa, "Leonhart Fuchs on the Importance of Pictures," *Journal of the History of Ideas* 58, no. 3 (1997): 403–27 (412).

37   Katherine Park suggested that the box would be for simples, and Faye Cook remembered the connection with the story of Cosmas and Damian's martyrdom.

38   See Saunders and O'Malley, *Illustrations*, 25. The 1641 statutes for the barber's guild specified: "Per evitare molti scandali che possono succedere, essendo che vi sono molti, quali vanno cavando sangue, & non sono dell'Arte, nè meno essaminati, nè approbati, tornando il tutto in danno de' corpi humani, però si prohibisce, che n'essuno ardisca, ne presuma, nè di cavar sangue, nè far altro essercito spettante al Barbiero, se prima non sarà approbato, & essaminato, & havuta licenza, conforme li Statuti." (*Statuti . . . de Barbieri di Roma*, 67). It is also mentioned in the statutes that the officers of the guild must know how to read and write, so assumptions about barbers being unlettered mean specifically someone not trained in a university or in Latin rather than technically illiterate.

39   Johannes Ketham, *Fasciculo di medicina vulgare*, Venice, 1493. On this book see Bylebyl, "Interpreting," (with more complete bibliography); he emphasizes the utility of this illustrated book for the non-academic, non-Latin literate anatomists although later editions (such as the one illustrated here) were printed in Latin. See also Andrea Carlino, *La fabbrica del corpo*, Turin: Einaudi, 1994.

40   Bylebyl, "Interpreting," explains this picture in relation to the earliest and clearest version of it, which was used for the 1493 edition. The version that appears here comes from a later edition in Latin: *Fasciculus medici[n]e*, Venice: Cesare Arrivabene, 1522. The original woodblock print shows the demonstrator using a stick as a pointer held in his right hand; in this version not only the pointer but the whole left hand is missing. In addition, the reader on the lectern has a book before him in later versions of the woodcut; in the original illustration he recites an implied text from memory. See also Carlino, *La fabbrica*, 87–9.

41   Vesalius, *De humani corporis*, fol. iii.

42   Vesalius, *De humani corporis*, fol. iii.

43   Saunders and O'Malley, *Illustrations*, 25 and 203–27. Several examples of the flap-anatomies are illustrated in M. Cazort, M. Kornell, and K. B. Roberts, *The Ingenious Machine of Nature: Four Centuries of Art and Anatomy*, Ottawa: National Gallery of Canada, 1996. Also see D. Hillman and C. Mazzio, *The Body in Parts*, New York and London: Routledge, 1997, xv–xvi, for the fragmentation, ordering, and disordering of the body in flap-books.

44   Andreas Vesalius, *Epitome*, Basel, 1543, for which see also L. R. Lind, *The Epitome of Andreas Vesalius*, Cambridge, Mass.: MIT Press, 1969. Cazort, Kornell, and Roberts, *Machine*, 133, mention that a copy of Vesalius's figures from the *Epitome* in which his instructions can be seen carried out is in the Karolinska Institute Library in Stockholm.

45   Juan Valverde de Amusco, *Historia de la composicion del corpo humano . . .* , Rome, 1556.

46   Glenn Harcourt, "Andreas Vesalius and the Anatomy of Antique Sculpture," *Representations* 17 (1987): 28–61, emphasizes the very pragmatic nature of Vesalius's use of the canon of Polykleitos compared to the "nuanced richness" of Galen's (42 and n. 46).

47   Magni, *Discorsi*, 1586, fol. 25r.

48   Magni, *Discorsi*, 1586, fol. 24r.

49   From the 1584 edition in the Countway Library of Harvard University. The jugular vein is treated in chapter X, "Delle vene della gola chiamate giugulari," 33–4.

50 For the Scultori family, see Stefania Massari, *Incisore Mantovani del Cinquecento*, Rome: De Luca, 1981; Paolo Bellini, *L'opera incisa di Adamo e Diana Scultori*, Vincenza: Neri Pozza, 1991; the important corrections and amplifications to Bellini by Pagani, "Adamo Scultori"; Evelyn Lincoln, *The Invention of the Italian Renaissance Printmaker*, New Haven and London: Yale University Press, 2000; and a short discussion by Christopher Witcombe, *Print Publishing in Sixteenth-Century Rome*, London: Harvey Miller, 2008, 336–7. The frontispiece was for *Rime de gli Academici Eterei dedicate alla serenissima madama Margherita di Vallois duchessa di Sauoia* Padua: Gli Eterei (1567?); the same plate was used with changes made to it for the frontispiece of Antonio Pagani, *Il discorso uniuersale della sacra legge canonica di F. Antonio Pagani Vinitiano minore osseruante, nel qual contiensi l'origine, et l'ordine di varie leggi a diuersi popoli date . . .*, Venice: Bolognino Zaltieri, 1570. See Bellini, *L'opera*, 112, cat. 99.

51 The abbreviation "sculp." for "sculpsit" was commonly used to designate the role of the engraver, and "exc." for "excudit" referred to the publisher, or owner of the plates, and could also mean the printer. For a rather complete list of the terms that usually appear in the inscriptions on plates, see Witcombe, *Print Publishing*, 15–18. Adamo appears in documents as a printer, publisher, and engraver in Rome. From about 1577 his name appears on prints with notations including his full name and "excudebat," for example: "Adam Schulptor Mantuanus exc. Romae"; see Bellini, *L'opera*, 156–61. His earlier prints, mostly engraved after designs by others, generally bear a monogram with a large capital "A" and a small "s" descending from its cross bar. One print, a *Nativity with Four Saints* after Giulio Romano, bears the inscription "Adamo Scultore Mant. Scul." in the third state; see Bellini, *L'opera*, 134.

52 *Herbario nuovo di Castore Durante* (Rome: Bartholomeo Bonfadino, and Tito Diani, 1585); see Masetti Zannini, 215 and 279.

53 The series of seventy-three figures from the Sistine ceiling (in some sets only seventy-two) appeared both bound and unbound, with a frontispiece of an escutcheon inscribed simply with the words: "Michael Angelus Bonarotus Pinxit Adam Sculptor Mantuanus Incidit"; see Bellini, *L'opera*, 64–104, cat. 21–92; and Pagani, "Adamo Scultori," 77. In the inventory of Scultori's plates made in 1613 the book, which was given to Cristofano Blanco (who republished it), is called *Libro de la volta di Michelangelo*.

54 Magni, *Discorsi*, 1586, chapter V, "Provisione à gli huomini timorosi, & pusillanimi, & astutie che in essi per sanguinargli, si possono usare, con alcuni rimedij per preservargli dalle sincopi," 16.

55 Magni, *Discorsi*, 1586, 40.

56 Magni, *Discorsi*, 1586, 3–6, 17–20.

57 Magni, *Discorsi*, 1586, 94r: "Mi pare necessario di raccontarvi quello che più volte mi è successo, essendo chiamato per far questa operatione à qualche donna vergognosa, laquale essendoli da signori Fisici state ordinate le sanguisughe da applicarsi à detta parte, non volendo per vergogna che io gli applicasse, ho consigliato i parenti, che si copra la faccia & tutta quanta si nasconda, accioche ne io lei, ne lei me veda & conosca: & così si è contentata, per havere un tanto soccorso. Poi che i parenti di lei sapevano, & essa ancora conosceva, che la speranza della sua vita dipendeva dal detto singolar rimedio. Questo dico, perche questa operatione da pochi debitamente è fatta. Et benche alcune don[n]e si trovino, che facciano professione di questa cosa, & altre simili, non però la fanno fare, ne meno in loro tutte le persone di deveno considare. Perche se molti Barbieri si trovano, che in questa applicatione fanno molti errori, quanto maggiormente li saranno le donne, che sono molto lontane dall'arte de i Barbieri?"

58 In the 1584 edition the chapters are numbered sequentially from the first page of the book, while in the 1586 edition the section on leeches, and on cupping, etc., are given chapter numbers within their own sections. Therefore, in the 1584 edition the chapter on bleeding the *fondamento* is XXXI, 89–92.

59 For the effect of the increased importance given to the licensing of medical practitioners between the twelfth and sixteenth centuries on women herbalists, midwives, and apothecaries, see Pomata, "Practicing," 121; Green, "Women's Medical Practice"; and M. Green, "From 'Diseases of Women' to 'Secrets of Women': The Transformation of

Gynecological Literature in the Later Middle Ages," *Journal of Medieval and Early Modern Studies* 30 (2000): 5–39, where she also speaks about the shame associated with medical inspection of a woman's private parts (8). See Park, *Secrets of Women*, 255–8, for an estimation of the influence and value of non-literate female practitioners in Italy who specialized in tending to women.

60   Pomata, "Practicing," 123–4; and Siraisi, *Medicine*, 19, on the non-uniformity of licensing in this period.

61   Green, "Women's Medical Practice," 456–68; Green, "Transformation," 10; Park, *Secrets of Women*, 92–103; also Elizabeth Cohen, "Miscarriages of Apothecary Justice: Un-separate Spaces of Work and Family in Early Modern Rome," in a special edition of *Renaissance Studies* 21, no. 4 (September 2007): 480–504, and the introduction to the same volume by David Gentilcore, "Spaces, Objects and Identities in Early Modern Italian Medicine," 473–9.

62   Green, "Women's Medical Practice," 471; Lingo, "Empirics," 593–6.

63   Magni, *Discorsi*, 1586, 19–20: "Ma mi pare che sarebbe il dovere, che tutte queste cose fossero considerate dal Medico, & che da lui fosse lasciato in scritto, non solamente, che l'amalato s'ha da sanguinare, ma di che vena, & se il taglio deve esser largo ò stretto, & quanta quantità di sangue s'ha da cavare: questo dico non per dar ordine alli Signori Fisici, che avvertiscano le sudette cose, ma per quelli che nol fanno, che in vero sono molti & molti, di che mi son assai doluto; perche queste inavertenze causano molti in convenienti, che ogni giorno si veggono: Et questo tanto maggiormente doverebbono i Fisici fare, quanto più si vede l'inesperienza & dapocaggine de'moderni Barbieri (sia sempre havuta da me riverenza à buoni, saggi, & prudenti; perche contra questi io non parlo) i quali à posta & chimera loro senza sapere l'ordine dei Medici, fanno le sanguigne, & le più volte per il contrario; perilche ne vien leso il paziente: sarà anco bene à farlo, per levar l'occasione à quelle donne, che procurano l'aborto, cioè il sconciamento; perche molte sono, che sanno la sanguigna della safena facilmente poter quello provocare." See Pomata, "Practicing,"

on the popular familiarity with the Hippocratic aphorism about the saphenous vein being bled to procure abortions.

64   For the construction of the scientist as medical hero in the sixteenth and seventeenth centuries, see John M. Steadman, "Beyond Hercules: Bacon and the Scientist as Hero," *Studies in the Literary Imagination*, 4 (1971): 3–47.

65   Magni, *Discorsi*, 1586, fol. 27r.

66   In discussing the illustrations of the *Fasciculo di medicina* of 1493 in the Italian edition, Bylebyl mentions that here, too, the only female figure in the illustrations is introduced when the subject matter is that part of the anatomy particular to her sex, the uterus: Bylebyl, "Interpreting," 295, 299, and 305.

67   Magni, *Discorsi*, 1586, fol. 65r. This prescription is traditional, and is also mentioned in the *Fasciculo*. See also Gianna Pomata, *Contracting a cure: Patients, Healers and the Law in Early Modern Bologna*, trans. by the author with Rosemary Foy and Anna Taraboletti-Segre, Baltimore, Md: John Hopkins University Press, 1998, 22; and Park, *Secrets of Women*, 106–9, for the relationship to Ketham, and throughout the book for the female body as the subject of anatomical viewing.

68   Magni, *Discorsi*, 1586, 78–9.

69   For the love-making euphemism, see Leo Steinberg, "The Metaphors of Love and Birth in Michelangelo's *Pietàs*," in T. Bowie and E. Christianson (eds.), *Studies in Erotic Art*, New York: Basic Books, 1970, 231–335. For images of birthing which were available not only in specialized midwifery manuals but as decorative motifs on maiolica and other household items, see Jacqueline Musacchio, *The Art and Ritual of Childbirth in Renaissance Italy*, New Haven and London: Yale University Press, 1999.

70   Magni, *Discorsi*, 1586, fol. 19r; see above n. 63.

71   Cazort, Kornell, and Roberts, *Machine* 139; and Bette Talvecchia, *Taking Positions: On the Erotic in Renaissance Culture*, Princeton: Princeton University Press, 1999, 161–87, with further bibliography.

72   Steadman, "Beyond Hercules," 11, on the role of invention in the concept of the hero.

73   Sawday, "Fate of Marsyas," 115, emphasizes the

importance of the link between criminality and anatomized body in anatomical illustration. The anatomy text of Mondino instructs the reader to begin with "the body of one dead from beheading or hanging": Bylebyl "Interpreting," 305, and 311–12 for mention of the Venetian statutes that direct anatomy subjects to be requested from the police in charge of capital crimes; see also Carlino, *La fabbrica*, 97–126.

74 Bylebyl, "Interpreting," 299–300 and 306, mentions comments by Bernardino Scardeone that the *Fasciculo di medicina* had been translated into Italian so that non-university trained surgeons (referred to by Bylebyl as the *chirurgi volgari*), could make use of the treatise even if they were lacking in Latin letters.

75 Magni, *Discorsi*, 1586, 4.

76 Bonfadino is described in documents as "Bartholomeo Bonfadino brescianensis," see Chapter 1, n. 54.

77 The printer's name is also found as Cristoforo Bianchi: Bury, *Print in Italy*, 233. The document appears to be lost, but a transcription is found in A. Bertolotti, *Artisti francesi in Roma nei secoli XV, XVI e XVII*, Mantua: G. Mondovi, 1886, 94. The plates for Magni's book do not appear in Bertolotti's transcription of Aurelia's dowry. Scholars have wondered why the dowry was issued so long after Aurelia's marriage, which seems to have taken place in 1593. See Bellini, *L'opera*, 29; the important corrections and amplifications to it by Pagani, "Adamo Scultori,"; and Witcombe, *Print Publishing*, 336–7.

78 *Discorsi di Pietro Paolo Magni piacentino sopra il modo di sanguinare attaccar le sanguisughe, et le ventose, far le fregagioni, et vessicatorij a corpi humani*, di nuovo ristampato ad istanza di Pietro Fetti libraro in Parione, Rome: Iacomo Mascardi, 1613: "Pietro Fetti Salute. Il giovare communemente à ciascuno, & particolarmente con quelle opere, che possano apportare utile à qualche arte, ò scienza: e stata sempre stimata cosa degna & laudabile. Però sendomi venuta alle mani la presente opera di M. Paolo Magno Barbiere sopr' l'Insanguinare: non ho voluto mancar di farla di nuovo stampare per beneficio commune, tanto per i professori di essa arte, quanto per gl'infermi. Et considerando che

meglio non la poteva indrizzare che à questo loro Collegio, la dedico volontieri alle SS. VV. si per esser materia di loro professione, si anco per maggior difesa dell'opera, à tale che vengo à far utile integrando la fama dell'istesso Autore, e giovamento à voi medesimi, e tanto più per essermi questa opera assai domandata. Assicurandomi ch'aggradiranno il mio buon animo, e si degnaranno ancora ricevarla con quella gratitudine ch'io la dedico, e presento."

79 Gaetano Moroni, *Dizionario di erudizione storico-ecclesiastica da S. Pietro sino ai nostri giorni*, Venice: Tipografia Emiliana, 1840–61, vol. 44, 136.

80 The book was first republished in Rome in 1613 with recut plates that were reused for an edition from the same printer in 1626. I used the 1626 edition at Countway Medical Library of Harvard University (see appendix).

81 Magni, *Discorsi*, 1586, fol. 11r.

82 Magni, *Discorsi*, 1586, fol. 65–6.

5   Courts and Other Theaters

1 Magino's request to grow mulberry trees in the Medici states was rejected by Duke Francesco I on 18 June 1587 (*MdP*, 270 DocID 16533). An earlier letter from the duke to Magino (*MdP*, 270 DocID 19330) says he is returning Magino's book because he is not interested in a financial venture (unspecified) that Magino made to him. Accessed on-line at the Medici Archive project, BIA (bia.medici.org), a password protected database, on 11 August 2013.

2 Magino Gabrielli, *Dialoghi di M. Magino Gabrielli Hebreo Venetiano sopra l'utili sue inventioni circa la seta. Ne'quali anche si dimostrono in vaghe Figure Historiati tutti gl'essercitij, & instrumenti, che nell'Arte della Seta si ricercano*, Rome: per gli Heredi di Giovanni Gigliotti, 1588. Brunet, vol. 3, 1295 [10259].

3 The history of the literature on silk is surveyed in Luca Molà, *The Silk Industry of Renaissance Venice*, Baltimore and London: Johns Hopkins University Press, 2000, 227–30; see also Daria Perocco, "La seta nella letteratura Italiana dal Duecento al Seicento," in L. Molà, R. C. Mueller,

and Claudio Zanier (eds.), *La seta in Italia dal Medioevo al Seicento*, Venice: Marsilio, 2000, 241–61.

4   Gabrielli, *Dialoghi*, fol. 4 and elsewhere.

5   Daniel Jütte, "Handel, Wissenstransfer und Netzwerke: Eine Fallstudie zu Grenzen und Möglichkeiten unternehmerischen Handelns unter Juden zwischen Reich, Italien und Levante um 1600," *Vierteljahrschrift für Sozial- und Wirtschaftsgeschichte* 95, no. 3 (July 2008): 263–90.

6   Luca Molà, "Le donne nell'industria serica veneziana del rinascimento," in Molà, Mueller, and Zanier, *La seta*, 423–60.

7   Gabrielli, *Dialoghi*, fol. 2r. The document authorizing the activity is ASR, 30 Not. Cap. uff. 24 vol. 95 (July 1587), fols. 735r–738v. See Molà, *Silk Industry*, 207; and Evelyn Lincoln "The Jew and the Worms: Portraits and Patronage in a 16th-Century How-To Manual," *Word & Image*, 19: 1/2 (2003): 86–99.

8   Molà, *Silk Industy*, 214.

9   Gabrielli, *Dialoghi*, fol. 43, Figura Settima: "F: Hebreo che insegna il modo di cibargli doppo ciascuna muta."

10 For the historical Magino (d. 1601) see most recently with accumulated bibliography: Dora Liscia Bemporad, *Maggino di Gabriello: "Hebreo Venetiano"; I dialoghi sopra l'utili sue inventioni circa la seta*, Florence: Edizioni Firenze, 2010; Jütte, "Handel"; Daniel Jütte, "Abramo Colorni, jüdischer Hofalchemist Herzog Friedrichs I, und die hebräische Handelskompanie des Maggino Gabrielli in Württemberg am Ende des 16. Jahrhunderts," *Ashkenas* 15, no. 2 (2005): 435–98 (an English-language translation of this book will be published by Yale University Press in 2013, with the title *The Age of Secrecy. Jews, Christians and the Economy of Secrets (1400–1800)*); and Jütte, *Das Zeitalter des Geheimnisses: Juden, Christen und die Ökonomie des Geheimen (1400–1800)*, Göttingen: Vandenhoeck and Ruprecht, 2011, 148–50, 302–3. Molà, *Silk Industy*, is the most complete about the context of silk manufacturing and Magino's mercantile activity, as well as the activity of the Jewish community in the silk business.

11 For Guidoboni and Magino, see Molà *Silk Industy*, 204–14.

12 Molà, who has studied the Venetian documents in the case, writes that Magino, "while having the power to present the petitions and receive the patents in his own or someone else's name . . . was obliged, if required, to have an official document drawn up in which he declared how the ownership of the inventions was actually shared." See Molà *Silk Industy*, 206.

13 Jütte, "Handel," discusses the technical aspects of Magino's glass process.

14 ASR, Not. RCA 1078, (1588) fols. 1005r–1008v, 10014r–v, "Appaltus super conficiendis mensuris pro Vino et alijs vitreis per xx anno Pro Magino hebreo Veneto." For the story of the rejection of the carafes by the *osti*, see Ermete Rossi, "La foglietta di Maggino Ebreo," *Roma: Rivista di studi e di vita romana* 5 (May 1928): 210–13 (for which reference I thank Daniel Jütte).

15 ASR, 30 Not. Cap. uff. 1, vol. 37 (1588), fols. 579r–v, 19 October 1588: "Obligatio pro Magistro Francesco de Solarijs Mediolanensis in Regione Campo Martij ex una et Maginus hebreus ex altera partibus [. . .] M. Francesco promette et si obliga fare a detto Magino p[rese]nte dodici lanternoni con Cherubini, et altre figure, et hornamento secondo il modello de doi lanternoni che detto M. Francesco ha nella bottegha, li quali lanternoni habbiano da essere palmi di longhezza cinque è mezzo cominciando dalla Cornice della Cuppula in qui et larghi palmi tre di bacocco, cio è l'armature debbiano essere di legno, et l'ornamenti di Carta Pesta, con fare l'incastre, dove habbiano à stare li Christalli li quali Christalli convengono che li debbia dare il detto magino, et il detto maestro Francesco sia obligato à metterli, et il simile habbia à fare delli fogli di latta. Item convengono che le detti lanternoni, detto Maestro Francesco sia obbligato ogni dodici giorni darne un finito à detto Magino, et cosi seguitare sino al fine de detta opera, et la detta Promessa il detto Maestro Francesco fa al detto Magino presente. Per prezzo, et nome di Prezzo di scudi tre per ciascheduno lanternone da pagarsi sicome detto Magino promette pagare al detto Maestro Francesco presente de volta in volta che li darrà il lavoro fatto al detto Magino qui in Roma [&c]."

16 For the privilege for the glass carafes and the

necessity to provide windows to the Holy See, see ASR, Not. RCA, vol. 1078, (1588) fols. 1005r–1008v, 1014r–v. The mendicant hospital at the foot of the Ponte Sisto is alternately described in documents as "ospedale" and "carcere," an important lack of distinction clarified for me by Carla Keyvanian. For the payment records for glass provided by Magino hebreo and Martino Briosa, vetraro, for the Palazzo apostolico Lateranense in 1598, see Anna Maria Corbo, *Fonti per la storia artistica Romana al tempo di Clemente VIII*, Rome: Archivio di Stato, 1975, 45, 53, and 76. I am grateful to Daniel Jütte for this last reference.

17   BAV, Urb. Lat 1057, Avvisi 1589: the announcement of the edict in January, fol. 42r: "È uscito in istampa il nuovo ordine dato agli Osti, o gli avari, et altri, che vendono liquori di adoprare carrafe di cristallo come si fà a Napolij per misurare il vino, invece delle misure di terra . . . et comprendosi Masino Hebreo l'inventore de q.ta novita sia obligato a pagarli à ragione di dui baiocchi la libra." The particulars of the measure caused an outcry among the *osti* when they appeared in March of that year, fol. 137v: "Finalmente hebbe essention l'uso delle misure d'osti di cristallo in questa città fin da Mercore passato con gusto universale fuorche delli professori di questo essercitio, che maledicono ad ogni hora l'inventore di questa novità di Masino hebreo, il quale paga all Camera 25 (scudi), con facoltà che egli solo per 25 anni possa fondere et vendere queste misure, non essendo giovato agli osti la offerta fatta di 12 scudi, per non essere obbligati a questa legge con pena 25 scudi per ogni volta, che saranno trovati à misurare con altri vasi." See also Rossi, "La foglietta". The document in which Magino contracts with Paolo Blado to print the *bando* itself in an edition of 50,000 on 30 January 1589: ASR, De Marchis, sec. and canc. RCA, vol. 1079, fol. 12r–v, in Masetti Zannini, 287–8.

18   Rossi, "La foglietta," 211–12. The tax was abolished in 1591.

19   ASR, Not. Cap. uff. 9 Garganus, vol. 7, (1588) fols. 235r–236r; and same notary, vol. 6, (1588–9) fols. 555r–v, 568r.

20   This office was abolished by his Jewish colleagues by 1593; however, Magino continued to benefit from Medici patronage and protection. Molà, *Silk Industry*, 213; Renzo Toaff, *La Nazione Ebrea a Livorno e a Pisa (1591–1700)*, Florence: Olschki, 1990, 42–9, 531–2. Magino had requested a copy of the *Lettere Patenti* of 30 June 1591, setting the terms for allowing the Jews to reside in Pisa and Livorno, from Duke Ferdinando before his arrival in Pisa in July 1591 (53–4). See also Lucia Frattarelli Fischer, "Reti locali e reti internazionali degli ebrei di Livorno nel Seicento," in Diogo Ramada Curto and Anthony Molho (eds.), *Commercial Networks in the Early Modern World*, Badia Fiesolana: European University Institute, 2002, 148–67 (148–9).

21   Toaff, *Nazione Ebrea*, 109–12. Magino seems to have been selling already woven lengths of cloth.

22   Toaff, *Nazione Ebrea*, 112; Corbo, *Fonti*, 76; and now Jütte, *The Age of Secrecy*.

23   See Gabrielli, *Dialoghi*, fol. 17, and elsewhere.

24   It is not known if this mandate was ever met. For the document about the edition size, see "Conventionis inter d. Contugho de Contughis et Mangino de Gabrielis Hebreo." ASR, 30 Not Cap uff. 28 Tondinus vol. 11 (1588) fols. 1046r–v.

25   See Erina Russo de Caro, "Iconografia di Sisto V nella pittura: Tre ritratti inediti," in P. dal Poggetto (ed.), *Le arti nelle Marche al tempo di Sisto V*, Milan: Silvana, 1992, 22–9.

26   Loren Partridge and Randolph Starn, *A Renaissance Likeness*, Berkeley: University of California Press, 1980, 49.

27   On the formation of individual identity in terms of groups, see Stephen Mennell, "The Formation of We-Images: A Process Theory", in C. Calhoun (ed.), *Social Theory and the Politics of Identity*, Cambridge, Mass.: Blackwell, 1994, 175–97.

28   See also Russo de Caro, "Iconografia," 27, which reproduces a similar portrait in a frame that carries Sistine *impresa* around the tondo portait.

29   A very similar version of this print, but with the portrait printed in reverse, served as the frontispiece for Giovanni Pinadello, *Dom invicti quinarii numeri series: quae summatim a superioribus pontificibus et maxime a Sixto Quinto; res praeclare quadriennio gestas adnumerat ad eundem Sixtum Quintum, Pont. Opt. Max*, Rome: F. Zanetti, 1589. For Sistine iconography, see Corinne Mandel, "Intro-

duzione all'iconologia della pittura a Roma in éta sistina" in Maria Luisa Madonna (ed.), *Roma di Sisto V: Le arti e la cultura*, Rome: De Luca, 1993, 3–16.

30 Gabrielli, *Dialoghi*, [fol.ii.r]: "Ho con molta mia spesa figurati nel presente libro tutti gl'instrumenti, & essercitij, che vi intervengono."

31 William Eamon, *Science and the Secrets of Nature*, Princeton: Princeton University Press, 1994, 137, for the standard claim that the book is an unburdening of secrets for the benefit of the world.

32 Gabrielli, *Dialoghi*, [fol. ii.r].

33 Attilio Milano, *Storia degli ebrei in Italia*, Turin: Einaudi, 1992, 258; Robert Bonfil, *Jewish Life in Renaissance Italy*, trans. A. Oldcorn, Berkeley: University of California Press, 1991, 72–3.

34 Jean Delumeau, *La vie économique et sociale de Rome dans la seconde moitié du XVIe siècle*, Paris: E. de Boccard, 1959, vol. 1, 404–14; and more recently Carla L. Keyvanian, "Charity, Architecture and Urban Development in post-Tridentine Rome: The Hospital of the SS.ma Trinità dei Pellegrini e Convalescenti (1548–1680)," PhD thesis, Massachusetts Institute of Technology, Department of Architecture, 2000, 141 and 150–8. Keyvanian mentions that Sixtus funded the hospital through taxes on wood unloaded at the river port, a tax on playing cards, and the concession of a patent on the production of glass, which was, of course, Magino's. See Charles Burroughs, "Opacity and Transparence: Networks and Enclaves in the Rome of Sixtus V," *RES: Anthropology and Aesthetics* 41 (Spring 2002): 56–71.

35 For the poverty of Rome's permanent residents see Delumeau, *La Vie économique*, vol. 1, 503; Jean Delumeau, *Rome au XVIe siècle*. Paris: Hachette, 1975, 123–4. Renata Ago, *Economia barocca, mercato e istituzioni nella Roma del Seicento*, Rome: Donzelli, 1998, 7–8, 12–13, dates attempts to introduce the Arte della Seta into Rome to 1568, and then in 1588 with Pietro Valentino of Pienza, and the Neapolitan Giovan Battista Corcione with a Milanese, Bernardo Crivelli. Delumeau, *Rome*, explains that Valentino was made a "prefect of the silk trade" in 1588, in charge of affirming that each of the Papal States was planted with mulberry trees (124); see also Delumeau, *La Vie*

économique, 502–10, for the silk and wool industries in late sixteenth-century Rome. A recent explanation of the Italian silk industry, with detailed explanation of Magino's enterprise, is in Molà, *Silk Industry*, 204–14.

36 Delumeau, *La vie économique*, 503–4; see also Ago, *Economia barocca*, 12–13, for immigrant artisans who came to Rome to practice specialized luxury-producing trades, arguing that the economy was suitably lively to support such enterprises. See also Katherine Rinne, "The Landscape of Laundry in Late Cinquecento Rome," *Studies in the Decorative Arts* 9 (2001–2): 34–60.

37 For example, Gabrielli, *Dialoghi*, fols. 12, 49 and elsewhere. Robert Bonfil attributes the voicing of a similar notion, that Jews were useful to Christian society because they were well suited to business, to the later publication by Simone Luzzato, *Discorso circa il stato de gl'hebrei*, Venice, 1638. See Bonfil, *Jewish Life*, 73–4.

38 Fol. iiv. "Sarfati" is Hebrew for "French"; see Toaff, *Nazione Ebrea*, 42. For help with the Hebrew I am grateful to Shaye Cohen and Saul Olyan from the Judaic Studies department at Brown University.

39 Gabrielli, *Dialoghi*, fols iiir–v. A Sebastiano Tellarini is noted as the author of *Nella morte di Alessandro Cardinale Farnese*, Rome: Tito e Paolo Diani, 1589; see LAIT (Libri antichi in Toscana 1501–1885, Catalogo cumulato di edizioni antiche conservate in biblioteche toscane, http://lait.signum.sns.it/Isis/servlet/Isis?Conf/usr/local/IsisGas/laitConf/lait.syst_ext.file&SrcWin1&Opt browse&Gizmono&Dsfr22744 (accessed 29 August 2011). The book is not listed in EDIT 16, and would seem to be one of many of the same title commemorating the death of Cardinal Farnese printed by that publisher the same year. I have been unable to trace the whereabouts of this publication or other publications by this person.

40 *Dialoghi*, fol. 87.

41 Gabrielli, *Dialoghi*, [fol. iii.v] "E sol pensando al giovamento humano / Trà boschi alpestri, e trà selvaggie fronde / Ho speculato tanto al monte, e al piano, / Che n'ho scoperto al fin virtù profonde, / Che de l'herbe, ch'io colgo con mia

mano / Ne fò mistura artificiosa: d'onde/ Vien oglio chiaro, e christallo, che tanti / Palagi splender fà, come diamanti."

42 For which see Lincoln, "The Jew and the Worms".

43 Vesalius (1543), xii; and Fontana, *Della trasportatione,* frontispiece.

44 The arms at the lower right show three (rather truncated) fleurs-de-lis over an eight-pointed star. An eight-pointed star and three rosettes appear on a crest used by the Sarfatti family in Padua (where Magino was from), according to Cecil Roth, "Stemmi di famiglie ebraiche italiane," in D. Carpi, A. Milano, and A. Rofé (eds.), *Scritti in memoria di Leone Carpi*, Jerusalem: Sally Meyer Foundation, 1967, 165–84. Roth describes the shield as containing "tre fiordalisi" but they are inexplicably pictured in his article as rosettes. He does not mention anything resembling the coat of arms on the left, which contains what looks to be a harp or a shield, possibly referring to David. For identification of the Sarfati coat of arms, see Jütte, "Handel."

45 For identifying signs and Jewish clothing, see Moses A. Shulvass, *The Jews in the World of the Renaissance*, Leiden: Brill, 1973, 348–9; Diane Owen Hughes, "Distinguishing Signs: Ear-Rings, Jews and Franciscan Rhetoric in the Italian Renaissance City," *Past and Present* 112 (1986): 3–59; and Benjamin Ravid, "From Yellow to Red: On the Distinguishing Head-Covering of the Jews of Venice," *Jewish History* 6, nos. 1–2 (1992): 179–210. According to Ravid, wearing a black hat was almost as definitive in assuming the identity of a Christian as a yellow or red hat was in identifying oneself as a Jew. Jews were allowed to wear the black hat and therefore to look indistinguishable from Christians only for specific reasons (187). In 1496 the Venetian Council of Ten replaced a law stating that all Jews should wear a yellow circle on their clothing with one commanding them to wear a yellow *baretta*, visible at all times (183).

46 From a charter of 1548: Ravid, "From Yellow to Red," 183–5 and 187.

47 Kenneth R. Stow, *The Jews in Rome, vol. 2: 1551–1557*. Leiden and New York: Brill, 1997; and K. R.

Stow, *Theater of Acculturation: The Roman Ghetto in the Sixteenth Century*, Seattle and London: University of Washington Press, 2001.

48 "Et di più che esso Maggino et suoi figlioli, ò figliole, che nasceranno in questa Città et nel stato ecclesiastico non siano obligati à contribuire con l'università delli hebrei di Roma, ò altri di detto stato à pesi, collette, ò altri carichi cosi imposti come da imporsi, cosi ordinarie, come straordinarie per qual si voglia et necessarijssima causa, della quale fosse necessario farne espressa mentione nelli presenti capitoli, oltra all somma di scudi vinticinque d'oro in oro l'anno quail habbino à servire per sovventione de i poveri ammalati et per il non sia sottoposto ad alcun bando o prohibitione ò suggettione da Sommi Pontefici et da detta Camera overo dall Illustrissimo Camerlengo ò Ilustrissimo Vicario ò Governatore et altri ministri di S.Sta. passati presenti et futuri fatte, ò che si facessero con li hebrei cosi di Roma come di detto stato per qual si voglia causa Reservandogli però tutti li privilegij et emolumenti et ogni'altra cosa concessi ò che si concedessero per l'avvenire dalla Santità di Nostro Signore et R. Camera et altri superiori à detti hebrei come di sopra stantia massime detta contributione delli scudi venticinque predetto l'anno, et promette detta Camera condederli patenti et ogni altra provisione sopra cio necessario et opportuna eccetto per il portare il segno solito portasi dalli altri Hebrei." ASR, Not. RCA vol. 1078, fol. 1014r.

49 Magino's signature (Mazino del q. Gabriel hebreo) using Venetian spelling appears on a document of 10 July 1587 in ASR, 30 Not. Cap. uff. 24, vol. 95 (1587), fol. 737v, in which Magino signs over part of his profits to Camilla Peretti. His name appears on the title page of the *Dialoghi* as "Magino Gabrielli Hebreo Venetiano" and in the lettering around the portrait as "Maggino Hebreo Venet." Three out of five cosigners of the document above refer to him as "Mazino hebreo", one as "Magino hebreo," and documents in the Roman archives deal with his unusual name in various ways, the most common of which was "Magino q. Gabrielli hebreo."

50 Jan van der Straet (Giovanni Stradano or Johannes

Stradanus), from *Vermis Sericus* see Marjolein Leesberg and Huigen Leeflang, "Johannes Stradanus, "in F. W. H. Hollstein, *The New Hollstein, Dutch & Flaming etchings, engravings and woodcuts, 1450–1700,* [vol. 71] Ouderkerk aan den Ijssel: Sound & Vision Publications, 2008, part 3, 33, cat. 346. see Alessandra Baroni Vannucci, *Jan Van Der Straet detto Giovanni Stradano flandrus pictor et inventor,* Milan: Jandi Sapi Editori, 1997, 410, cat. 699. The book is *Vermis Sericus,* Antwerp, 1583. Published by Philip Galle and engraved by Karel van Mallery for Costanza Alamanni, wife of Raffaello de' Medici.

51   For a discussion of Magino and Jewish access to the secrets of nature, see Jütte "Handel." (2005), and Ariel Toaff, *Il prestigiatore di Dio: Avventure e miracoli di un alchimista ebreo nelle corti del Rinascimento,* Milan: Rizzoli, 2010.

52   Gabrielli, *Dialoghi,* frontispiece: "Del cerchio eccelso nella destra parte / Come nasca sì cibi, opri la seta / Il Verme apprende, e sotto con qual'arte / Si caccia integra da' bucciuoli, e lieta / La naspan, tingon, tessono, e fan parte / Gl'Angeli di bel drappo in l'altra meta; Sopra, per la corona, e giusto motto / Sai che à questa ogn'altra arte stà di sotto."

53   Archivio Segreto Vaticano, Sec. Brev., Registro 129, fol. 294r, 4 July 1587.

54   Gabrielli, *Dialoghi,* [fol. v. r]: "Magino de Gabriele Hebraeo in Civitate Venetiarum commoranti, viam veritatis agnoscere, & agnitam custodire." C. Vittorio Massimo, *Notizie istoriche della Villa Massimo alle Terme,* Rome: Salvivcci 1836, 122, also notes this, giving the usual beginning as "Salutem et Apostolicam benedictionem," although other variants occur. The printing privilege for Diana Mantuana began, "Cum sicut accepimus dilecta in Christo filia Diana Mantuana uxor dilecti filii Francisci Cipriani Architecti." See Lincoln, *The Invention of the Italian Renaissance Printmaker* (2000), 189.

55   Gabrielli, *Dialoghi,* proemio [fol. v. r].

56   ASR, 30 Not. Cap. uff. 16, vol. 10 (1588), fols. 153r–v, 31 January 1588.

57   Written in pen at the bottom of the official notice that the children of Leonardo Parasole, who died intestate, will be considered his rightful heirs: "Io Curtio trombetto ha publicato il presente bandi nella sola Campidoglio essendo publica audientia et per roma ali lochi soliti" (ASR, 30 Not. Cap. uff. 21, vol. 84 (1612) fol. 320r.

58   Almost the entire output of the Gigliotti press, both before and after the printer's death when it was being run by his widow and son, was printed in quarto and octavo. The only publication in folio besides Magino's registered in the ICCU (http://edit16.iccu.sbn.it/web_iccu/ihome.htm) was published in the same year, Giuseppe Castiglione's unillustrated *De victoria britannica Iosephi Castalionis ode,* Rome: Haeredes Ioannis Giliotti, 1588, which consisted of a single sheet. There are only five other illustrated books from a total output of sixty-four listed in the EDIT16 database catalogue of sixteenth-century printed books.

59   For the vestibule as described by Carlo Sigonio in his treatise on dialogues, see Jon R. Snyder, *Writing the Scene of Speaking: Theories of Dialogue in the Late Italian Renaissance,* Stanford: Stanford University Press, 1989, 62.

60   Gabrielli, *Dialoghi,* fol. 58.

61   Gabrielli, *Dialoghi,* fol. 2: "H: Che piu? già si è stampata l'inventione historiata di bellissime figure, le quali rappresentano ciò che distintamentesi deve fare in detta materia, e sono universalmente publicati gl'utilissimi suoi secreti in volgare, e latino, acciò si possano intendere ancora di là da Monti, e credo certo che ne porterà hoggi un libro à mia consorte, perche ella gli hà detto che non ritorni più senza portarneli."

62   See Kenneth R. Stow, "The Church and the Jews," in David Abulafia (ed.), *The New Cambridge Medieval History,* vol. 5, Cambridge: Cambridge University Press, 1999, 204–19 (206); and K. R. Stow, *Catholic Thought and Papal Jewry Policy, 1555–1593,* New York: Jewish Theological Seminary of America, 1977. See also Stefanie B. Siegmund, *The Medici State and the Ghetto of Florence,* Stanford: Stanford University Press, 2006, xvii–xix, 19, and 53. For the role of Jews in the papal economy, see Bonfil, *Jewish Life,* 71–7; and Bernard D. Cooperman, "Ethnicity and Constitution Building among Jews in Early Modern Rome," *AJS Review* 30, no. 1 (2006): 119–45.

63   "l'inventore al mio parere riuscirebbe delli più

grandi, e riguardevoli Hebrei che siano per avventura nati dal savio Rè Salamone in quà; eccettuandone però gl'huomini Santi, e diletto di Dio, che da quell Popolo ne son venuti": Gabrielli, *Dialoghi*, fol. i.

64 Gabrielli, *Dialoghi*, fols. 1–2.

65 Gabrielli, *Dialoghi*, privilege: "acciò che I detti nostri sudditi possano più facilmente imparare l'esercitio di queste nuove inventioni, coi ci havete promesso di fare che in tutti I giorni di fiere, nelle piazze & luoghi piublici si ritrovaranno & potranno leggere in stampa edificij d'esercitar tal'invention nuova, dalla quale i predetti sudditi aiutati, facilmente possino essercitare l'istessa arte, con poche spese, & gran giovamento, pur che noi con voi vogliamo usare di qualche liberalità."

66 David Gentilcore, *Medical Charlatanism in Early Modern Italy*, Oxford and New York: Oxford University Press, 2006, 2. The discussion of charlatans here depends heavily on Gentilcore's excellent book. See also W. Eamon, *The Professor of Secrets. Mystery, Medicine and Alchemy in Renaissance Italy*, Washington, DC: *National Geographic,* 2010, especially chapter 19.

67 Gentilcore, *Charlatanism*, 5. Also, 163 for merchant charlatans' preference for obtaining privileges, 302 for their use of spectacle.

68 For charlatans' use of printed and oral forms of publication, as well as legal-looking *bandi* and decrees to advertise their wares, and broadsheets as published protection of their intellectual property, see Gentilcore, *Charlatanism*, 335–50; and Eamon, *Science*, chapter 7.

69 Gentilcore, *Charlatanism*, 311.

70 H. R. F. Brown, *The Venetian Printing Press*, London: J. Nimmo, 1891, 106.

71 Gabrielli, *Dialoghi*, fol. 6r: "Non è dubbio ch'ella è la più necessaria cosa, & il maggior dono che possa l'huomo ricevere dal supremo Creatore, peroche mediante quella speculiamo le cose divine, ci pacifichiamo & uniamo con Dio, onde ne conseguiamo poi, e ciò che s'appartiene al viver, e comodo nostro, alla giustitia, e sapienza, & à gl'essercitij manuali, e finalmente la felicità eterna."

72 The author is certainly profiting from printed versions of antique discussions of the origin of inventions, such as Polydore Vergil's *De rerum inventoribus*, printed first in 1499 but also Pliny's *Natural History*, which had been printed in Venice in 1561. See also George W. McClure, *The Culture of Profession in Late Renaissance Italy*, Toronto: University of Toronto Press, 2004, 72.

73 For Colorni and Garzoni, see McClure, *Culture*, 83; for the partnership between Magino and Colorni in 1598, see Jütte, *The Age of Secrecy.*

74 See Virginia Cox, *The Renaissance Dialogue: Literary Dialogue in its Social and Political Contexts, Castiglione to Galileo*, Cambridge and New York, Cambridge University Press, 1992, 73–5.

75 Gabrielli, *Dialoghi*, 11–16 (note: the pages are misnumbered in all copies, these are consecutive pages):

"M: Buon giorno alle Signorie vostre, me perdonino di gratia che me è bisognato tardare alla Stampa sino ad hora che non hò potuto haver prima spedite le figure.

"H: Siate per mille volte il ben venuto, e non potevate arrivare più a proposito d'adesso, perche il Signor Cesare, & io habbiamo ragionato quasi due hore delle cose vostre, e pur quando veniste dicevamo che se i vermicelli havessero spirito, ò intendimento non si dovrebbono satiar giamai di honorar, e celebrar' voi lor si grande illustratore, e noi tutti appresso che habbiamo, la Dio mercè, intelletto dovemo tenervi in ammiratione, e stima, come benefattor publico di tutto il Christianesimo, & in particolare di molti fideli, che impiegarete con lor notabile utile ne' traffichi vostri, non è egli vero Signor Cesare?

"C: E verrissimo, M. MAGINO, anzi il Signor Horatio ha discorso sopra molte belle inventioni, & à lungo sopra diversi inventori, & ha concluso che l'inventioni vostre siano (16, which is next page) mirabili, come quelle che apporteranno utili à tutti perche i Principi aumentaranno l'entrate de' datij, i Popoli saranno più denarosi non estraendo denari per cavar seta di levante, e d'infideli; i mercanti, e quei che attendono all'arti cresceranno i lor traffichi, e molte private persone riceveranno inviamento mediante l'opere vostre oltre che le povere donne si guadagnaranno meglio il vivere con questo che con altro essercitio, e quei c'haveranno celsi ne raddoppieranno il frutto per vendergli nella seconda raccolta; per le quai cose dice e lo confermo io che voi siate un inventore più che mediocre fra i grandi, & huomo di sublime ingegno; e vi stavamo ambedue aspettando con quella voluntà che voi aspettate il Messia."

76   See Adriano Prosperi, "L'Inquisizione Romana e gli ebrei," in M. Luzzati (ed.), *L'Inquisizione e gli ebrei in Italia*, Rome: Laterza, 1994, 67–120. Prosperi makes the point that in Rome the Inquisition was formed to eradicate heresy among Christians, not to persecute Jews, who could not be heretics as they were not Christians who believed in false doctrine, nor were they infidels as they shared a history with the Christians.

77   Wayne C. Booth, "The Self-Conscious Narrator in Comic Fiction before Tristram Shandy," *PMLA* 67, no. 2 (1952): 163–85, and reprised in his *The Rhetoric of Fiction*, 2nd ed., Chicago: University of Chicago Press, 1983, 221–40. *Don Quixote* was published in 1605, *Tom Jones* in 1749/50, and *Tristram Shandy* in 1759.

78   Booth, *Rhetoric*, 225.

79   The third dialogue contains six figures as opposed to nine in the second dialogue. In the third dialogue the headings and some of the instructions in the key have been changed.

80   Gabrielli, *Dialoghi*, fol. 17r, "Dialogo Secondo. Nel quale M. Magino Inventore insegna alla Signor Isabella il vero modo di essercitar l'arte della seta, & I secreti suoi; e di più gli dimostra in varie figure vagamente disegnato tutto ciò, che à detto esercitio si appartiene."

81   Gabrielli, *Dialoghi*, fol. 18r: "M.: in questa figura vederete ritratto dal naturale tutto ciò che vi hò detto sin quì, & il medesimo ordine osservaremo in tutti gl'altri magisterij che vi occorreranno; e sotto à ciascuna figura sarà la dichiaratione delle lettere dell'alphabeto, e del contenuto in essa: però sommariamente."

82   Gabrielli, *Dialoghi*, fol. 21r: "I: Resto molto sodisfatta d'haver visto questa figura, e mi par bellis-

sima inventione il metter cosi diligentemente ogni cosa in disegno: circa l'accomodare i bocciuoli secondo, la vostra regola conosco chiaro ch'è la migliore che si possa fare: mà vorrei sapere quali bocciuoli son più buoni per questo effetto, perche ce ne sono di gialli, di bianchi, di rancetti, e di verdi chiari come sapete"

83   Gabrielli, *Dialoghi*, fol. 22r: "M: mà sarà buono adesso che voi guardiate bene questo secondo quadro, acciò che non vi scordasse il modo di preparare il seme, e dapoi seguitaremo à dire come dovete accomodarlo per farlo nascere. I: Sarà ben fatto, date quà che io vederò voluntieri la figura per intrattenimento ancora, non tanto perche io m'habbia scordato l'ordine vostro. M: Eccovela."

84   *Directions for the breeding and management of silkworms: Extracted from the treatises of the Abbé Boissier de Sauvages, and Pullein. With a preface, giving some account of the rise and progress of the scheme for encouraging the culture of silk, in Pennsylvania, and the adjacent colonies*, Philadelphia, 1770.

85   Marco Girolamo Vida, *Silkworms: A Poem in Two Books. Written originally in Latin by Marc. Hier. Vida, Bishop of Alba, and now translated into English*, trans. D. O. M. London: J. Peele, 1723, 5–7.

86   Gabrielli, *Dialoghi*, fol. 25.

87   Gabrielli, *Dialoghi*, fol. 26: "I: Horsù io vi capito benissimo lasciatemi solo dar un'occhiata à questo terzo quadro & apparecchiatevi in tanto à mostrarmi, come si debbano governare i vermi dapoi che sono venuti nella luce di questo Mondo."

88   Fontana, *Della trasportatione*.

89   Gabrielli, *Dialoghi*, fol. 29.

90   Gabrielli, *Dialoghi*, fol. 29.

91   I have seen three copies of the book: in the Vatican Library, the Biblioteca Nazionale Centrale di Firenze, and the Biblioteca Nazionale di Napoli. I have seen photographs of the copy at the Biblioteca Marciana, Venice. I have not seen the copies at the National Library of Israel, or the Bodleian Library. A copy in the Library of Congress, Washington DC, may be seen on-line but lacks the proemio and first frontispiece, as well as the engraving of Sixtus V. The second frontispiece, missing in the Vatican copy, is placed at the beginning of the book. The Vatican copy also lacks fols.

43–9, including Figura Settima, set in Naples.

92 Gabrielli, *Dialoghi*, fol. 49:

"I: Il disegno è si ben fatto, e da voi così sufficientemente dichiarato, che io semplice donna non saprei mai che cosa mi vi apporre, ne di che più intorno à cio domandarvi: però anco in questo ottavo si come nel terzo un sol difetto ci sò conoscere, & è che l'intagliatore s'ha scordato d'intagliarvi le lettere dell'Alfabeto, secondo che sono nel sommario in dichiaratione e della figura; verò è che per esservi poche persone figurate, & apparentemente espresso ciò che quelle fanno, si potrà in ogni modo senza esse lettere intendere il suggetto: dovendone poi far stampare dell'altre, fatevele pur segnare, acciò non s'imputasse l'errore à voi per la negligenza, ò per la troppa fretta: ma scopritemi un poco questo bel secreto, che havete, da porre à lavorare la seta i vermi senza frasche?

"M: Havendole da stampar di nuovo, come crederò di essere sforzato, perche nel Regno di Napoli, e nell'altre restanti parti d'Italia siano ancora adoperate, son resoluto di megliorarle in più di quattro luoghi, ne vi dovete maravigliare se in questa prima impressione, non vi si scorge tutta quella eccellenza che desiderareste, considerando che la stagione ci sia sopraggiunta prima che io habbia voluto conferir punto i secreti miei con persona (fuor che nell'isperienza ad alcuni Principi dimostrata) per non haver finiti d'espedir tutti i privilegi, i quali havendo ultimamente ottenuti, hanno causata furia grande d'intagliar, & stampare le figure mostratevi senza matura revisione, di maniera che la mattina s'è in un subito fatto il disegno, e la dichiaratione d'esse, & il giorno si son date alla stampa, non con buona voluntà di quel che n'ha hauta la cura: mà forzatamente per non perdere anco questo anno; oltre che per voler far una cosa perfetta convien tornarvi più d'una volta."

93 See Martine Boiteux, "Les juifs dans le carnaval de la Rome moderne," *Mélanges de L'École Française de Rome Moyen Âge et Temps Modernes*, 88, no.

2 (1976): 745–87; Barbara Wisch, "Violent Passions. Plays, Pawnbrokers and the Jews of Rome, 1539," in A. Terry-Fritsch and E. F. Labbie (eds), *Beholding Violence in Medieval and Early Modern Europe*, Aldershot: Ashgate, 2012, 197–213.

94 For Sommi (*c.*1525–*c.*1590), see Leon de' Sommi, *Quattro dialoghi in materia di rappresentazione sceniche*, ed. Ferruccio Marotti, Milan: Il Polifilo, 1968; and Donald Beecher, "Leone de' Sommi and Jewish Theater in Renaissance Mantua," *Renaissance and Reformation* 17, no. 2 (1993): 5–19. See also Jütte, "Abramo Colorni," 457–8.

95 For the benefits of strongly rhythmic music at low volumes in the raising of leeches, see John Colapinto, "Bloodsuckers," *The New Yorker* (25 July 2005): 72.

96 Gabrielli, *Dialoghi*, fol. 72:

"C: Mà nel vostro disegno ci hò veduta quella donna signficata per A. che par che si levi del petto il seme si come facea nel primo raccolto: il che voi havete detto che non si deve fare: e similmente è scritto nella dichiaratione del A sotto alla figura; e perche non ogn'uno sà leggere, ma tutti quegli che non sono ciechi veggono, potrebbe esser che alcuno reggendosi secondo il ritratto delle fallisse, di modo che più sicuro era il mutare il disegno.

"M: La brevità del tempo è causa di questo poco disordine, che oltre l'improprietà della vista non è di momento nessuno, perche deviamo presupporre che almeno uno per casa sappia leggere, e se non; come potrà succedere in qualche strana villa; lo sapra intendere il vicino, & uno l'insegna à molti."

See Eamon, *Science*, 127, on the growth of literacy among previously illiterate people (artisans and women) for scientific books in the sixteenth century, and Molà, *Silk Industry*, 428, for many actual silk masters being illiterate.

97 Gabrielli, *Dialoghi*, fol. 49: "non teniate i cavalieri piu netti, e politi, che sia possibile, e vedendone alcuno di color giallo, e spiacevole, e non simile a gl'altri, lo portarete all'aria à ripiglar lena [le là?] per un quarto d'hora, avvertendo pero, che il

sole non lo percuota, che altramente invece di
medicina gli dareste il veleno."

98   ASR, 30 Not. Cap. uff. 28, vol. 10, (1588) fols.
717r–720v

99   A Francesco Solaro appears among the *intagliatori*
paid for working on the façade of St. Peter's
under Carlo Maderno throughout 1613; see
Oskar Pollak, "Ausgewählte Akten zur Geschichte
der römischen Peterskirche (1535–1621)," *Jahrbuch
der Königlich Preussischen Kunstsammlungen* 36
(1915): 21–117 (81–100).

100 The document is published in Masetti Zannini,
287: "Che detto messer Paulo sia tenuto et obligato come promette et si obliga di far stampare
a tutte sue spese eccettuata la carta qual gli darà
detto messer Maggino in carta reale, cinquanta
milla bandi sopra le caraffe conforme il Motu
proprio emesso da Nostro Signore al detto messer
Maggino, per prezzo di giuli 9 il cento." The
population of Rome in 1601 was 101,912; see
Eugenio Sonnino, "Between the Home and the
Hospice," in John Henderson and Richard Wall
(eds.), *Poor Women and Children in the European
Past*, London: Routledge, 1994, 96.

101 ASR, 30 Not. Cap. uff. 28, Tondinus, vol. 11
(1588), fols. 1046r–v:

> "Conventiones inter D. Contugum de Contu-
> ghis et Maginum de Gabrielis Hebreum Mag-
> nificus dominus Contugus de Contughis
> Romano sponte ac omnibus et Promisit D.
> Manginò de Gabrielis Veneto Hebreo presente
> et finire de Compore et corregere alla stampa
> Tre dialogi sopra li Inventione di far lo seta dui
> Volte li Anno, et Migliorarla Intitulata all santità
> D. N. Ste. PP Sixte V che sonno Intutto Libri
> mille et cinquecento che al presente si stam-
> pono nella stamparia delle Herede del q. M.
> Giovanni Gigliotti, et tutto volta che sarra
> bisognio, et da esso M. Mangi sono recerca et
> debba dare adetto stampa per compore, et cor-
> regere dette opre sinò al fine di detto opre di
> Tre dialog(i) da detti Libri 1500 che già comin-
> siati, et Il tutto fare con quella diligentià che
> in simili opre far si suole. . . .
>
> "*Die quinta men. Juilj 1588*
> "D. Maginus de Gabrielis ante. d. sponte &

quietavit d. Contugum de Contughis presen-
tem & de operis con laborerijs per ipsum in
Supradetto Instrumento fueri promissis Magi-
norum asseruit et asserit esse Juxtae dictas con-
ventiones factarum ulteriusque idem Contughus
promisit facere revidere unam declarationem
seu dialogum versuum lingue latine vulgaris, et
hebraici ac ab dicto Magino presente & lecita
duo decim cum di medio fosse fuit habuisse et
sunt ad computum scutorum triginta quinque
ipsi debitorum ex causa de qua in supradetta
Instrumento de quibus sine previdicto residui
quod esse calcularunt scutorum viginti duorum
cum dimedio quietavit cum pacto & et tacto
calamo et respec[. . .]s scripteris Jurarunt supra
quibus & Actum Romae in offitio me & presentibus d. Lenonti Rentio Sabinense et Mercurio accursio testibus.

102 Gabrielli, *Dialoghi*, 6: "E di tutte queste cose poi
potrete ricevere sodisfattion maggiore da Contugo
Contughi carissimo amico mio, giovane come in
parte sapete, tanto desideroso di virtù, quanto
tenuto basso dalla sorte, dal quale anch'io ho
cavata la piu parte delle mie cose."

103 Gabrielli, *Dialoghi*, 13: "H: ma certo se si trovasse
alcuno, che come costui fà à noi medesimi vedere
le bellezze, ò brutezze del volto nostro così ne
facesse la malitia, ò bontà conoscere del core
altrui, fuor di dubbio molto più nobile inventione
sarebbe, e quella sola, che conservaria il Mondo
in grandissima tranquillità, e pace, come suol dire
il nostro Contugo che da questo spererebbe ricevere almeno qualche honorato inviamento nella
servitù d'alcun magnanimo Prelato."

104 McClure, *Culture*, 76–7, in which he also discusses
Tomaso Garzoni as a "compositori de' libri."

105 ASR, 30 Not. Cap. uff. 28, Tondinus, vol. 11
(1588), fol. 1046r. 24 May 1588: "et il tutto fare
con quella diligentià che in simil' opre far si
vuole."

106 "Questo adunque che s'è scritto, finche più particolare relatione ne sopragiunga, dovrà bastare se
non a sodisfare compiramente la nobile curiosità,
& perfetto discorso d V.S. Ill. & di quelli, che
desiderando veder quello, che n'era pervenuto a
mia notitia à ricercarne; almeno spero che sup-

plirà a me per usare dell obligo in che il desidero suo, & loro mi haveva posto, massime considerando la brevità del tempo, nel quale si può dire siano abbozzate queste carte sopra material difficile, & non ben chiara, & la poca commodità mia." Contugo Contughi, "Relatione della gran città del Quinsay, & del Rè della China." The version of the esssay I read is dated 1584 and was printed in Giulio Belli's *Praxis prudentiae politicae*, Frankfurt, 1611, 439–55, but a manuscript version exists dated 1583 and is mentioned in Boleslaw Szezesniak, "The City of Utopia: Quinsay," *Literature East & West* 4, 2/3 (1957) 5. On the collections of travel descriptions and ambassadorial letters known collectively as *Tesori politici*, see Simone Testa, "From the 'Bibliographical Nightmare' to a Critical Bibliography: *Tesori politici* in the British Library, and Elsewhere in Britain," *Electronic British Library Journal* (eBLJ) 1 (2008), http://www.bl.uk/eblj/2008articles/article1.html, accessed 6 August 2008. According to Testa, "This report is another edition of Marco Polo's description of China, with additions taken from other travellers" (11 n. 29). Contughi's *Relatione* appears in printed form as early as 1601 in *La seconda parte del Thesoro politico nella quale si contengono Trattati, Discorsi, Relationi, Ragguagli, Instruttioni di molta importanza per li maneggi, interessi, pretensioni, di pendenze, e disegni dei principi.* Milan: Girolamo Bordone e Pietro Martire Locarni, 1601. His description was indeed eventually replaced in succeeding editions by a newer description of China, *Iacobi Pantogia de amplissimo Sinarum regno*; see Testa, "*Tesori politici*" 28 n. 74.

107 Gabrielli, *Dialoghi*, fols. 12–13.

108 I am grateful to Daniel Jütte for pointing out the references to payment to "Magino hebreo" in 1598 for glass windows for the "novo palazzo apostolico de San Giovanni Laterano per tutto l'apartamento delle stanzie nobile dove havrà da habitare Nostro Signore." See Corbo, *Fonti*, 76.

6 Talking Pictures

1 Pamela Long's work is indispensable for discussions of the role of literature in raising the status of crafts. See particularly her "Power, Patronage, and the Authorship of *Ars*."

2 Magino Gabrielli, *Dialoghi di M. Magino Gabrielli Hebreo Venetiano . . .* , Rome: per gli Heredi di Giovanni Gigliotti, 1588. The pages are misnumbered at this juncture. This should, if numbered correctly, be folio 11, but it is the first of two folios numbered 13:

"C: Però dunque si chiama Musica dalle Muse: e fine qui parmi che habbiate ragionato delle lettere à sufficienza, solo vorrei sapere quei che da prima si servirono de Dialoghi, e brevemente com'esser debbano: peròche questo discorso, che insieme facciamo haverà forsi qualche forma di Dialogo. H: A mio giuditio non debbono intervenir nel Dialogo più di quattro persone, e forse quattro in un dialogo solo porne à discorrere è vitioso; più di tanti fariano una comedia, e non un dialogo; due, ò tre è il più usato & lodato modo; puossi sopra ogni materia fare, eccetto pastorale, che muta il nome in egloga; dev'essere di cosa presupposta vere, e non finta; non hà da esser troppo breve che si cangiarebbe in iscena, e principalmente vi si deve osservare il decoro, e la proprietà de gl'interventori." For the number of people in a dialogue see Jon R. Snyder, *Writing the Scene of Speaking: Theories of Dialogue in the Late Italian Renaissance*, Stanford: Stanford University Press, 1989, 99.

3 Jean-Louis Fournel, *Les dialogues de Sperone Speroni: Libertés de la parole et règles de l'écriture*, Marburg: Hitzeroth, 1990, 160–2.

4 Snyder, "Writing," 86–133; and Wayne A. Rebhorn (ed. and trans.), "Sperone Speroni," in *Renaissance Debates on Rhetoric*, Ithaca: Cornell University Press, 2000, 111. Sperone's second sojourn in Rome lasted from 1573 to 1578.

5 Virgina Cox, *The Renaissance Dialogue: Literary Dialogue in its Social and Political Contexts, Castiglione to Galileo*, Cambridge and New York: Cambridge University Press, 1992, 70–6.

6 Cox, *Renaissance Dialogue*, 71.

7 Cox, *Renaissance Dialogue*, 72.

8 Snyder, *Writing*, 100.

9 Snyder, *Writing*, 96–102.

10   Cox, *Renaissance Dialogue*, 178 n. 27: "Speroni analyses the components of the writer's 'decoro particolare': he must write as is fitting to his 'grado', his 'costumi', his 'professione' and 'le leggi della città'." Speroni's probable motivations and intentions for moving from a concern with the decorum of the interlocutors, in the first part of his book, to a concern for the comportment of the writer by the end, are discussed most recently in Cox, *Renaissance Dialogue*, 76–7.

11   Cox, *Renaissance Dialogue*, 75–6.

12   For Sigonio, see Snyder, *Writing*, 55–8. Speroni's *Dialogo della rhetorica* appeared in 1546.

13   On Sforza Pallavicino, *Trattato dello stile e del dialogo*, Rome, 1662, see Cox, *Renaissance Dialogue*, 3 and 79.

14   Eugenio Gentilini, *Il perfeto bombardiero et real instruttione di artiglieri: sperimentata, & composta da Eugenio Gentilini . . .* , Venice: Gio. Antonio & Giacomo de'Franceschi, 1606; Angelo Viggiani, *Lo schermo d'Angelo Viggiani dal Montone da Bologna: nel quale per uia di dialogo si discorre intorno all'eccellenza dell'armi, & delle lettere . . .* , Venice: Giorgio Angelieri, 1575.

15   Geminiano Montanari, *Le forze d'Eolo: dialogo fisico-matematico sopra gli effetti del vortice, ò sia turbine, detto negli stati Veneti la Bisciabuova. Che il giorno 29 lvglio 1686 hà scorso, e flagellato molte ville, e luoghi de' territorj di Mantova, Padova, Verona, &c. / Opera postuma del sig. dottore Geminiano Montanari modanese, astonomo, e meteorista dello studio di Padova*, Parma: Andrea Poletti, 1694.

16   Sperone Speroni, *Dialogo della rhetorica*, trans. in Rebhorn, *Renaissance Debates*, 113–14 and 120.

17   Mikhail Bakhtin, *Problems of Dostoevsky's Poetics*, trans. Caryl Emerson, Minneapolis: University of Minnesota Press, 1984, 110. Virginia Cox provides a more historical discussion of this aspect of the dialogue in *Renaissance Dialogue*, 76–8.

18   Cox, *Renaissance Dialogue*, throughout, but particularly 62–3, 79; Snyder, *Writing*, 103; and Speroni on rhetoric as in n. 16, above.

19   Cox, *Renaissance Dialogue*, 5.

20   Cox, *Renaissance Dialogue*, 104–5, also mentions the use of diagrams in dialogues by Tartaglia and Galileo, and on p. 113 points to his *Dialogo dei due massimi sistemi* as a surprising example of a modern scientific treatise in dialogue form.

21   Gabrielli *Dialoghi*, fol. 18.

22   Galileo Galilei, *Dialogo sopra I due massimi sistemi el mondo Tolomaico e Copernicano*, ed. Ottavio Besomi and Masio Helbing, 2 vols., Padua: Antenore, 1998, 6: "Ho poi pensato tornare molto a proposito lo spiegare questi concetti in forma di dialogo, che, per non esser ristretto alla rigorosa osservanza delle leggi matematiche, porge campo ancora a digressioni, tal ora non meno curiose del principale argomento." Translation from *Dialogue Concerning the Two Chief World Systems*, trans. Stillman Drake, Berkeley: University of California Press, 1953, 6.

23   Snyder, *Writing*, 8 and 103; J. L. Heilbron, *Galileo*, Oxford and New York: Oxford University Press, 2010, 128.

24   Lorraine J. Daston, "Galilean Analogies: Imagination at the Bounds of Sense," *Isis* 75, no. 2 (1984): 302–10. See also Isabelle Pantin, "Une 'École d'Athènes' des Astronomes? La representation de l'astronomie antique dans les frontispieces de la Renaissance," in Emmanuèle Baumgartner and Laurence Harf-Lancner (eds.), *Images de l'Antiquité dans la littérature française: Le texte et son illustration*, Paris: Editions Rue d'Ulm, 1993, 87–99.

25   François Rabelais, *The Complete Works*, trans. Donald M. Frame, Berkeley: University of California Press, 1981, 193–201.

26   Rabelais, *Works*, 195.

27   Rabelais, *Works*, 197. See Jeffrey C. Persels, "Bragueta Humanistica, or Humanism's Codpiece," *Sixteenth Century Journal*, 28, no. 1 (Spring 1997): 79–99, who also clarifies the representation of effeminacy in the Renaissance to signify powerlessness (see particularly 91 n. 24).

28   Renaissance anti-academicism, particularly at court, has been much written about: see Mario Biagioli, "Galileo the Emblem Maker," *Isis* 81, no. 2 (June 1990): 230–58 (237); Warren W. Wooden, "Anti-Scholastic Satire in Sir Thomas More's *Utopia*," *Sixteenth Century Journal* 8, no. 2 (July, 1977): 29–45; Paul Oskar Kristeller, "Erasmus from an Italian Perspective," *Renaissance Quarterly* 23, no. 1 (Spring 1970): 1–14.

29   Alan Perreiah, "Humanist Critiques of Scholastic Dialectic," *Sixteenth Century Journal* 13, no. 3 (Autumn, 1982): 4; and J. Austin Gavin and

Thomas M. Walsh, "The Praise of Folly in Context: The Commentary of Girardus Listrius," *Renaissance Quarterly* 24, no. 2 (Summer 1971): 193–209.

30 Franco Giacomo, "Rabelais et Annibal Caro: Traditions, filiations et traductions littéraires," *Revue d'histoire littéraire de la France* 99, no. 5 (1999): 963–73, for Caro's burlesque and carnivalesque *La Nasea* (Rome, 1539) in the context of the Roman Accademia della Virtù.

31 *Opere di Galileo*, ed. Antonio Favaro, vol. 14, Florence, Tipografia Berbero, 1904, 297: "fui pregato da una mano di gentilhuomini di garbo e litterati di spiegarli i principii della geometria." See Joseph Connors, "St. Ivo alla Sapienza: The First Three Minutes," *JSAH* 55, no. 1 (March 1996): 38–57 (51).

32 Galileo Galilei, *Dialogo dei due massimi sistemi . . .* , Florence: Gio. Batista Landini, 1632.

33 Pantin, "La representation," 95: "L'astronome antique des pages de titre, après avoir été un simple emblème de la discipline, aus service d'une conception traditionelle du savoir, s'est donc trouvé engagé dans ses conflits, parfois à son corps défendant." See also L. Tongiorgio Tomasi, "Image, Symbol and Word on the Title Pages and Frontispieces of Scientific Books from the Sixteenth and Seventeenth Centuries," *Word & Image* 4 (1988): 372–82.

34 Mario Biagioli, *Galileo, Courtier*, Chicago: University of Chicago Press, 1993, 207; Marta Spranzi Zuber, "Dialectic, Dialogue and Controversy: The Case of Galileo," *Science in Context* 11, no. 2 (1988): 181–203, and her "Galileo's 'Dialogue on the Two Chief World Systems': Rhetoric and Dialogue," in Fernand Hallyn and Lyndia Roveda (eds.), *La rhetorique des textes scientifiques au XVIIe siècle*, Turnhout: Brepols, 2005, 97–114.

35 Volker Remmert, *Widmung, Welterklärung und Wissenschaftslegitimierung: Titelbilder und ihre Funktionen in der Wissenschaftlichen Revolution*, Wiesbaden: Harrassowitz in Kommission, 2005, 64–8. See also Roberto Paolo Ciardi and Lucia Tongiorgi Tomasi, "La 'scienza illustrata': Osservazioni su frontespizi delle opere di Athanasius Kircher e di Galileo Galilei," *Annali dell'Istituto Storico Italo-Germanico in Trento* 11 (1985): 68–79.

36 Miles Chappell, "Cigoli, Galileo, and Invidia," *The Art Bulletin* 62 (1975): 97–8 (91) n. 4. A fundamental text for Galileo's relationship to visual culture is Erwin Panofsky, "Galileo as a Critic of the Arts: Aesthetic Attitude and Scientific Thought," *Isis* 47, no. 1 (March 1956): 3–15. Galileo's own drawing is discussed in several places: Samuel Edgerton, "Galileo, Florentine 'Disegno,' and the 'Strange Spottednesse' of the Moon," *Art Journal* 44 (1984): 225–32; Mary Winkler and Albert Van Helden, "Representing the Heavens: Galileo and Visual Astronomy," *Isis* 83, no. 2 (June 1992): 215, for Galileo and his involvement in the Accademia del Disegno as a gentleman amateur. Eileen Reeves, *Painting the Heavens: Art and Science in the Age of Galileo*, Princeton: Princeton University Press, 1997, 6–9, for Galileo's awareness of optical effects in art and relationship to art and painting, a discussion sustained throughout the book. This is discussed in depth also, with particular relationship to the drawings of sunspots, by Horst Bredekamp, *Galilei der Künstler: Der Mond, die Sonne, die Hand*, Berlin: Akademie Verlag, 2007, especially 34–8, 78–82, and on his relationship to copperplate printing in particular, 189–208; see also Alexander Marr, *Between Raphael and Galileo: Mutio Oddi and the Mathematical Culture of the Late Renaissance*, Chicago: University of Chicago Press, 2011, particularly 167–9. David Freedberg, *The Eye of the Lynx: Galileo, his Friends, and the Beginning of Modern Natural History*, Chicago: University of Chicago Press, 2002, 349–50, discusses Galileo's dissatisfaction with the efficacy of pictures "to define what was essential about the things in nature."

37 Della Bella left Florence to train in Rome in 1633. Phyllis Dearborn Masser, "Presenting Stefano della Bella," *The Metropolitan Museum of Art Bulletin*, new series 27, no. 3 (November 1968): 159–76; P. D. Masser, "Costume Drawings by Stefano della Bella for the Florentine Theater," *Master Drawings* 8, no. 3 (Autumn 1970): 243–66, 297–317. Most of della Bella's theatrical designs in Paris and Florence were later than this early frontispiece design. See also Françoise Viatte, "Allegorical and Burlesque Subjects by Stefano della Bella," *Master Drawings* 15, no. 4 (Winter 1977):

347–65, 425–44. For della Bella's early work with theatrical subjects, see Ulrike Ilg, "Stefano della Bella and Melchior Lorck: The Practical Use of an Artist's Model Book," *Master Drawings* 41, no. 1 (Spring 2003): 30–43.

38 Galileo, *Dialogo Sopra*, 5: "Giudicai, come pienamente instrutto di quella prudentissima determinazione, comparir publicamente nel teatro del mondo, come testimonio di sincera verità."

39 Thanks to Sara Schechner for the identification of Copernicus's model. Remmert calls it a tellurium.

40 Galileo, *Dialogo Sopra*, 5: "A questo fine ho presa nel discorso la parte Copernicana, procedendo in pura ipotesi matematica, cercando per ogni strada artifiziosa di rappresentarla superiore, non a quella della fermezza della Terra assolutamente, ma secondo che si difende da alcune che, di professione Peripatetici, ne ritengono solo il nome, contenti, senza passaggio, di adorar l'ombre, non filosofando con l'avvertenza propria, ma con solo la memoria di quattro principii mal intesi."

41 Zuber, "Dialectic," 198; Remmert, *Widmung*, 66.

42 Anne Reynolds, "Galileo Galilei's Poem 'Against Wearing the Toga'," *Italica* 59, no. 4 (Winter 1982): 330–41. The poem was probably written while Galileo was in Pisa, 1589–92; professors were fined for not wearing this item; see Heilbron, *Galileo*, 60.

43 Galileo, *Dialogo Sopra*, 142: "Giornata seconda: Salviati: Prima che proceder più oltre, devo dire al Sig. Sagredo che in questi nostri discorsi fo da Copernichista, e lo imito quasi su maschera; ma quello che internamente abbiano in me operato le ragioni che par ch'io produca in suo favore, non voglio che voi lo giudichiate dal mio parlare mentre siamo nel fervor della rappresentazione della favola, me dopo che avrò deposto l'abito, che forse mi troverete diverso da quello che mi vedete in scena." Translation from *Dialogue*, Drake, 131. See also Heilbron, *Galileo*, 230.

44 Cox, *Renaissance Dialogue*, 72; see also Zuber, "Dialectic," 100–4.

45 Remmert, *Widmung*, 68–70.

46 Galileo, *Dialogo Sopra*, 113; translation from *Dialogue*, Drake, 104.

47 Galileo, *Dialogo Sopra*, 113; translation from *Dia-logue*, Drake, 104.

48 Galileo, *Dialogo Sopra*, 388.

49 Galileo, *Dialogo Sopra*, 357.

50 Daston, "Galilean Analogies," and François De Candt, "Galileo Furioso? Evidence and Conviction in the 'Dialogo,'" *Philosophy & Rhetoric* 32, no. 3 (1999): 197–209 (201). François de Candt, comparing Galileo's writing to that of his favorite poet, Ariosto, says that Galileo wrote of "how difficult it is sometimes to reconcile what we believe with what we see. . . . The Aristotelian refuses what he can nevertheless see."

51 Pantin, "La representation," 95: "L'astronome antique des pages de titre, après avoir été un simple emblème de la discipline, aux service d'une conception traditionelle du savoir, s'est donc trouvé engagé dans ses conflits, parfois à son corps défendant."

52 Reeves, *Painting the Heavens*; Panofsky, "Galileo as a Critic," 3–15.

53 Heilbron, *Galileo*, 50; the unpublished work is *De motu antiquiora* (c.1590).

54 Galileo, *Dialogo Sopra*, 200–2; translation from *Dialogue*, Drake, 186–7.

55 For Landini and the publishing arrangement with Galileo, see Gustavo Bertoli, "Un episodio inedito della vita di Galileo: Una lite giudizaria con Giovanbattista Landini, l'editore del *Dialogo*," *Galileana* 2 (2005): 219–27.

56 Remmert, *Widmung*, 16: *Le opere*, 369: Filippo Magalotti to Mario Giuducci, 7 August 1632: "Questo fù che, con molta segretezza, mi significò che era stata fatta molta reflessione sopra l'impresa, che io credo che sia nel frontespizio del libro, se male non mi ricordo (dico questo, perchè non ci ho fatto mai molta reflessione ancor io, e di presente non ho il libro appresso di me) e sono: s'io non m'inganno, quei tre delfini, che l'uno tiene in bocca la coda dell'altro, con non so che motto. A questo non potetti tenermi di non ridere e far atti di maraviglia, perchè io credevo di poter assicurare che il Sr. Galileo non pensava a queste bassezze e minuzie, con le quali volesse coprire gran misteri, avendo detto le cose assai chiare; e credevo risolutamente poter fermare che fosse dello stampatore." Freedberg, *Eye of the Lynx*,

144–5, mentions that the Barberini crest with its trigon of bees was added to the frontispiece of Galileo's *Il Saggiatore* (1623) along with the crest of the Accademia dei Lincei, and that it was ever more evident in Linean publications. Landini also published music prints in Florence; see Tim Carter, "Music-Printing in Late Sixteenth- and Early Seventeenth-Century Florence: Giorgio Marescotti, Cristofano Marescotti and Zanobi Pignoni," *Early Music History* 9 (1990): 27–72.

57 Cox, *Renaissance Dialogue*, 73; and Snyder, *Writing*, 100.

58 Bakhtin, *Problems*, 164: "Carnival laughter does not permit a single one of these aspects of change to be absolutized or to congeal in one-sided seriousness." Also, Cox, *Renaissance Dialogue*, 76: "The carnival of dialogue, [Speroni] perceived, was over, its mask of sophism put into storage and the careless idol of 'gioco', under whose aegis it had held its revels, had been ousted by the Lenten face of 'utile' and 'truth'."

Bibliography

Primary Sources

Agrippa, Camillo. *Trattato di scientia d'arme, con un dialogo di filosofia di Camillo Agrippa Milanese*. Rome: Antonio Blado, 1553.

——. *Trattato di scienza d'arme et un dialogo in detta materia*. Venice: Antonio Pinargenti, 1568.

——. *Trattato di Camillo Agrippa Milanese di trasportar la guglia in su la piazza di San Pietro*. Rome: Francesco Zannetti, 1583.

——. *Dialogo . . . sopra la generatione de venti, baleni, tuoni, fiumi, laghi, valli, & montagne*. Rome: Bartolomeo Bonfadino and Tito Diani, 1584.

——. *Dialogo di Camillo Agrippa Milanese del modo di mettere in battaglia presto & con facilità il popolo di qual si voglia luogo con ordinanze & battaglie diverse*. Rome: Bartolomeo Bonfadino, 1585.

——. *Nuove invenzioni di Camillo Agrippa sopra il modo di navigare*. Rome: Domenico Gigliotti, 1595.

——. *La virtù, dialogo di Camillo Agrippa Milanese sopra la dichiarazione de la causa de' Moti, tolti de la parole scritte nel Dialogo de'Venti*. Rome: Stefano Paolini, 1598.

——. *Fencing: A Renaissance Treatise*. Trans. and ed. Ken Mondschein. Rev. ed. New York: Italica Press, 2013.

Ahmet, Pasha. *Antiquités Romaines expliquées dans les memoires du Comte de B****. *Contenant ses avantures, un grand nombre d'histoires & anecdotes du tems très-curieuses, ses recherches & ses découvertes sur les antiquités de la ville de Rome & autres curiosités de l'Italie*. The Hague: Jean Neaulme, 1750.

Boissard, Jean Jacques. *Topographiae Urbis Romae*. Frankfurt, 1597.

Caro, Annibale. *Lettere famigliari*. Ed. Aulo Greco. 3 vols., Florence: F. Le Monnier, 1957–61.

Cassian, John. *Collationes partum XXIV*. Available in English as *John Cassian: The Conferences*. Trans. and ed. Boniface Ramsay. New York: Paulist Press, 1997.

Cesariano, Cesare. *Di Lucio Vitruuio Pollione De architectura libri dece, traducti de latino in vulgare affigurati, comentati, & con virando ordine insigniti: per il quale facilmente potrai trouare la multitudine de laibstrusi & reconditi vocabuli a li soi loci & in epsa tabula con summo studio expositi & enucleati ad immensa utilitate de ciascuno studioso & beniuolo di epsa opera . . .* Como: Gottardo da Ponte for Agostino Gallo and Aloisio Pirovano, 15 July 1521.

Coccius, Huldreich. *Opera D. Gregorii Papae huius nominis primi*. Basel, 1564.

Contughi, Contugo. "Relatione della gran città del Quinsay, & del Rè della China." In Giulio Belli, *Praxis prudentiae politicae*. Frankfurt, 1611, 439–55.

The Dialogues of Saint Gregory, surnamed the Greate: Pope of Rome & the first of that name. Divided into Four Books, wherein he intreateth of the Lives and Miracles of the Saintes in Italie . . . Trans. P. W. Paris: Charles Boscard, 1608. Early English Books Online: http://gateway.proquest.com/openurl?ctx_ver=Z39.882003&res_id=xri:eebo&rft_id=xri:eebo:image:21746:28, accessed 9 May 2011.

Directions for the breeding and management of silkworms: Extracted from the treatises of the Abbé Boissier de Sauvages, and Pullein. With a preface, giving some account of the rise and progress of the scheme for encouraging the culture of silk, in Pennsylvania, and the adjacent colonies. Philadelphia, 1770.

Durante, Castore. *Herbario nouvo di Castore Durante medico, & cittadino romano. Con figure, che rappresantano le vive piante, che nascono in tutta Europa, & nelle'Indie orientali & occidentali* . . . Rome: per Iacomo Bericchia & Iacomo Tornierij, 1585 (In Rome: nella stamperia di Bartholomeo Bonfadino, and Tito Diani, 1585).

Erasmus, Desiderius. *The Praise of Folly.* Trans. Clarence H. Miller. New Haven and London: Yale University Press, 2003.

Fabrizi, Principio. *Delle allusioni, imprese, et emblemi del sig. Principio Fabricij da Teramo sopra la vita, opere, et attioni di Gregorio XIII pontefice massimo libri VI nei quali sotto l'allegoria del drago, arme del detto pontefice, si descriue anco la uera forma d'un principe christiano; et altre cose, la somma delle quali si legge doppo la dedicatione dell'opera all'ill.mo et ecc.mo s. duca di Sora.* Rome: Bartolomeo Grassi, 1588 (Romae: apud Iacobum Ruffinellum).

Falimirz, Stefan. *O ziolach i o moczy ich* . . . Crakow, 1534.

Fontana, Domenico. *Della trasportatione dell'obelisco vaticano.* Rome: Domenico Basa, 1590.

Gabrielli, Magino. *Dialoghi di M. Magino Gabrielli Hebreo Venetiano sopra l'utili sue inventioni circa la seta. Ne'quali anche si dimostrono in vaghe Figure Historiati tutti gl'essercitij, & instrumenti, che nell'Arte della Seta si ricercano.* Rome: Heredi di Giovanni Gigliotti, 1588.

Galilei, Galileo. *Dialogo di Galileo Galilei Linceo matematico sopraordinario dello studio di Pisa. E filosofo, e matematico primario del serenissimo gr. duca di Toscana. Doue ne i congressi di quattro giornate si discorre sopra i due massimi sistemi del mondo tolemaico, e copernicano; proponendo indeterminatamente le ragioni filosofiche, e*

naturali tanto per l'una, quanto per l'altra parte. Florence: Gio. Batista Landini, 1632.

—. *Systema cosmicum: in quo dialogis IV. De duobus maximis mundi systematibus, Ptolomaico & Copernicano, rationibus utrinque propositis indefinitè disseritur.* Lyon: I. A. Huegutan, 1641.

—. *Le opere di Galileo Galilei.* Ed. Antonio Favaro. Vol. 14. Florence: Tipografia Barbera, 1904.

—. *Dialogue Concerning the Two Chief World Systems.* Trans. Stillman Drake. Berkeley: University of California Press, 1953.

—. *Dialogo sopra I due massimi sistemi el mondo Tolomaico e Copernicano.* Ed. Ottavio Besomi and Mario Helbing. 2 vols. Padua: Antenore, 1998.

—. *The Essential Galileo.* Ed. and trans. Maurice Finocchiaro. Indianapolis: Hackett, 2008.

Garimberto, Girolamo. *Della fortuna libri sei di Girolamo Garimberto.* Venice: Michele Tramezzino, 1547.

—. *Il capitano generale.* Venice: Giordano Zietti, 1556.

Garzoni, Tomaso. *La piazza universale di tutte le professioni del mundo.* Ed. G. B. Bronzini. 2 vols. Florence: Olschki, 1996.

Gentilini, Eugenio. *Il perfeto bombardiero et real instruttione di artiglieri: sperimentata, & composta da Eugenio Gentilini* . . . Venice: Gio. Antonio & Giacomo de'Franceschi, 1606.

Gersdorff, Hans von. *Feldtbuch der Wundartzney.* Strasbourg, 1540.

Gregory, St. *Vita et miracula sanctissimi patris Benedicti ex libro ii dialogorum beati Gregorii papae et monachi collecta, et ad instantiam devotorum monachorum congregationis eiusdem sancti Benedicti Hispaniarum aeneis typis accuratissimè delineata.* Rome: 1579.

—. *Vita et miracula sanctissimi patris Benedicti, ex libro ii dialogorum beati Gregorii papae et monachi collecta, et ad instantiam devotorum monachorum congregationis eiusdem S. Benedicti Hispaniarum aeneis typis accuratissime delineata.* Rome: Sumptu Paullini Arnolfini Lucen[sis], 1596.

—. *Vita et miracula sanctissimi patris Benedicti ex libro ii dialogorum beati Gregorii papa, et monachi collecta per Thomam Triterum, et e Latina in Hispanicum lingua conversa per D. Franciscum Cabrera. Adiuncta vita, et effigie eiusdem. / S. Benedicti. Ad Philippum Hispanioarum et Indiarum principem / Sumptu Paullini Arnolfini Lucen[sis].* Rome: 1597.

—. *Vita beatissimi patris Benedicti monachorum patriarchae sanctissimi.* Rome: A. Vaccario successor, 1611.

—. *The Second Book of the Dialogues, 1638.* English Recusant Literature 1558–1640, Vol. 294. Ed. D. M. Rogers. Ilkley and London: Scolar Press, 1976.

I statuti, ordini e constitutioni della Università de Barbieri, & Stufaroli dell'Alma Citta di Roma. Rome: Antonio Blado, 1559.

Ketham, Johannes. *Fasciculo di medicina vulgare.* Venice, 1493.

—. *Fasciculus medicine,* Venice: Joannem & Gregori de Gregorijs fratres, 1500.

—. *Fasciculus medici(n)e.* Venice: Cesare Arrivabene, 1522.

Leonardo da Vinci. *Libro di Pittura. Codice Urbinate lat. 1270 nella Biblioteca Apostolica Vaticana.* Ed. Carlo Pedretti. 2 vols. Florence: Giunti, 1995.

Lomazzo, Giovanni Paolo. *Rime . . . divise in sette libri . . .* Milano: per Paolo Gottardo Pontio, 1587.

Magni, Pietro Paolo. *Discorsi di Pietro Paolo Magni piacentino intorno al sanguinar i corpi humani, il modo di ataccare le sanguisuche e ventose è far frittioni è vessicatorii: con buoni et utili avertimenti.* Rome: appresso Bartolomeo Bonfadini, and Tito Diani, 1584.

—. *Discorsi di Pietro Paolo Magni piacentino sopra il modo di sanguinare attaccar le sanguisughe, & le ventose far le fregagioni & vessicatorij a corpi humani. Di nuouo stampati, corretti & ampliati di utili auuertimenti dal proprio autore.* Rome: per Bartholomeo Bonfadino nel Pellegrino, 1586.

—. *Discorso di Pietro Paolo Magni sopra il modo di fare I cautarii ò rottorii à corpo humani, nel quale si tratta de siti, ove si hanno da fare, de ferri che usar vi si debbono, del modo di tenergli aperti, delle legature, & delle palline, & dell'utilità che da essi ne vengono, cose utilissime non solo à barbieri, ma à tutte le person, che n'hanno bisogno.* Rome: Bartolomeo Bonfadino, 1588.

—. *Discorsi di Pietro Paolo Magni piacentino sopra il modo di sanguinare attaccar le sanguisughe, et le ventose, far le fregagioni, et vessicatorij a corpi humani. Di nuouo ristampato ad istanza di Pietro Fetti libraro in Parione.* Rome: Iacomo Mascardi, 1613.

—. *Discorsi di Pietro Paolo Magni piacentino sopra il modo di sanguinare, attaccar le sanguisughe & le ventose, far le fregagioni & vessicatorij a corpi humani. Di nuouo stampati, corretti & ampliati di utili auuertimenti dal proprio autore.* Brescia: Bartholomeo Fontana, 1618.

—. *Discorsi di Pietro Paolo Magni Piacentino sopra il modo di sanguinare attaccar le sanguisughe, & le ventose, far le fregagioni, & vessicatorii a corpi humani. Di nuouo stampati, corretti, & ampliati di vtili auuertimenti dal proprio autore.* Rome: ad istanza di Iacomo Marcucci in Piazza Nauona: per Iacomo Mascardi, 1626.

—. *Discorsi di Pietro Paolo Magni piacentino sopra il modo di sanguinare attaccar le sanguisughe, e le ventose, far le fregagioni, & vessicatorij a corpi humani.* Bologna: Gio. Recaldini, 1674.

—. *Discorsi di Pietro-Paolo Magni piacentino sopra il modo di sanguinare, attaccar ventose, e sanguisugue [!], far le fregagioni, e vessicatorij a' corpi humani, nouamente ristampato con sue figure.* Bologna: nella stamperia del Longhi, 1703.

Marozzo, Achille. *Opera nuova chiamato duello, o vero, fiore dell'armi de singulari abattimento, offensivi, & diffensivi, composta per Achille Marozzo gladiatore Bolognese . . .* Modena: Antonio Bergola, 1536.

Mercati, Michele. *De gli obelischi di Roma.* Rome: Domenico Basa, 1589.

Meyer, Cornelis. *L'arte di restituire a Roma la tralasciata navigatione del suo Tevere.* Rome: Lazzari Varese, 1685.

Montanari, Geminiano. *Le forze d'Eolo: dialogo fisicomatematico sopra gli effetti del vortice, ò sia turbine, detto negli stati Veneti la Bisciabuova. Che il giorno 29 lvglio 1686 hà scorso, e flagellato molte ville, e luoghi de' territorj di Mantova, Padova, Verona, &c. / Opera postuma del sig. dottore Geminiano Montanari modanese, astonomo, e meteorista dello studio di Padova.* Parma: Andrea Poletti, 1694.

Nadal, Jerome. *Evangelicae historiae imagines: ex ordine Euangeliorum, quae toto anno in Missae sacrificio recitantur, in ordinem temporis vitae Christi digestae.* Antwerp: Martin Nutius, 1593.

—. *Adnotationes et meditationes in Evangelia quae in sacrosancto Missae sacrificio toto anno leguntur: cum Evangeliorum concordantia historiae integritati sufficienti . . .* Antwerp: Martinus Nutius, 1594 (2d ed. 1595).

—. *Annotations and Meditations on the Gospels, Vol. I.* Trans. and ed. Frederick A. Homann, S.J. Philadelphia: Saint Joseph's University Press, 2003.

Pacioli, Luca. *De divina proportione.* Venice: Paganius, 1509.

Pinadello, Giovanni. *Dom invicti quinarii numeri series: quae summatim a superioribus pontificibus et maxime a Sixto Quinto; res praeclare quadriennio gestas adnumerat ad eundem Sixtum Quintum, Pont. Opt. Max.* Rome: F. Zanetti, 1589.

Rabelais, François. *The Complete Works.* Trans. Donald M. Frame. Berkeley: University of California Press, 1981.

Ramazzini, Bernardino. *De morbis artificum.* Modena, 1700. English translation: *Diseases of Workers.* Trans. W. C. Wright. Ontario: OH&S Press, 1993.

Rossini, Giovanni Pietro. *Il Mercurio Errante . . .* Rome: Zenobj, 1715.

Rufinus of Aquileia. *Historia monachorum sive de vita sanctorum partum.* Ed. Eva Schulz-Flügel. Berlin: W. de Gruyter, 1990.

Sangrino, Angelo, *Speculum, & exemplar christicolarum. Vita beatissimi patris Benedicti monachorum patriarchae sanctissimi. Per R.P.D. Angelum Sangrinum abbatem Congregationis Casinensis carmine conscripta.* Rome: Bartolomeo Bonfadino, 1587.

Santonis, Petri Maillardi. *Petri Maillardi Santonis Epigrammatum libri duo . . .* Turin: Haeredes Nicolai Bevilaqua, 1576.

Sommi, Leon de'. *Quattro dialoghi in materia di rappresentazione sceniche.* Ed. Ferruccio Marotti. Milan: Il Polifilo, 1968.

Statuti, ordini e costitutioni della Venerabil Compagnia & Università de Barbieri di Roma. Rome: Stamperia della Rev. Camera Apostolica, 1641.

Straet, Jan van der. *Vermis Sericus.* Antwerp, 1583.

Titi, Filippo. *Descrizione delle pitture, sculture e architetture esposte in Roma.* Rome: Marco Pagliarini, 1763.

Valverde di Hamusco, Juan. *Historia de la composicion del corpo humano . . .* Rome, 1556.

Vasari, Giorgio. *Le vite de' più eccellenti pittori scultori e architettori: nelle redazioni del 1550 e 1568.* Ed. R. Bettarini and P. Barocchi. Florence: Sansoni, 1966–.

Vasi, Giuseppe. *Indice istorico del gran prospetto di Roma . . . ovvero Itinerario istruttivo di Roma.* Rome: Marco Pagliarini, 1763.

Vesalius, Andreas. *De humani corporis fabrica.* Basel: Iohannis Oporinus, 1543.

—. *Epitome.* Basel, 1543.

—. *On the Fabric of the Human Body, Vol. 1.* Trans. W. F. Richardson and J. B. Carman. San Francisco: Norman Publishing, 1998.

Vida, Marco Girolamo, *Silkworms: A Poem in Two Books. Written originally in Latin by Marc. Hier. Vida, Bishop of Alba, and now translated into English.* Trans. D. O. M. London: J. Peele, 1723.

Viggiani, Angelo. *Lo schermo d'Angelo Viggiani dal Montone da Bologna: nel quale per via di dialogo si discorre intorno all'eccellenza dell'armi, & delle lettere . . .* Venice: Giorgio Angelieri, 1575.

Zuccaro, Federico. *Scritti d'arte di Federico Zuccaro.* Ed. D. Heikamp. Florence: Olschki, 1961.

Secondary Sources

Ago, Renata. *Economia barocca, mercato e istituzioni nella Roma del Seicento.* Rome: Donzelli, 1998.

—. *Gusto for Things.* Trans. B. Bouley, C. Tazzara, and P. Findlen. Chicago: Chicago University Press, 2013.

Andrade, E. N. da C. "Galileo." *Notes and Records of the Royal Society of London* 19, no. 2 (December 1964): 120–30.

Andretta, Elisa. "Mercati, Michele." *DBI* Vol. 73, Rome, 2009, 606–11.

Annas, Julia. "Moral Knowledge as Practical Knowledge." In *The Philosophy of Expertise.* Ed. Evan Selinger and Robert P. Crease. New York: Columbia University Press, 2006, 280–301.

Arbizzoni, Guido. "Immagini per le vite dei Santi," in *Visibile Teologia. Il libro sacro figurato tra Cinquecento e Seicento in Italia.* Ed. Erminia Ardissino and Elisabetta Selmi. Rome: Edizioni di Storia e Letteratura, pp. 83–113.

Arcari, Elisabetta. "Polemica Caro Castelvetro: Ragione del Castelvetro contro il Caro con postile autografe dell'autore." In *Vincenzo Borghini: Filologia e invenzione nella Firenze di Cosimo I.* Ed. Gino Belloni and Riccardo Drusi. Florence: Olschki, 2002, 315–18.

Arikha, Noga. *Passions and Tempers: A History of the Humours*. New York: HarperCollins, 2007.

Ash, Eric H. *Power, Knowledge and Expertise in Elizabethan England*. Baltimore and London: Johns Hopkins University Press, 2004.

Atwood, Philip. *Italian Medals c.1530–1600 in British Public Collections*. London: British Museum, 2003.

Bakhtin, Mikhail. *Problems of Dostoevsky's Poetics*. Trans. Caryl Emerson. Minneapolis: University of Minnesota Press, 1984.

Balsamo, Luigi. "The Origins of Printing in Italy and England." *Journal of the Printing Historical Society* 11 (1976–7): 48–63.

Barberi, Francesco. *Mostra del libro illustrato romano del Cinquecento*. Rome: Biblioteca Angelica, 1950.

—. "Blado, Antonio." *DBI*. Vol. 10, Rome, 1968, 753–57.

Barnhart, Robert K., ed. *The Barnhart Dictionary of Etymology*. New York: H. W. Wilson Co., 1988.

Bascetta, Carlo. *Sporti e giuochi: Trattati e scritti dal xv al xvi secolo*. Milan: Edizioni il Polifilo, 1978.

Becker, Rotraud. "Cristoforo Madruzzo" and "Giovanni Ludovico Madruzzo." *DBI*. Vol. 67. Rome, 2007, 175–80, 181–6.

Beecher, Donald. "Leone de' Sommi and Jewish Theater in Renaissance Mantua." *Renaissance and Reformation* 17, no. 2 (1993): 5–19.

Bell, Janis Callen. "Alberti." In *Grove Art Online*. Oxford Art Online, http://www.oxfordartonline.com/subscriber/article/grove/art/T001520pg2, accessed 1 October 2011.

Bellini, Paolo. *L'opera incisa di Adamo e Diana Scultori*. Vincenza: Neri Pozza, 1991.

Bemporad, Dora Liscia. *Maggino di Gabriello: "Hebreo Venetiano"; I Dialoghi sopra l'utili sue inventioni circa la seta*. Florence: Edizioni Firenze, 2010.

Benigni, U. "Mint." In *The Catholic Encyclopedia*, Vol. 10. New York: Robert Appleton Company, 1911, 335.

Bernstein, Jane A. *Print Culture and Music in Sixteenth-Century Venice*. New York and Oxford: Oxford University Press, 2001.

Bertoli, Gustavo. "Un episodio inedito della vita di Galileo: Una lite giudizaria con Giovanbattista Landini, l'editore del *Dialogo*." *Galileana* 2 (2005): 219–27.

Bertolotti, Achille. "Le tipografie orientali e gli orientalisti a Roma nei secoli XVI e XVII." *Rivista europea* 9:2 (1878): 217–68.

—. *Artisti lombardi a Roma nei secoli XV, XVI, e XVII*, Vol. 2. Milan: U. Hoepli, 1881.

—. *Artisti francesi in Roma nei secoli XV, XVI e XVII*. Mantua: G. Mondovi, 1886.

Biagioli, Mario. "The Social Status of Italian Mathematicians 1450–1600." *History of Science* 27, no. 1 (1989): 41–95.

—. "Galileo the Emblem Maker." *Isis* 81, no. 307 (June 1990): 230–58.

—. *Galileo, Courtier*. Chicago: University of Chicago Press, 1993.

Bierens de Haan, Johan Catharinus Justus. *L'oeuvre gravé de Cornelis Cort, graveur hollandaise, 1533–1578*. La Haye: M. Nijhof, 1948.

Birali, Alessandro, and Paolo Morachiello. *Immagini dell'ingegnere tra Quattro e Settecento: Filosofo, soldato, politecnico*. Milan: Franco Angeli, 1985.

Blair, Ann M. "Mosaic Physics and the Search for a Pious Natural Philosophy in the Late Renaissance." *Isis* 91, no. 1 (March 2000): 32–58.

—. *Too Much to Know: Managing Scholarly Information Before the Modern Age*. New Haven and London: Yale University Press, 2010.

Boiteux, Martine. "Les Juifs dans le carnaval de la Rome moderne." *Mélanges de L'École Française de Rome Moyen Âge et Temps Modernes* 88, no. 2 (1976): 745–87.

Bolzoni, Lina. *La stanza della memoria*. Turin: Einaudi, 1995.

Bonfil, Robert. *Jewish Life in Renaissance Italy*. Trans. A. Oldcorn. Berkeley: University of California Press, 1991.

Booth, Wayne C. "The Self-Conscious Narrator in Comic Fiction before Tristram Shandy." *PMLA* 67, no. 2 (1952): 163–85.

—. *The Rhetoric of Fiction*. 2nd ed. Chicago: University of Chicago Press, 1983.

Bora, Giulio. "Note Cremonesi, II: L'eredità di Camillo e i Campi." *Paragone Arte* 28, no. 327 (May 1997): 55–88.

Borghini, Giuseppe. "Salasso e l'opera di Pietro Paolo Magni Piacentino." *Piacenza sanitaria* 4, no. 9 (1956): 3–30.

Börner, Lore. *Die italienische Medallien der Renaissance und des Barock (1450 bis 1750)*. Berlin: Gebr. Mann Verlag, 1997.

Brach, Carla Cassetti. "Diani, Tito." *DBI*. Vol. 39, 1991, 650–52.

Brakke, David. "Ethiopian Demons: Male Sexuality, the Black-Skinned Other, and the Monastic Self." *Journal of the History of Sexuality* 10, no. 3/4 (July–October 2001): 501–35.

—. *Demons and the Making of the Monk.* Cambridge, Mass.: Harvard University Press, 2006.

Brandt, Kathleen Weil-Garris. "Michelangelo's Pietà for the Cappella del Re di Francia." In *Michelangelo: Selected Scholarship in English*. Ed. William E. Wallace. Hamden, Conn.: Garland, 1995, 217–60.

Bredekamp, Horst. *Galilei der Künstler: Der Mond, die Sonne, die Hand.* Berlin: Akademie Verlag, 2007.

Bridges, Thomas W. "Simone Verovio." In *The New Grove Dictionary of Music and Musicians*. Ed. Stanley Sadie. 2nd ed. London and New York: Macmillan, 2001, Vol. 26, 489–90.

Broise, Henri, and Vincent Jolivet. "Pincio (Jardins de Lucullus)." *Mélanges de l'École française de Rome – Antiquité* 110, no. 1 (1998): 492–5.

Brown, Clifford M. *Our Accustomed Discourse on the Antique: Cesare Gonzaga and Gerolamo Garimberto, Two Renaissance Collectors of Greco-Roman Art.* New York: Garland, 1993.

Brown, Clifford M., and Anna Maria Lorenzoni. "Major and Minor Collections of Antiquities in Documents of the Later Sixteenth Century." *The Art Bulletin* 66, no. 3 (September 1984): 496–507.

Brown, Horatio R. F. *The Venetian Printing Press.* London: J. Nimmo, 1891.

Bruschi, Arnaldo. *Bramante architetto.* Bari: Laterza, 1969.

Bundy, Murray Wright. *The Theory of Imagination in Classical and Mediaeval Thought.* Urbana: University of Illinois, 1927.

Burnham, John C. "How the Concept of Profession Evolved in the Work of Historians of Medicine." *Bulletin of the History of Medicine* 70, no. 1 (1996): 1–24.

Burroughs, Charles. "Opacity and Transparency: Networks and Enclaves in the Rome of Sixtus V." *RES: Anthropology and Aesthetics* 41 (Spring 2002): 56–71.

Bury, Michael. *The Print in Italy, 1550–1620.* London: British Museum, 2001.

Bylebyl, Jerome J. "Interpreting the *Fasciculo* Anatomy Scene." *Journal of the History of Medicine* 45 (1990): 285–316.

Camasasca, Ettore, ed., *Lettere sull'arte di Pietro Aretino, Vol. 2: 1543–1555.* Milan: Edizione di Milione, 1957.

Camille, Michael. Review of Hans Beltung, *Bild und Kulte: Eine Geschichte des Bildes vor dem Zeitalter der Kunst. The Art Bulletin* 74, no. 3 (September 1992): 517.

Campbell, Thomas Joseph, SJ, *The Jesuits, 1534–1921: A History of the Society of Jesus from its Foundation to the Present Time*, New York: The Encyclopedia Press, 1921.

Candt, François de. "Galileo Furioso? Evidence and Conviction in the 'Dialogo'." *Philosophy & Rhetoric* 32, no. 3 (1999): 197–209.

Capalbi, Monica. "Le fortificazioni del Mediterraneo negli *Atlanti* di Matteo Neroni." *Ricerche di storia dell'arte* 86 (2005): 39–42.

Carlino, Andrea. *La fabbrica del corpo.* Turin: Einaudi, 1994.

Carruthers, Mary. *The Book of Memory.* Cambridge and New York: Cambridge University Press, 1990.

—. *The Craft of Thought.* Cambridge and New York: Cambridge University Press, 1998.

Carter, Tim. "Music-Printing in Late Sixteenth- and Early Seventeenth-Century Florence: Giorgio Marescotti, Cristofano Marescotti and Zanobi Pignoni," *Early Music History* 9 (1990): 27–72.

Casali, Scipione. *Annali della tipografia veneziana di Francesco Marcolini da Forli.* Forli: M. Casali, 1861.

Castelli, Patrizia, ed. *Iconografia di San Benedetto nella pittura della Toscana: Immagini e aspetti culturali fino al XVI secolo.* Florence: Centro d'incontro della Certosa di Firenze, 1982.

Cavallo, Sandra. *Artisans of the Body in Early Modern Italy.* Manchester: Manchester University Press, 2007.

Cazort, M., M. Kornell, and K. B. Roberts. *The Ingenious Machine of Nature: Four Centuries of Art and Anatomy*. Ottawa: National Gallery of Canada, 1996.

Ceresa, Massimo. *Una stamperia nella Roma del primo Seicento. Annali tipografici di Guglielmo Facciotti ed eredi (1592–1640)*. Rome: Bulzoni, 2000.

Certeau, Michel de. *The Practice of Everyday Life*. Trans. S. Rendell, Berkeley: University of California Press, 1984.

Chambers, D. S. "The Economic Predicament of Renaissance Cardinals." In *Studies in Medieval and Renaissance History*, Vol. 3. Ed. William M. Bowsky. Lincoln: University of Nebraska Press, 1966, 289–313.

Chappell, Miles. "Cigoli, Galileo, and Invidia." *The Art Bulletin* 62 (1975): 91–8.

Chartier, Roger. *The Order of Books*. Trans. L. G. Cochrane. Stanford: Stanford University Press, 1994.

Christian, Kathleen Wren. "The De' Rossi Collection of Ancient Sculptures, Leo X, and Raphael," *JWCI* 65 (2002): 132–200.

Ciardi, Roberto Paolo, and Lucia Tongiorgi Tomasi. "'La scienza illustrata': Osservazioni su frontespizi delle opere di Athanasius Kircher e di Galileo Galilei." *Annali dell'Istituto Storico Italo-Germanico in Trento* 11 (1985): 68–79.

Clark, Elizabeth A. *Reading Renunciation: Asceticism and Scripture in Early Christianity*. Princeton: Princeton University Press, 1999.

—. "Foucault, the Fathers and Sex." In *Michel Foucault and Theology: The Politics of Religious Experience*. Ed. James W. Bernauer and Jeremy R. Carrett. Burlington, Vt.: Ashgate, 2004, 39–56.

Clark, Francis. *The "Gregorian" Dialogues and the Origins of Benedictine Monasticism*. Leiden and Boston: Brill, 2003.

Clarke, W. Mitchell. "The History of Bleeding, and its Disuse in Modern Practice." *British Medical Journal* 2, no. 759 (17 July 1875): 67–70.

Coffin, David R. *The Villa D'Este at Tivoli*. Princeton: Princeton University Press, 1960.

—. *The Villa in the Life of Renaissance Rome*. Princeton: Princeton University Press, 1997.

Cohen, Elizabeth. "Miscarriages of Apothecary Justice: Un-Separate Spaces of Work and Family in Early Modern Rome." *Renaissance Studies* 21, no. 4 (September 2007): 480–504.

Colapinto, John. "Bloodsuckers." *The New Yorker* (25 July 2005): 72–81.

Collett, Barry. *Italian Benedictine Scholars and the Reformation: The Congregation of Santa Giustina of Padua*. Oxford and New York: Clarendon Press, 1985.

Conforti, Maria, and Silvia de Renzi. "Sapere anatomico negli ospedali romani: Formazione dei chirurghi e pratiche sperimentali (1620–1720)." In *Rome et la science moderne: Entre Renaissance et Lumières*. Ed. Antonella Romano. Rome: École française de Rome, 2008, 433–72.

Connors, Joseph. "St. Ivo: The First Three Minutes." *JSAH* 55, no. 1 (March 1996): 38–57.

Cooperman, Bernard D. "Ethnicity and Constitution Building among Jews in Early Modern Rome." *AJS Review* 30, no. 1 (2006): 119–45.

Corbo, Anna Maria. *Fonti per la storia artistica Romana al tempo di Clemente VIII*. Rome: Archivio di Stato, 1975.

Corns, Thomas N. "The Early Modern Search Engine: Indices, Title Pages, Marginalia and Contents." In *The Renaissance Computer*. Ed. Neil Rhodes and Jonathan Sawday. London and New York: Routledge, 2000, 95–105.

Costamagni, Philippe. "Il ritrattista." In *Francesco Salviati (1510–1563) o la Bella Maniera*. Ed. Catherine M. Goguel. Milan: Electa, 1998, 47–52.

Cox, Virginia. *The Renaissance Dialogue: Literary Dialogue in its Social and Political Contexts, Castiglione to Galileo*. Cambridge and New York: Cambridge University Press, 1992.

Crillo, Giuseppe. *Carlo Urbino da Crema, disegni e dipinti*. Parma: Grafiche STEP Editrice, 2005.

Curran, Brian A., A. Grafton, P. O. Long, and B. Weiss. *Obelisk: A History*, Cambridge, Mass.: Burndy Library, 2009.

Darnton, Robert. "What is the History of Books?" *Daedalus* 3, no. 3 (Summer 1982): 65–83.

Daston, Lorraine J. "Galilean Analogies: Imagination at the Bounds of Sense." *Isis* 75, no. 2 (June, 1984): 302–10.

Davis, Charles. "Ammanati, Michelangelo and the

Tomb of Francesco del Nero." *Burlington Magazine* 118, no. 80 (1976): 472–84.

Debuchy, Paul. "Spiritual Exercises of Saint Ignatius." In *The Catholic Encyclopedia,* Vol. 14. New York: Robert Appleton Company, 1912. http://www.newadvent.org/cathen/14224b.htm, accessed 9 December 2012.

Delumeau, Jean. *La vie économique et sociale de Rome dans la seconde moitié du XVIe siècle.* 2 vols. Paris: E. de Boccard, 1959.

—. *Rome au XVIe siècle.* Paris: Hachette, 1975.

Dibner, Bern. *Moving the Obelisks.* Cambridge, Mass.: Burndy Library, 1991.

Dolza, Luisa M. "Reframing the Language of Inventions: The First Theatre of Machines." In *The Power of Images in Early Modern Science.* Ed. Wolfgang Lefèvre, J. Renn, and U. Schoepflin. Basel and Boston: Birkhäuser, 2003, 89–207.

Dominicus, Claudio de. *Membri del Senato della Roma Pontificia: Senatori, Conservatori, Caporioni e loro Priori e Lista d'oro delle famiglie dirigenti (secc. X–XIX).* Rome: Fondazione Marco Besso, 2009.

Eamon, William. *Science and the Secrets of Nature.* Princeton: Princeton University Press, 1994.

—. "The Professor of Secrets. Mystery, Medicine and Alchemy in Renaissance Italy.: Washington, DC: *National Geographic,* 2010.

Edgerton, Samuel. "Galileo, Florentine 'Disegno,' and the 'Strange Spottednesse' of the Moon." *Art Journal* 44 (1984): 225–32.

—. *The Heritage of Giotto's Geometry: Art and Science on the Eve of the Scientific Revolution.* Ithaca: Cornell University Press, 1991.

Enzo, Carlo. *Le storie di San Benedetto a Monteoliveto Maggiore.* Milan: Silvana, 1980.

Esposito, Anna. "Stufe e bagni pubblici a Roma nel Rinascimento." In *Taverne, locande e stufe nel rinascimento.* Ed. Massimo Miglio et al, Rome: Roma nel Rinascimento, 1999, 77–91.

Fani, Sara and Margherita Farina, eds. *Le vie delle lettere: la Tipografia medicea tra Roma e l'Oriente.* Florence: Mondragon, 2012.

Fara, Amelio. *Il sistema e la città: Architettura fortificata dell'Europa moderna dai trattati alle realizzazioni 1464–1794.* Genoa: SAGEP, 1989.

Farago, Claire. "The Defense of Art and the Art of Defense." *Achademia Leonardi Vinci* 10 (1997): 13–22.

Farenga, Paola, ed. *Editori ed edizioni a Roma nel Rinascimento.* Rome: Roma nel Rinascimento, 2005.

Field, J. V. "Piero della Francesca's Mathematics." In *The Cambridge Companion to Piero della Francesca.* Ed. Jeryldene M. Wood. Cambridge and New York: Cambridge University Press 2003, 152–70.

Finocchiaro, Maurice. *Galileo and the Art of Reasoning: Rhetorical Foundations of Logic and Scientific Method.* Boston: Kluwer, 1980.

Fiocca, Alessandra, D. Lamberini, and C. Maffioli. *Arte e scienza della Acque nel Rinascimento.* Venice: Marsilio, 2003.

Fischer, Lucia Frattarelli, "Reti locali e reti internazionali degli ebrei di Livorno nel Seicento." In *Commercial Networks in the Early Modern World.* Ed. Diogo Ramada Curto and Anthony Molho. Badia Fiesolana: European University Institute, 2002, 148–67.

Fournel, Jean-Louis. *Les dialogues de Sperone Speroni: Libertés de la parole et règles de l'écriture.* Marburg: Hitzeroth, 1990.

Fragnito, Gigliola, ed. *Church, Censorship and Culture in Early Modern Italy.* Cambridge: Cambridge University Press, 2001.

Freedberg, David. *The Power of Images: Studies in the History and Theory of Response.* Chicago: University of Chicago Press, 1989.

—. *The Eye of the Lynx: Galileo, his Friends, and the Beginning of Modern Natural History.* Chicago: University of Chicago Press, 2002.

Gallo, Valentina. "Longo, Alberigo." *DBI.* Vol. 65, Rome: 2005, 686–7.

Gavin, J. Austin, and Thomas M. Walsh. "The Praise of Folly in Context: The Commentary of Girardus Listrius." *Renaissance Quarterly* 24, no. 2 (Summer 1971): 193–209.

Gennette, Gérard. *Narrative Discourse: An Essay on Method.* Ithaca: Cornell University Press, 1980.

Gentilcore, David. "Charlatans: Regulation of the Marketplace and the Treatment of Venereal Disease in Italy." In *Sins of the Flesh: Responding to Sexual Disease in Early Modern Europe.* Ed. Kevin Patrick

Siena. Toronto: Centre for Renaissance and Refor-mation Studies, 2005, 57–80.

—. *Medical Charlatanism in Early Modern Italy.* Oxford and New York: Oxford University Press, 2006.

—. "Spaces, Objects and Identities in Early Modern Italian Medicine." *Renaissance Studies* 21, no. 4 (September 2007): 473–9.

Giacomo, Franco. "Rabelais et Annibal Caro: Tra-ditions, filiations et traductions littéraires." *Revue d'histoire littéraire de la France* 99, no. 5 (1999): 963–73.

Ginzburg, Carlo, *Formaggio e i vermi.* Turin: Einaudi, 1976. English translation: *The Cheese and the Worms.* Trans. John and Anne Tedeschi. Baltimore: Johns Hopkins University Press, 1980.

Goldman, William. *The Princess Bride.* New York: Ballantine, 1973.

González de Zarate, Jesús María. "Aportaciones del Coro Alto de San Benito de Valladolid a la Iconografía de San Benito." *Boletín del Seminario de Estudios de Arte y Arqueología* 52 (1986): 357–68.

Grafton, Anthony. *Commerce with the Classics.* Ann Arbor: University of Michigan Press, 1997.

—. *Bring Out Your Dead: The Past as Revelation.* Cambridge, Mass.: Harvard University Press, 2001.

Greco, Aulo. *Annibal Caro, cultura e poesia.* Rome: Edizioni di storia e letteratura, 1950.

Green, Monica. "Women's Medical Practice and Health Care in Medieval Europe." *Signs* 14, no. 2 (Winter 1989): 434–73.

—. "From 'Diseases of Women' to 'Secrets of Women': The Transformation of Gynecological Lit-erature in the Later Middle Ages." *Journal of Medieval and Early Modern Studies* 30 (2000): 5–39.

Grendler, Paul. *The Universities of the Italian Renais-sance.* Baltimore: Johns Hopkins University Press, 2002.

Harcourt, Glenn. "Andreas Vesalius and the Anatomy of Antique Sculpture." *Representations* 17 (1987): 28–61.

Harkness, Deborah E. *The Jewel House: Elizabethan London and the Scientific Revolution.* New Haven and London: Yale University Press, 2007.

Heilbron, J. L. *Galileo.* Oxford and New York: Oxford University Press, 2010.

Henkel, Willi. *Die Druckerei der Propaganda Fide: Eine Dokumentation.* Munich: Schöningh, 1977.

Henneberg, Josephine von. *L'oratorio dell'Arcicon-fraternita del Santissimo Crocifisso di San Marcello.* Rome: Bulzoni, 1974.

Herzlich, Claudine. "Modern Medicine and the Quest for Meaning: Illness as a Social Signifier." In *The Meaning of Illness: Anthropology, History and Society.* Ed. Marc Augé and Claudine Herzlich. London: Harwood Academic Publishers, 1995, 151–74.

Hillman, David, and C. Mazzio. *The Body in Parts.* New York and London: Routledge, 1997.

Hoffman, Eve R. "The Author Portrait in Thirteenth-Century Arabic Manuscripts: A New Islamic Context for a Late-Antique Tradition." *Muqumas* 10 (1993): 6–20.

Hoffmann, Paula. *Il Monte Pincio e la Casina Vala-dier.* Rome: Edizioni del Mondo, 1967.

Hollingsworth, Mary and Carol M. Richardson, eds. *The Possessions of a Cardinal: Politics, Piety and Art, 1450–1700.* University Park: Pennsylvania State Uni-versity Press, 2010.

Hollstein, F. W. H. *The New Hollstein: Dutch and Flemish Etchings, Engravings and Woodcuts 1450–1700,* Amsterdam: Rijksmuseum, 1993.

Howe, Eunice D., trans. and ed. *The Churches of Rome.* Binghamton: Center for Medieval and Early Renaissance Studies, State University of New York, 1991.

Hughes, Diane Owen. "Distinguishing Signs: Ear-Rings, Jews and Franciscan Rhetoric in the Italian Renaissance City." *Past and Present* 112 (1986): 3–59.

Ilg, Ulrike. "Stefano della Bella and Melchior Lorck: The Practical Use of an Artist's Model Book." *Master Drawings* 41, no. 1 (Spring 2003): 30–43.

Jay, Martin. *Songs of Experience.* Berkeley: Univer-sity of California Press, 2005.

Jayawardene, S. A. "Rafael Bombelli, Engineer-Architect: Some Unpublished Documents of the Apostolic Camera." *Isis* 56, no. 3 (Autumn, 1965): 298–306.

—. "The Influence of Practical Arithmetics on the Algebra of Rafael Bombelli." *Isis* 64, no. 4 (Decem-ber 1973): 510–23.

Johns, Adrian. "Science and the Book in Modern

Cultural Historiography." *Studies in the History of Science* 29, no. 2 (1997): 167–94.

—. *The Nature of the Book*. Chicago: University of Chicago Press, 1998.

—. "How to Acknowledge a Revolution." *American Historical Review* 107, no. 1 (February 2002): 106–25.

Johnson, A. F. "A Catalogue of Italian Writing Books of the Sixteenth Century." *Signature*, new series 10 (1950): 22–48.

Jones, Peter Murray. "Image, Word and Medicine in the Middle Ages." In *Visualizing Medieval Medicine and Natural History 1200–1550*. Ed. J. A. Givens, K. M. Reeds, and A. Tonwaide. Aldershot: Ashgate, 2006, 1–24.

Jones, Robert, "The Medici Oriental Press (Rome 1584–1614) and the Impact of its Arabic Publications on Northern Europe." In *The "Arabick" Interest of the Natural Philosophers in Seventeenth-Century England*. Ed. G. A. Russell. Leiden: Brill, 1994, 88–108.

Jütte, Daniel. "Abramo Colorni, jüdischer Hofalchemist Herzog Friedrichs I, und die hebräische Handelskompanie des Maggino Gabrielli in Württemberg am Ende des 16. Jahrhunderts." *Ashkenas* 15, no. 2 (2005): 435–98.

—. "Handel, Wissenstransfer und Netzwerke: Eine Fallstudie zu Grenzen und Möglichkeiten unternehmerischen Handelns unter Juden zwischen Reich, Italien und Levante um 1600." In *Vierteljahrschrift für Sozial- und Wirtschaftsgeschichte* 95, no. 3 (July 2008): 263–90.

—. *Das Zeitalter des Geheimnisses: Juden, Christen und die Ökonomie des Geheimen (1400–1800)*. Göttingen: Vandenhoek & Ruprecht, 2011.

—. "Trading in Secrets: Jews and the Early Modern Quest for Clandestine Knowledge." *Isis* 103:4 (December 2012): 668–86.

Kagan, Richard. *Lawsuits and Litigants in Castile 1500–1700*. Chapel Hill: University of North Carolina Press, 1981.

Kardong, Terrence G. *Benedict's Rule: A Translation and Commentary*. Collegeville, Minn.: Liturgical Press, 1996.

Keele, Kenneth. *Leonardo da Vinci's Elements of the Science of Man*. New York and London: Academic Press, 1983.

Kemp, Martin. "Leonardo and the Visual Pyramid." *JWCI* 40 (1977): 128–49.

—. *The Science of Art: Optical Themes in Western Art from Brunelleschi to Seurat*. New Haven and London: Yale University Press, 1990.

Keyvanian, Carla L. "Charity, Architecture and Urban Development in post-Tridentine Rome: The Hospital of the SS.ma Trinità dei Pellegrini e Convalescenti (1548–1680)." PhD thesis, Massachusetts Institute of Technology, Department of Architecture, 2000.

Kristeller, Paul Oskar. "Erasmus from an Italian Perspective." *Renaissance Quarterly* 23, no. 1 (Spring 1970): 1–14.

Kusukawa, Sachiko. "Leonhart Fuchs on the Importance of Pictures." *Journal of the History of Ideas* 58, no. 3 (1997): 403–27.

—. *Picturing the Book of Nature: Image, Text, and Argument in Sixteenth-century Human Anatomy and Medical Botany*, Chicago: University of Chicago Press, 2012.

Lamberini, Daniela. "Collezionismo e patronato dei Medici a Firenze nell'opera di Matteo Neroni, 'cosmografo del granduca'." In *Il disegno di architettura*. Ed. P. Carpeggiani and L. Patetta. Milan: Guerrini e Associati, 1989, 33–8.

Landrus, Matthew. *Leonardo da Vinci's Giant Crossbow*. Berlin: Springer-Verlag, 2010.

LeBlanc, Charles. *Le graveur en taille-douce*. Leipzig: R. Weigel, 1847.

Leclercq, Jean. "Lectio divina," *Worship* 58, no. 3 (1984) 239–48.

Leesberg, Marjolein, and Huigen Leeflang. "Johannes Stradanus." In Hollstein, F. W. H. *The New Hollstein, Dutch & Flemish etchings, engravings and woodcuts, 1450–1700*. 3 vols. Ouderkerk aan den Ijssel: Sound & Vision Publications, 2008.

Lewis, R. and M., and S. Boorsch. *The Engravings of Giorgio Ghisi*. New York: Metropolitan Museum of Art, 1985.

Lincoln, Evelyn. *The Invention of the Italian Renaissance Printmaker*. New Haven and London: Yale University Press, 2000.

—. "Curating the Renaissance Body." *Word & Image* 17, nos. 1 and 2 (January–June 2001): 42–61.

—. "The Devil's Hem: Allegorical Reading in a Sixteenth-century Illustrated Life of St. Benedict." In *Early Modern Visual Allegory: Embodying Meaning.* Aldershot: Ashgate, 2007, 135–53.

—. "Invention and Authorship in Early Modern Italian Visual Culture." *DePaul Law Review* 52:4 (Summer 2003): 1093–119.

—. "The Jew and the Worms: Portraits and Patronage in a 16th Century How-To Manual." *Word & Image* 19, nos. 1 and 2 (January–June 2003): 86–99.

—. "Invention, Origin, and Dedication: Republishing Women's Prints in Early Modern Italy." In *Making and Unmaking Intellectual Property: Creative Production in Legal and Cultural Perspective.* Ed. M. Biagioli, P. Jaszi, and M. Woodmansee. Chicago: University of Chicago Press, 2011, 339–57.

—. "Camillo Graffico, Printmaker and Fountaineer at the Villa Farnesina." *Print Quarterly* 29, no. 3 (September 2012): 259–80.

Lind, L. R. *The Epitome of Andreas Vesalius.* Cambridge, Mass.: MIT Press, 1969.

Lindberg, David C. *Theories of Vision from al-Kindi to Kepler.* Chicago: University of Chicago Press, 1976.

—. "Science and the Early Church." In *God and Nature: Historical Essays on the Encounter Between Christianity and Science.* Berkeley: University of California Press, 1986, 18–48.

Lingo, Alison Klairmont. "Empirics and Charlatans in Early Modern France: The Genesis of the Classification of the 'Other' in Medical Practice." *Journal of Social History* 19 (Summer 1986): 583–603.

Lingo, Estelle. "The Evolution of Michelangelo's Magnifici Tomb: Program vs. Process in the Iconography of the Medici Chapel." *Artibus et Historia* 16:32 (1995): 91–100.

Lombardi, Leonardo. "Camillo Agrippa's Hydraulic Inventions on the Pincian Hill (1574–1578)." *Waters of Rome Occasional Journal* 5 (2008): 1–10.

Long, Pamela O. "Power, Patronage and the Authorship of *Ars*: From Mechanical Know-How to Mechanical Knowledge in the Last Scribal Age." *Isis* 88, no. 1 (March 1997): 1–41.

—. *Openness, Secrecy, Authorship: Technical Arts and the Culture of Knowledge from Antiquity to the Renaissance.* Baltimore and London: Johns Hopkins University Press, 2001.

—. *Artisan/Practitioners and the Rise of the New Sciences, 1400–1600.* Corvallis: Oregon State University Press, 2011.

Lowry, Martin. *The World of Aldus Manutius: Business and Scholarship in Renaissance Venice.* Ithaca: Cornell University Press, 1979.

Lukehart, Peter M., ed. *The Accademia Seminars: The Accademia di San Luca in Rome, c.1590–1635.* Washington, DC, National Gallery of Art, 2009.

Lynch, Michael, and John Law, "Pictures, Texts and Objects: The Literary Language Game of Bird-Watching." In *The Science Studies Reader.* Ed. Mario Biagioli. New York and London: Routledge, 1999, 317–41.

McClure, George W. *The Culture of Profession in Late Renaissance Italy.* Toronto: University of Toronto Press, 2004.

MacDougall, Elisabeth. *Fons Sapientiae: Renaissance Garden Fountains,* Washington, DC: Dumbarton Oaks, 1978.

McVaugh, Michael R. "Bedside Manners in the Middle Ages." *Bulletin of the History of Medicine* 71, no. 2 (1997): 201–23.

Maffioli, Cesare. *La Via delle Acque (1500–1700): Appropriazione delle arti e trasformazione delle matematiche.* Florence: Olschki, 2010.

Mandel, Corinne. "Felix Culpa and Felix Roma: On the Program of the Sixtine Staircase at the Vatican." *The Art Bulletin* 75, no. 1 (1993): 65–90.

—. "Introduzione all'iconologia della pittura a Roma in éta sistina." In *Roma di Sisto V: Le arti e la cultura.* Ed. Maria Luisa Madonna. Rome: De Luca, 1993, 3–16.

Marr, Alexander. *Between Raphael and Galileo: Mutio Oddi and the Mathematical Culture of the Late Renaissance.* Chicago: University of Chicago Press, 2011.

Marshall, Amy. *Mirabilia urbis Romae: Five Centuries of Guidebooks and Views.* Toronto: University of Toronto Library, 2002.

Martini, Antonio. *Arti, mestieri e fede nella Roma dei papi.* Bologna: Cappelli, 1965.

Massari, Stefania. *Incisori Mantovani del Cinquecento.* Rome: De Luca, 1981.

Masser, Phyllis Dearborn. "Presenting Stefano della Bella." *The Metropolitan Museum of Art Bulletin* new series 27, no. 3 (November 1968): 159–76.

—. "Costume Drawings by Stefano della Bella for the Florentine Theater." *Master Drawings* 8, no. 3 (Autumn 1970): 243–66, 297–317.

Massimo, C. Vittorio. *Notizie istoriche della Villa Massimo alle Terme.* Rome: Salviucci, 1836.

Matsen, Herbert S. "Students' 'Arts' Disputations at Bologna around 1500." *Renaissance Quarterly* 47, no. 3 (Autumn 1994): 533–55.

Melion, Walter. "Introduction." In Jerome Nadal, *Annotations and Meditations on the Gospels*, Vol. 1. Trans. and ed. Frederick A. Homann, S.J. Philadelphia: Saint Joseph's University Press, 2003.

Menato, Marco, E. Sandal, and G. Zappella, eds. *Dizionario dei tipografi e degli editori italiani: Il Cinquecento.* Milan: Editrice Bibliografica, 1997.

Mennell, Stephen. "The Formation of We-Images: A Process Theory." In *Social Theory and the Politics of Identity.* Ed. C. Calhoun. Cambridge, Mass.: Blackwell, 1994, 175–97.

Miglio, Massimo, and O. Rossini, eds. *Gutenberg e Roma,* Naples: Electa Napoli: 1997.

Milano, Attilio. *Storia degli ebrei in Italia.* Turin: Einaudi, 1992.

Missini, Melchior. *Memorie per servire alla storia della Romana Accademia di S. Luca.* Rome: de Romanis, 1823.

Modigliani, Anna. "Tipografi a Roma (1467–1477)." In *Gutenberg e Roma.* Ed. M. Miglio and O. Rossini. Naples: Electa Napoli, 1997, 41–66.

—. "Printing in Rome in the XVth Century: Economics and the Circulation of Books." In *Editori ed edizioni a Roma nel Rinascimento.* Ed. Paola Farenga. Rome: Roma nel Rinascimento, 2005, 65–76.

—. " Massimo, Pietro." *DBI.* Vol. 72, 2008, 15–16.

Molà, Luca. "Le donne nell'industria serica veneziana del rinascimento." In *La seta in Italia dal Medioevo al Seicento.* Ed. L. Molà, R. C. Mueller, and Claudio Zanier. Venice: Marsilio, 2000, 423–60.

—. *The Silk Industry of Renaissance Venice.* Baltimore and London: Johns Hopkins University Press, 2000.

Monfasani, John. "Aristotelians, Platonists and the Missing Ockhamists: Philosophical Liberty in Pre-Reformation Italy." *Renaissance Quarterly* 46, no. 2 (1993): 247–76.

Morison, Stanley. *Early Italian Writing Books: Renaissance to Baroque.* Boston: Godine, 1990.

Moroni, Gaetano. *Dizionario di erudizione storico-ecclesiastica da S. Pietro sino ai nostri giorni.* Venice: Tipografia Emiliana, 1840–61, Vol. 44, 136.

Musacchio, Jacqueline. *The Art and Ritual of Childbirth in Renaissance Italy.* New Haven and London: Yale University Press, 1999.

Nenci, Elio. "Camillo Agrippa: Un ingegniere rinascimentale di fronte ai problemi della filosofia naturale." *Physis* 29 (1992): 71–119.

—. *Bernardino Baldi (1553–1617), studioso rinascimentale: Poesia, storia, linguistica, meccanica, architettura.* Milan: Franco Angeli, 2005.

Norton, Frederick John. *Italian Printers, 1501–1520: An Annotated List.* London: Bowes and Bowes, 1958.

Nova, Alessandro, and Alessandro Cecchi. "Francesco Salviati e gli editori." In *Francesco Salviati (1510–1563) o la Bella Maniera.* Ed. Catherine M. Goguel. Milan: Electa, 1998, 66–74.

Olivieri, Iolanda. "Il corvo e il pane: Iconografia degli episodi legati al cibo nei cicli figurativi." In *Il cibo e la regola.* Ed. Angela Adriana Cavarna et al. Rome: Biblioteca Casanatense, 1996, 129–42.

O'Malley, John W. *The First Jesuits,* Cambridge, Mass.: Harvard University Press, 1993, 37–50.

Orbaan, J. F. *Documenti sul barocco in Roma.* Rome: Società romana di storia patria, 1920.

Orgel, Stephen. "Textual Icons: Reading Early Modern Illustrations." In *The Renaissance Computer.* Ed. Neil Rhodes and Jonathan Sawday. London and New York: Routledge, 2000, 59–94.

Pagani, Valeria. "Adamo Scultori and Diana Mantovana." *Print Quarterly* 9, no. 1 (1992): 72–87.

—. "The Dispersal of Lafreri's Inheritance 1581–89 – III: The De'Nobili-Arbotti-Clodio Partnership," *Print Quarterly* 28, no. 2 (2011): 119–35.

Panofsky, Erwin. "Galileo as a Critic of the Arts: Aesthetic Attitude and Scientific Thought." *Isis* 47, no. 1 (March 1956): 3–15.

Pantin, Isabelle. "Une 'École d'Athènes' des Astronomes? La representation de l'astronomie antique dans les frontispieces de la Renaissance." In *Images de l'Antiquite dans la littérature française: Le texte et son illustration*. Ed. Emmanuèle Baumgartner and Laurence Harf-Lancner. Paris: Editions Rue d'Ulm, 1993, 87–99.

Park, Katherine. *Doctors and Medicine in Early Renaissance Florence*. Princeton: Princeton University Press, 1985.

—. "The Organic Soul." In *The Cambridge History of Renaissance Philosophy*. Ed. Charles Schmitt, Q. Skinner, and E. Kessler. Cambridge and New York: Cambridge University Press, 1988, 464–84.

—. "Nature in the Person: Medieval and Renaissance Allegories and Emblems." In *The Moral Authority of Nature*. Ed. L. Daston and F. Vidal. Chicago: University of Chicago Press, 2004, 50–73.

—. *Secrets of Women: Gender, Generation, and the Origins of Human Dissection*. New York: Zone Books, 2006.

Parshall, Peter. "Antonio Lafreri's *Speculum Romanae Magnificentiae*." *Print Quarterly* 23, no. 1 (2006): 3–27.

Partridge, Loren. "The Sala d'Ercole in the Villa Farnese at Caprarola, Part I." *The Art Bulletin* 53, no. 4 (December 1971): 467–86.

—. "Discourse on Asceticism in Bertoja's Room of Penitence in the Villa Farnese at Caprarola." *Memoirs of the American Academy in Rome* 40 (1995):145–74.

—. "The Farnese Circular Courtyard at Caprarola: God, Geopolitics, Geneology, and Gender." *The Art Bulletin* 83, no. 2 (June 2001): 259–93.

Partridge, Loren, and Randolph Starn. *A Renaissance Likeness*. Berkeley: University of California Press, 1980.

Passamani, Bruno. "Caprioli, Aliprando." *DBI.* Vol. 19, 1976, 209–10.

Pastor, Ludwig von. *The History of the Popes, from the Close of the Middle Ages*. London: J. Hodges, 1891.

Peay, Rev. Bede, ed. *Saint Benedict, Life and Miracles*. Trans. Mary Jean Lutz-Bujdos. Subiaco: Errebigrafica, 1993.

Pedretti, Carlo. "Introduction." In Leonardo da Vinci, *Libro di pittura: Codice Urbinate lat. 1270 nella Biblioteca Apostolica Vaticana*. Ed. Carlo Pedretti. 2 vols. Florence: Giunti, 1995, 11–82.

Pedretti, Carlo, and Sergio Marinelli. "The Author of the Codex Huygens." *JWCI* 44 (1981): 214–20.

Pepper, Simon. "Artisans, Architects and Aristocrats: Professionalism and Renaissance Military Engineering,." In *The Chivalric Ethos and the Development of Military Professionalism*. Ed. David J. B. Trim. Leiden and Boston: Brill, 2003, 117–48.

Perocco, Daria. "La seta nella letteratura Italiana dal Duecento al Seicento." In *La seta in Italia dal Medioevo al Seicento*. Ed. L. Molà, R. C. Mueller, and Claudio Zanier. Venice: Marsilio, 2000, 241–61.

Perreiah, Alan. "Humanist Critiques of Scholastic Dialectic." *Sixteenth Century Journal* 13, no. 3 (Autumn, 1982): 3–22.

Persels, Jeffrey C. "Bragueta Humanistica, or Humanism's Codpiece." *Sixteenth Century Journal* 28, no. 1 (Spring 1997): 79–99.

Petrucci, Armando. *Writers and Readers in Medieval Italy*. New Haven and London: Yale University Press, 1995.

Pettegree, Andrew. *The Book in the Renaissance*. New Haven and London: Yale University Press, 2010.

Pinkus, Karen. *Picturing Silence*. Ann Arbor: University of Michigan Press, 1996.

Plaisance, Michel. "Culture et politique à Florence de 1542 à 1551." In *Les écrivains et le pouvoir en Italie à l'époque de la Renaissance*. Ed. A. Rochon. Paris: Université de la Sorbonne Nouvelle, 1973, 148–242.

Pollak, Oskar. "Ausgewählte Akten zur Geschichte der römischen Peterskirche (1535–1621)." *Jahrbuch der Königlich Preussischen Kunstsammlungen* 36 (1915): 21–117.

Pollard, J. Graham. *Italian Renaissance Medals in the Museo Nazionale of Bargello, Vol. 3: 1513–1640*. Florence: Associazione Amici del Bargello, 1984–5.

Pomata, Gianna. "Barbari e comari." In *Medicina, erbe e magia*. Ed. G. Adani and G. Tamagnini. Milan: Silvana editore, 1981, 161–83.

—. *Contracting a Cure: Patients, Healers and the Law in Early Modern Bologna*. Trans. by the author with Rosemarie Foy and Anna Taraboletti-Segre. Baltimore, Md.: Johns Hopkins University Press, 1998.

—. "Practicing between Earth and Heaven: Women Healers in Seventeenth-Century Bologna." *Dynamis* 19 (1999): 119–43.

Popplow, Marcus. "Why Draw Pictures of Machines? The Social Contexts of Early Modern Machine Drawings." In *Picturing Machines*. Ed. Wolfgang Lefèvre. Cambridge, Mass.: MIT Press, 2004, 17–50.

Portoghesi, Paolo, and Adriano Carugo, *Della trasportatione dell'obelisco vaticano, 1590*. Milan: Il Polifilo, 1978.

Prosperi, Adriano. "L'Inquisizione Romana e gli ebrei." In *L'Inquisizione e gli ebrei in Italia*. Ed. M. Luzzati. Rome: Laterza, 1994, 67–120.

Ravid, Benjamin. "From Yellow to Red: On the Distinguishing Head-Covering of the Jews of Venice." *Jewish History* 6, nos. 1–2 (1992): 179–210.

Rebhorn, Wayne A., ed. and trans. *Renaissance Debates on Rhetoric*. Ithaca: Cornell University Press, 2000.

Reeves, Eileen. *Painting the Heavens: Art and Science in the Age of Galileo*. Princeton: Princeton University Press, 1997.

Remmert, Volker. *Widmung, Welterklärung und Wissenschaftslegitimierung: Titelbilder und ihre Funktionen in der Wissenschaftlichen Revolution*. Wiesbaden: Harrassowitz in Kommission, 2005.

Reynolds, Anne. "Galileo Galilei's Poem 'Against Wearing the Toga'." *Italica* 59, no. 4 (Winter 1982): 330–41.

Ricci, Antonio. "Lorenzo Torrentino and the Cultural Programme of Cosimo I de' Medici." In *The Cultural Politics of Duke Cosimo I de' Medici*. Ed. K. Eisenbichler. Aldershot: Ashgate, 2001, 103–19.

Richardson, Brian. *Printing, Writers and Readers in Renaissance Italy*. Cambridge and New York: Cambridge University Press, 1999.

Richter, Jean Paul. *The Notebooks of Leonardo da Vinci*. 2 vols. New York: Dover, 1970.

Ringbom, Sixten. "Action and Report: The Problem of Indirect Narration in the Academic Theory of Painting." *JWCI* 52 (1989): 34–51.

Rinne, Katherine. "The Landscape of Laundry in Late Cinquecento Rome." *Studies in the Decorative Arts* 9 (2001–2): 34–60.

Robertson, Clare. "Annibale Caro as Iconographer." *JWCI* 45 (1982): 160–81.

—. *The Invention of Annibale Carracci*, Milan: Silvana, 2008.

Robertson, Clare and Roberto Zapperi. "Odoardo Farnese." *DBI*. Vol. 45. Rome, 1995, 112–19.

Rossetti, Sergio. *Rome: A Bibliography from the Invention of Printing through 1899*. Vol. I: *The Guide Books*. 4 vols. Florence: Olschki, 2000.

Rossi, Ermete. "La foglietta di Maggino Ebreo." *Roma: Rivista di studi e di vita romana* 5 (May 1928): 210–13.

Roth, Cecil. "Stemmi di famiglie ebraiche italiane." In *Scritti in memoria di Leone Carpi*. Ed. D. Carpi, A. Milano, and A. Rofé. Jerusalem: Sally Meyer Foundation, 1967, 165–84.

Rubin, Patricia. "The Private Chapel of Cardinal Alessandro Farnese in the Cancelleria, Rome." *JWCI* 50 (1987): 82–112.

Ruffini, Marco. *Le imprese del drago: Politica, emblematica e scienze naturali alla corte di Gregorio XIII (1572–1585)*. Rome: Bulzoni, 2005.

—. *Art Without an Author: Vasari's Lives and Michelangelo's Death*. New York: Fordham University Press, 2011.

Ruggeri, Ugo. "Carlo Urbini e il Codice Huygens." *Critica d'arte* 11 (1978): 167–76.

Russo de Caro, Erina. "Iconografia di Sisto V nella pittura: Tre ritratti inediti." In *Le arti nelle Marche al tempo di Sisto V*. Ed. P. dal Poggetto. Milan: Silvana, 1992, 22–9.

Saenger, Paul. *The Space Between Words*. Stanford: Stanford University Press, 1997.

Said, Edward. "Opponents, Audiences, Constituencies, and Community." In *The Philosophy of Expertise*. Ed. Evan Selinger and Robert P. Crease. New York: Columbia University Press, 2006, 370–94.

Saltini, Guglielmo Enrico. "Della Stamperia Orientale Medicea e Giovan Battista Raimondi." *Giornale Storico degli Archivi Toscani* 4 (October–December 1860): 237–308.

San Juan, Rose Marie. *Rome: A City Out of Print*. Minneapolis: University of Minnesota Press, 2001.

Santoni, Barbara Tellini, and Alberto Mondadori. *Libri e cultura nella Roma di Borromini*. Rome: Retablo, 2000.

Saunders, J. B. de C. M., and C. D. O'Malley. *The*

Illustrations from the Works of Andreas Vesalius of Brussels. New York: Dover, 1950.

Sawday, Jonathan. "The Fate of Marsyas: Dissecting the Renaissance Body." In *Renaissance Bodies: The Human Figure in English Culture c.1540–1660*. London: Reaktion Books, 1990, 111–35.

Saxl, Fritz. "Macrocosm and Microcosm in Mediaeval Pictures." In *Lectures*, Vol. 1, London: The Warburg Institute, 1957, 58–72.

Schmitt, Jean Claude. "L'imagination efficace." In *Imagination und Wirklichkeit*. Ed. Klaus Krüger and Alessandro Nova. Mainz: Philipp von Zabern, 2000, 13–20.

Schudt, Ludwig. *Le guide di Roma: Materialien zu einer geschichte der römischen topographie*. Vienna: B. Filzer, 1930.

Scott, Katie. "Authorship, the Académie, and the Market in Early Modern France." *Oxford Art Journal* 21, no. 1 (1998): 27–41.

Sennet, Richard. *The Craftsman*. New Haven and London: Yale University Press, 2008.

Serrai, Alfredo. *Bernardino Baldi: La vita, le opere, la biblioteca*. Milan: Edizioni Sylvestre Bonnard, 2002.

Shapin, Steven. "'A Scholar and a Gentleman': The Problematic Identity of the Scientific Practitioner in Early Modern England." *History of Science* 29 (1991): 279–327.

—. *A Social History of Truth: Civility and Science in Seventeenth-Century England*. Chicago: University of Chicago Press, 1994.

Shell, Janice. "Bergognone, Ambrogio." In *Grove Art Online. Oxford Art Online*. http://www.oxford-artonline.com/subscriber/article/grove/art/T008121, accessed 20 August 2011.

Shulvass, Moses A. *The Jews in the World of the Renaissance*. Leiden: Brill, 1973.

Siegmund, Stefanie B. *The Medici State and the Ghetto of Florence*. Stanford: Stanford University Press, 2006.

Siraisi, Nancy. *Medieval and Early Renaissance Medicine*. Chicago: University of Chicago Press, 1990.

Smith, Pamela H. *The Body of the Artisan: Art and Experience in the Scientific Revolution*. Chicago: University of Chicago Press, 2006.

Snyder, Jon R. *Writing the Scene of Speaking: Theories of Dialogue in the Late Italian Renaissance*. Stanford: Stanford University Press, 1989.

Sonnino, Eugenio. "Between the Home and the Hospice." In *Poor Women and Children in the European Past*. Ed. John Henderson and Richard Wall. London: Routledge, 1994, 94–116.

Stallybrass, Peter. "Books and Scrolls: Navigating the Bible." In *Books and Readers in Early Modern England*. Ed. J. Anderson and E. Sauer. Philadelphia: University of Pennsylvania Press, 2002, 42–79.

Stallybrass, Peter, and Allon White. *The Poetics and Politics of Transgression*. Ithaca: Cornell University Press, 1986.

Steadman, John M. "Beyond Hercules: Bacon and the Scientist as Hero." *Studies in the Literary Imagination* 4 (1971): 3–47.

Steinberg, Leo. "The Metaphors of Love and Birth in Michelangelo's Pietàs." In *Studies in Erotic Art*. Ed. T. Bowie and E. Christianson. New York: Basic Books, 1970, 231–335.

Steinberg, S. H. *Five Hundred Years of Printing*. Revised J. Trevitt. London: British Library, 1996.

Stimilli, Davide. "Le forze del destino riflesse nel simbolismo all'antica, 1924." In *Aby Warburg: La dialettica dell' imagine, aut aut* 321/322, Milan: Il Saggiatore, 2004 18–20.

Stock, Brian. *After Augustine: The Meditative Reader and the Text*. Philadelphia: University of Pennsylvania Press, 2001.

Stow, Kenneth R. *Catholic Thought and Papal Jewry Policy, 1555–1593*. New York: Jewish Theological Seminary of America, 1977.

—. *The Jews in Rome, Vol. 2: 1551–1557*. Leiden and New York: Brill, 1997.

—. "The Church and the Jews." In *The New Cambridge Medieval History, Vol. 5*. Ed. David Abulafia. Cambridge: Cambridge University Press, 1999, 204–19.

—. *Theater of Acculturation: The Roman Ghetto in the Sixteenth Century*. Seattle and London: University of Washington Press, 2001.

Suster, Guido. "Dell'incisore Trentino Aliprando Caprioli." *Archivio Trentino* 18 (1903): 144–206.

Symonds, John Addington. *Renaissance in Italy, Vol. 4: Italian Literature*. New York: Scribner, 1914.

Szczesniak, Boleslaw. "The City of Utopia: Quinsay." *Literature East & West* 4 nos. 2/3 (1957): 5.

Tafuri, Manfredo. "Capriani, Francesco, detto Francesco da Volterra." *DBI*. Vol. 19, 1976, 189–95.

Talvecchia, Bette. *Taking Positions: On the Erotic in Renaissance Culture*. Princeton: Princeton University Press, 1999.

Testa, Simone. "From the 'Bibliographical Nightmare' to a Critical Bibliography: *Tesori politici* in the British Library, and Elsewhere in Britain." *Electronic British Library Journal* (eBLJ) 1 (2008). http://www.bl.uk/eblj/2008articles/article1.html, accessed 6 August 2008.

Tietze-Conrat, Erica. "Neglected Contemporary Sources Relating to Michelangelo and Titian." *The Art Bulletin* 25, no. 2 (June 1943): 154–9.

Tinto, Alberto. *La Tipografia Orientale Medicea*. Lucca: Maria Pacini Fazzi, 1987.

Toaff, Ariel. *Il prestigiatore di Dio: Avventure e miracoli di un alchimista ebreo nelle corti del Rinascimento*. Milan: Rizzoli, 2010.

Toaff, Renzo. *La Nazione Ebrea a Livorno e a Pisa (1591–1700)*. Florence: Olschki, 1990.

Tomasi, L. Tongiorgio. "Image, Symbol and Word on the Title Pages and Frontispieces of Scientific Books from the Sixteenth and Seventeenth Centuries." *Word & Image* 4 (1988): 372–82.

Toomer, G. F. *Eastern Wisedome and Learning: The Study of Arabic in Seventeenth-Century England*, Oxford: Clarendon Press, 1996.

Vannucci, Alessandra Baroni. *Jan Van Der Straet detto Giovanni Stradano flandrus pictor et inventor*. Milan: Jandi Sapi Editori, 1997.

Vasoli, Cesare. "The Renaissance Concept of Philosophy." In *The Cambridge History of Renaissance Philosophy*. Gen. ed. C. B. Schmitt. Cambridge and New York: Cambridge University Press, 1988, 55–74.

Verellen, Till. "Cosmas and Damian in the New Sacristy." *JWCI* 42 (1979): 274–7.

Vescovini, G. Federici. "Les *Vite di matematici Arabi* de Bernardino Baldi." In *Between Demonstration and Imagination*. Ed. L. Nauta and A. Vanderjagt. Leiden: Brill, 1999, 395–408.

Viatte, Françoise. "Allegorical and Burlesque Subjects by Stefano della Bella." *Master Drawings* 15, no. 4 (Winter 1977): 347–65, 425–44.

Vincent, Catherine. "Discipline du corps et de l'esprit chez les Flagellants au Moyen Âge," *Revue Historique* 302:615 (July–September 2000): 593–614.

Vogüé, Adalbert de. *The Rule of Saint Benedict*. Trans. J. B. Hasbrouck. Kalamazoo, Mich., Cistercian Publications, 1983.

Wallace, William A. "The Culture of Science in Renaissance Thought." *History of Philosophy Quarterly* 3, no. 3 (July 1986): 281–91.

Warburg, Aby. "Postscriptum alla conferenza di Alfred Doren 'Fortuna nel Medioevo e nel Rinascimento,' 1923." *aut aut* 321/322 (2004): 17.

Warnock, Mary. *Imagination*. Berkeley and Los Angeles, University of California Press, 1976.

—. *Memory*. London: Faber, 1987.

Whitman, Jon. *Allegory: The Dynamics of an Ancient and Medieval Technique*. Cambridge, Mass.: Harvard University Press, 1987.

Wisch, Barbara. "Violent Passions: Plays, Pawnbrokers and the Jews of Rome." In *Beholding Violence in Medieval and Early Modern Europe*. A. Terry-Fritsch and E. F. Labbie (eds). Aldershot: Ashgate, 2012, 197–213.

Winkler, Mary, and Albert Van Helden. "Representing the Heavens: Galileo and Visual Astronomy." *Isis* 83, no. 2 (June 1992): 195–217.

Witcombe, Christopher L. C. E. 'Cherubino Alberti and the Ownership of Engraved Plates.' *Print Quarterly* 6, no. 2 (1989): 160–9.

—. "Passeri, Bernardino." In *The Dictionary of Art, Vol. 24*. Ed. Jane Turner. New York: Grove, 1996, 234.

—. *Copyright in the Renaissance*. Leiden: Brill, 2004.

—. *Print Publishing in Sixteenth-Century Rome*. London: Harvey Miller, 2008.

Wittkower, Rudolph. *Art and Architecture in Italy, 1600 to 1750*. Harmondsworth: Penguin, 1958.

Wooden, Warren W. "Anti-Scholastic Satire in Sir Thomas More's Utopia." *Sixteenth Century Journal* 8, no. 2 (July, 1977): 29–45.

Zardin, Danilo. "Le *Adnotationes et Meditationes* illustrate de Nadal sui Vangeli del ciclo liturgigico: il modello e li riuso." In *Visible Teologia. Il libro sacro figurato tra Cinquecento e Seicento in Italia*. Ed. Erminia

Ardissino and Elisabetta Selmi. Rome: Edizione di Storia e Letteratura, 2012, 3–23.

Zorach, Rebecca. *The Virtual Tourist in Renaissance Rome: Printing and Collecting the Speculum Romanae Magnificentiae*. Chicago: University of Chicago Press, 2008.

Zuber, Marta Spranzi. "Dialectic, Dialogue and Controversy: The Case of Galileo." *Science in Context* 11, no. 2 (1988): 181–203.

——. "Galileo's 'Dialogue on the Two Chief World Systems': Rhetoric and Dialogue." In *La rhetorique des textes scientifiques au XVIIe siècle*. Ed. Fernand Hallyn and Lyndia Roveda. Turnhout: Brepols, 2005, 97–114.

Index

Illustration Credits

In most cases illustrative material has been provided by the owners or custodians of the work. Those for which further credit is due are listed below:

By permission of the Houghton Library, Harvard University: 1, 3, 7a, 7d, 8, 9, 10, 13, 15, 16, 17, 19, 20, 21, 22, 23, 24, 30, 36, 37, 38, 42, 47, 48, 49, 54, 55, 58, 60, 61, 62, 63, 65, 81, 84, 125; © Trustees of the British Museum: 2, 35, 43, 67, 86, 87, 93; Avery Library, Columbia University: 4, 7b, 7c; Author photographs: 5, 28, 68, 71, 72, 73, 74, 75, 76, 78, 79, 80, 88, 89, 90, 91, 103; Scala/Ministero per i Beni e le Attività culturali/Art Resource, NY: 11, 12, 14, 25; Scala/Art Resource, NY: 18; © Comune di Milano: 26, 27; By permission of the Ministero per i Beni e le Attività Culturali-Italia © Biblioteca Casanatense, Rome: 29; By permission of the Folger Shakespeare Library: 31, 50, 51, 64; Courtesy of the John Carter Brown Library at Brown University: 33; By permission of the Ministero per i Beni e le Attività Culturali-Italia © Biblioteca Angelica, Rome: 34, 66; © 2013 Biblioteca Apostolica Vaticana, Vatican City, Rome: 39, 45, 46, 99, 100, 106, 120b; Rare Books Division, New York Public Library, Astor, Lenox, and Tilden Foundations: 40, 41; Vanni/Art Resource, NY: 44; John Hay Library, Brown University: 52, 53, 85, 92, 94, 115, 120c, 120d, 120e, 120f, 120g, 120h, 121, 122, 124, 126, 127, 128; The Lessing J. Rosenwald Collection, Rare Book and Special Collections Division, Library of Congress, Washington, DC: 56, 57, 109, 117, 120a; © The Pierpont Morgan Library, NY Purchased in 1938: 59; Boston Medical Library in the Francis A. Countway Library of Medicine: 69, 82, 97, 98; Biblioteca del Senato della Repubblica "Giovanni Spadolini": 70; Digital Production Services, Brown University Library: 88; Pilcher Collection, Taubman Health Sciences Library, University of Michigan: 95; Biblioteca d'arte e di storia di San Giorgio in Poggiale, Bologna: 96; © The Metropolitan Museum of Art, The Elisha Whittelsey Collection, The Elisha Whittelsey Fund, 1949: 101; By permission of the Ministero per i Beni e le Attività Culturali-Italia © Biblioteca Nazionale di Napoli: 102, 104, 107, 108, 110, 111, 112, 113, 114, 118; By permission of the Ministero per i Beni e le Attività Culturali-Italia: 119; © Cornell University Library: 123.